Springer
Proceedings in Physics 79

Springer Proceedings in Physics

Managing Editor: H. K. V. Lotsch

46 *Cellular Automata and Modeling of Complex Physical Systems* Editors: P. Manneville, N. Boccara, G. Y. Vichniac, and R. Bidaux

47 *Number Theory and Physics* Editors: J.-M. Luck, P. Moussa, and M. Waldschmidt

48 *Many-Atom Interactions in Solids* Editors: R. M. Nieminen, M. J. Puska, and M. J. Manninen

49 *Ultrafast Phenomena in Spectroscopy* Editors: E. Klose and B. Wilhelmi

50 *Magnetic Properties of Low-Dimensional Systems II: New Developments* Editors: L. M. Falicov, F. Mejía-Lira, and J. L. Morán-López

51 *The Physics and Chemistry of Organic Superconductors* Editors: G. Saito and S. Kagoshima

52 *Dynamics and Patterns in Complex Fluids: New Aspects of the Physics–Chemistry Interface* Editors: A. Onuki and K. Kawasaki

53 *Computer Simulation Studies in Condensed-Matter Physics III* Editors: D. P. Landau, K. K. Mon, and H.-B. Schüttler

54 *Polycrystalline Semiconductors II* Editors: J. H. Werner and H. P. Strunk

55 *Nonlinear Dynamics and Quantum Phenomena in Optical Systems* Editors: R. Vilaseca and R. Corbalán

56 *Amorphous and Crystalline Silicon Carbide III, and Other Group IV-IV Materials* Editors: G. L. Harris, M. G. Spencer, and C. Y. Yang

57 *Evolutionary Trends in the Physical Sciences* Editors: M. Suzuki and R. Kubo

58 *New Trends in Nuclear Collective Dynamics* Editors: Y. Abe, H. Horiuchi, and K. Matsuyanagi

59 *Exotic Atoms in Condensed Matter* Editors: G. Benedek and H. Schneuwly

60 *The Physics and Chemistry of Oxide Superconductors* Editors: Y. Iye and H. Yasuoka

61 *Surface X-Ray and Neutron Scattering* Editors: H. Zabel and I. K. Robinson

62 *Surface Science: Lectures on Basic Concepts and Applications* Editors: F. A. Ponce and M. Cardona

63 *Coherent Raman Spectroscopy: Recent Advances* Editors: G. Marowsky and V. V. Smirnov

64 *Superconducting Devices and Their Applications* Editors: H. Koch and H. Lübbig

65 *Present and Future of High-Energy Physics* Editors. K.-I. Aoki and M. Kobayashi

66 *The Structure and Conformation of Amphiphilic Membranes* Editors: R. Lipowsky, D. Richter, and K. Kremer

67 *Nonlinearity with Disorder* Editors: F. Abdullaev, A. R. Bishop, and S. Pnevmatikos

68 *Time-Resolved Vibrational Spectroscopy V* Editor: H. Takahashi

69 *Evolution of Dynamical Structures in Complex Systems* Editors: R. Friedrich and A. Wunderlin

70 *Computational Approaches in Condensed-Matter Physics* Editors: S. Miyashita, M. Imada, and H. Takayama

71 *Amorphous and Crystalline Silicon Carbide IV* Editors: C. Y. Yang, M. M. Rahman, and G. L. Harris

72 *Computer Simulation Studies in Condensed-Matter Physics IV* Editors: D. P. Landau, K. K. Mon, and H.-B. Schüttler

73 *Surface Science: Principles and Applications* Editors: R. F. Howe, R. N. Lamb, and K. Wandelt

74 *Time-Resolved Vibrational Spectroscopy VI* Editors: A. Lau, F. Siebert, and W. Werncke

75 *Computer Simulation Studies in Condensed-Matter Physics V* Editors: D. P. Landau, K. K. Mon, and H.-B. Schüttler

76 *Computer Simulation Studies in Condensed-Matter Physics VI* Editors: D. P. Landau, K. K. Mon, and H.-B. Schüttler

77 *Quantum Optics VI* Editors: D. F. Walls and J. D. Harvey

78 *Computer Simulation Studies in Condensed-Matter Physics VII* Editors: D. P. Landau, K. K. Mon, and H.-B. Schüttler

79 *Nonlinear Dynamics and Pattern Formation in Semiconductors and Devices* Editor: F.-J. Niedernostheide

Volumes 1–45 are listed at the end of the book

F.-J. Niedernostheide (Ed.)

Nonlinear Dynamics and Pattern Formation in Semiconductors and Devices

Proceedings of a Symposium Organized Along with the International Conference on Nonlinear Dynamics and Pattern Formation in the Natural Environment
Noordwijkerhout, The Netherlands, July 4-7, 1994

With 168 Figures

Springer

Dr. Franz-Josef Niedernostheide

Institut für Angewandte Physik
Corrensstraße 2/4
D-48149 Münster, Germany

ISBN 3-540-58833-7 Springer-Verlag Berlin Heidelberg New York

CIP data applied for

Printed in Germany

Typesetting: Camera ready copy from the editor
SPIN: 10479837 54/3144 - 5 4 3 2 1 0 - Printed on acid-free paper

Preface

This volume contains the contributions given at the symposium about *Nonlinear Dynamics and Pattern Formation in Semiconductors and Devices* in the frame of the *International Conference on Nonlinear Dynamics and Pattern Formation in the Natural Environment* (ICPF'94) held at Noordwijkerhout, the Netherlands July 4-7, 1994. As the conference aimed at the communication of new results and the exploration of new ideas concerning the theory of nonlinear dynamics and the study of pattern generating phenomena in diverse biological, chemical and physical systems, the central issues discussed during the semiconductor-symposium involved fundamental self-organization processes and the physics of electrical instabilities in semiconductors and semiconductor devices.

One of today's most challenging questions concerns the understanding of evolution of life. This in turn entails the question of how and on what conditions patterns may arise from spatially uniform distributions and which minimal ingredients are necessary to achieve an aspired degree of complexity. In the last decades great progress in investigating this matter has been made in many fields of sciences, and fundamental concepts and basic mechanisms leading to the spontaneous appearance of spatial, temporal and spatio-temporal patterns have been developed. It turned out that many of the basic mechanisms appear in entirely different systems and, consequently, a concept worked out for the description of pattern formation in one system very often can be applied to a whole class of systems including systems of other fields. However, many questions are still unsolved and sometimes we are even far from a complete understanding of systems appearing very simple on a first view.

In comparison with biological or chemical systems, physical systems have a great advantage concerning preparation, handling, and parameter controlling. Among the physical systems, semiconductors and semiconductor devices are most suitable for investigations concerning self-organization and pattern formation because, on the one hand, a mature technology for their production is available and many theoretical works concerning the understanding of the microscopic processes have been done. On the other hand, two basic presuppositions for the evolution of dissipative structures, the existence of nonlinearities and the possibility to force the system into a state far from equilibrium, can be achieved in semiconductors in manifold ways, leading to a huge variety of spatiotemporal patterns reaching from simple stationary kinks, solitary structures or periodic patterns over structures showing regular dynamic behaviour as the propagation of fronts, various oscillation modes of localized structures, target or spiral patterns to complex spatiotemporal patterns as chaotic oscillations of localized structures or irregular wavepatterns. The contributions included in this volume take up many of these phenomena and discuss various mechanisms for pattern formation being of importance not only for semiconductors and devices.

The articles are organized in the following way. In the first four chapters the main attention is focussed on the development of models describing pattern formation in semiconductors and devices. Specific models concerning the dynamics of electric field domains in superlattices (Chapt. 1 and Chapt. 2), current filaments in heterostructures (Chapt. 2) and nonlinear waves in extrinsic semiconductors (Chapt. 3) are presented and a survey of the properties of current filaments and field domains is given (Chapt. 4). In the next five chapters, the emphasis is on experiments. The contributions deal with complex spatiotemporal luminescence patterns in a two-dimensional excitable medium consisting of a *II-VI* compound (Chapt. 5), nonlinear behaviour in organic and anorganic crystals (Chapt. 6) and current filaments in gallium arsenide (Chapt. 7). Furthermore, stationary and dynamic patterns of current filaments and various bifurcation sequences in silicon *p-n-p-n* (Chapt. 8) and *p-i-n* devices (Chapt. 9) are studied and corresponding models are developed. The last two chapters focus on possible applications of nonlinear phenomena. By means of low frequency oscillations appearing spontaneously, e. g., in *III-V* compounds, deep levels in semiconductors are analyzed (Chapt. 10) and the application of maximum and minimum detectors based on the winner-takes-all mechanism is proposed for optical pattern recognition and optical fuzzy logic computing (Chapt. 11).

I would like to thank the organizers of the ICPF'94, Prof. A. van Harten and Dr. A. Doelman, for their invitation to organize the semiconductor-symposium as part of the ICPF'94 and Dr. H. Lotsch from Springer-Verlag for his excellent cooperation.

Münster,
December 1994

F.-J. Niedernostheide

Table of Contents

*contribution not presented at the ICPF'94

1 Dynamics of Electric Field Domains in Superlattices

L.L. Bonilla

Universidad Carlos III de Madrid, Escuela Politécnica Superior, Butarque 15
28911 Leganés, Spain

Abstract. A discrete drift model of resonant tunneling transport in weakly-coupled GaAs quantum-well structures under laser illumination is introduced and analyzed. Results include explanations (via formation of electric field domains) of the oscillatory shape of the *I-V* diagram leading to multistability and hysteresis between stationary electric-field profiles for both doped and undoped superlattices under strong laser illumination. Moreover, the dynamics of electric-field domains and domain walls in our model account for damped and undamped time-dependent oscillations of the current in a dc voltage bias situation, with the laser photoexcitation acting as a damping factor. Our results agree with time-resolved photoluminescence and photocurrent experiments. An asymptotic analysis of the continuum limit of our discrete model shows that these current oscillations are due to the formation, motion, annihilation and regeneration of negatively charged domain walls on the superlattice. The situation is reminiscent of the classical Gunn-effect oscillations in bulk semiconductors due to dipole-domain dynamics, and in fact our present asymptotic analysis is an extension and adaptation of previous work of ours on the Gunn effect.

1.1 Introduction

In this paper I will discuss the formation and dynamics of electric field domains in semiconductor superlattices (SL). This is an old subject that goes back to Esaki and Chang in 1974 [1.1]. A SL is a succession of N quantum wells made out of alternating slabs of two (doped or undoped) semiconductors. The canonical example is the identical repetition of one superperiod composed of several tens of Å of GaAs (the "quantum valley") and several tens of Å of AlAs (the "quantum barrier"). When one such SL is subject to a large dc voltage bias across its growth direction, electric field domains [1.2] and time-dependent oscillations of the current may appear [1.3]. These phenomena are of potential importance in understanding the physics and technology of nanometer devices [1.4].

To understand the mechanism behind the formation of electric field domains in SL, let me consider the following oversimplified view of the Kazarinov and Suris's tight-binding results for conduction in a SL [1.5]. At zero applied electric field the dominant transport mechanism is miniband conduction: the energy levels of all the quantum wells (QW) are aligned along the growth direction of the SL and the electrons tunnel coherently from well to well across the SL. In

Springer Proceedings in Physics, Vol. 79 Nonlinear Dynamics and Pattern Formation in Semiconductors and Devices
Editor: F.-J. Niedernostheide

reality the energy levels of the QWs are subbands (consider wells that extend infinitely on the transversal direction) and have a finite width, but we will continue speaking of QW levels and thinking in terms of one space dimension for simplicity. When an electric field is applied, the levels of different wells cease to be aligned and there appears a localized state (Wannier-Stark localization). As a consequence the electric current through the structure drops to a small value. When the applied electric field E is large enough, a new conduction mechanism appears: sequential resonant tunneling (SRT). Let

$$E \sim \frac{\mathcal{E}_2 - \mathcal{E}_1}{e\tilde{l}}, \tag{1.1}$$

where $\mathcal{E}_2$ and $\mathcal{E}_1$ are the energies of the first (e_1) and the second (e_2) levels of one well, $\tilde{l}$ the SL period and e the electron charge. Then the level e_1 of one QW is aligned with the level e_2 of the following QW, and one electron may tunnel coherently between them. After this, scattering events (mainly scattering with phonons) send down the electron to the level e_1 again where it can proceed to the next QW in the same way. This SRT mechanism (which we denote by $e_1 \to e_2$) gives rise to a peak in the current-voltage diagram of the SL. For applied voltages (minus built-in potentials) which are smaller than $N(\mathcal{E}_2 - \mathcal{E}_1)/e$, electric field domains may form: part of the SL is at zero field (domain I) and the rest (domain II) at the field (1.1), so that the voltage across the SL equals the applied bias. For larger applied voltages other SRT mechanisms (between levels e_1 and e_3 for example) may give rise to domains with higher electric field. In doped SL domain formation was already experimentally demonstrated by Esaki and Chang [1.1]. Recent theories of domain formation include those of Laikhtman [1.6, 7], Prengel et al. [1.8] (for doped SL) and ours [1.9] (for undoped SL). I now comment the relevant experimental results in undoped SL under laser illumination by Kwok et al. [1.3], then I extend our model based upon them to doped and undoped SL under laser illumination and finally I explain our results.

In a typical experiment, an undoped 40 period 90 Å GaAs/40 Å AlAs SL was mounted in a *p-i-n* diode and continuously illuminated by laser light at 4 K [1.3]. When the laser power was in a certain interval and the applied *dc* voltage bias was large enough, damped time-dependent oscillations of the photocurrent (PC) and the peaks in the photoluminescence (PL) spectrum were observed. The applied fields in the experiment [1.3] are high, the potential barriers are wide, and the minibands correspondingly narrow so that the coherence length is comparable or smaller than the width of one quantum well. Then the quantum wells in the superlattice are weakly coupled and formation of electric field domains may appear [1.2, 10, 11]. In Kwok et al. experiments, the damped time-dependent oscillations of the PC were observed in regions of the current-voltage diagram where domains type II and III (with electric fields corresponding to $e_1 \to e_2$ and $e_1 \to e_3$ SRTs) coexist [1.3]. Similar but undamped time-dependent oscillations of the current have been observed in doped SL by the Grahn's group at Berlin [1.12] (The n-doped SL formed part of a n^+-n-n^+ diode as described in [1.11]). Under laser illumination, the amplitude of the oscillations decreased and they eventually became damped for strong enough laser power, [1.12]. Our theory aims to explain this dynamic behavior while other theories considered coexistence of zero field and $e_1 \to e_2$ domains [1.7, 8]. In these SL with wide QW, there are several important times that we have to consider in order to construct a reasonable model. First of all, the characteristic time scale of the PC oscillations in [1.3] is of 10-100 ns (the order of the recombination time). The time scale for carrier thermalization is of 0.1 ps while the carriers reach thermal equilibrium with the lattice after a time that ranges from 1 to 100 ps, smaller than the typical tunneling time of 500 ps (see [1.9] and references cited therein). This means that in time scales of the order of nanoseconds (the experimental time scale), we may consider the holes and electrons to be at local equilibrium within each QW j at the lattice temperature, and with given values of their densities, $\tilde{p}_j$ and $\tilde{n}_j$. The process of reaching a stationary state might be seen as the attempt of reaching a "global equilibrium" starting from "local equilibrium" through tunneling processes that communicate different QWs, self-consistency of the electric field and scattering and interband processes. In this spirit we consider the QWs as entities characterized by average values of the electric field, $\tilde{E}_j$ for the j-th well, and of the densities of holes and electrons, $\tilde{p}_j$ and $\tilde{n}_j$, respectively. We then propose the following transport equations to describe the dynamics of the SL [1.9]:

$$\tilde{E}_j - \tilde{E}_{j-1} = \frac{e\tilde{l}}{\epsilon}(\tilde{n}_j - \tilde{p}_j - \tilde{N}_D), \tag{1.2}$$

$$\epsilon\frac{d\tilde{E}_j}{d\tilde{t}} + e\,\tilde{v}(\tilde{E}_j)\,\tilde{n}_j = \tilde{J}, \tag{1.3}$$

$$\frac{d\tilde{p}_j}{d\tilde{t}} = \tilde{\gamma} - \tilde{r}\,\tilde{n}_j\,\tilde{p}_j, \tag{1.4}$$

where $j = 1, \ldots, N$. In this model (1.2) and (1.3) are, respectively, Poisson equation (averaged over one SL period) and Ampère's law; equation (1.4) is the hole rate equation containing the photogeneration rate $\tilde{\gamma}$ (proportional to the laser power), and the recombination constant $\tilde{r}$. The average permittivity is denoted by ϵ and $\tilde{N}_D$ is the average donor concentration (zero for undoped SL). $\tilde{v}(\tilde{E})$ is an effective electron velocity and $\tilde{J}$ is the total current density; the contribution of the heavier holes to the current is ignored.

In the calculation, we take $\tilde{v}(\tilde{E})$ as a datum and assume it does not vary with density. Qualitatively our results do not depend on the precise shape of $\tilde{v}(\tilde{E})$ provided it exhibits maxima at the resonant fields; the results reported here are based on the curve shown in the inset of Fig. 1.1. In the $3N$ equations (1.2)-(1.4) there are $3N+2$ unknowns, $\tilde{J}$, $\tilde{E}_0, \tilde{E}_j, \tilde{n}_j, \tilde{p}_j$, with $j = 1, \ldots, N$. One additional equation is the bias condition:

$$\frac{1}{N}\sum_{j=1}^{N}\tilde{E}_j = \frac{\tilde{\Phi}}{N\tilde{l}}. \tag{1.5}$$

Here $\tilde{\Phi}$ is the difference between the applied voltage and the built-in potential due to the doped regions outside the SL (1.5 Volts in [1.3]). $\frac{\tilde{\Phi}}{N\tilde{l}}$ is the average applied electric field on the SL, which we will henceforth call the *bias*. The missing condition is a boundary condition for the field at the zeroth QW, $\tilde{E}_0$, (*before* the SL). We do not have direct experimental evidence for what $\tilde{E}_0$ should be. Thus our choice for $\tilde{E}_0$ has to be validated *a posteriori* by comparing the results of our analysis with experiments and with the consequences of a different choice. We shall use throughout this paper the following boundary condition:

$$\tilde{E}_1(\tilde{t}) - \tilde{E}_0(\tilde{t}) = \frac{\tilde{c}\,\tilde{l}}{\epsilon}. \tag{1.6}$$

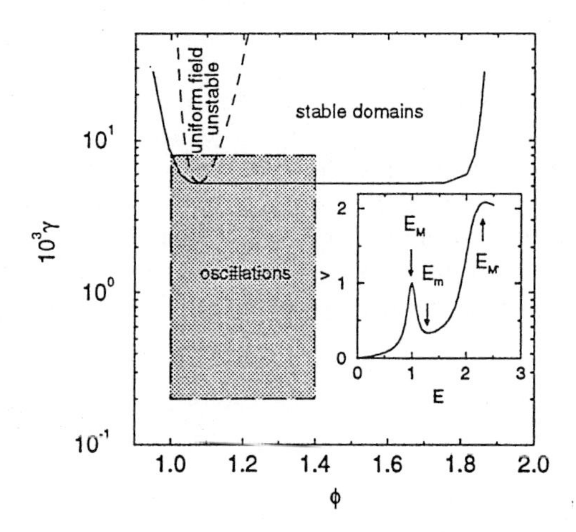

Fig. 1.1. Phase diagram of the undoped model: Photogeneration rate (γ) vs. applied voltage bias (ϕ) in units of, respectively, $\tilde{n}/\tau_p$ and $\tilde{E}_M N\tilde{l}$; see text. The shaded region denotes the range of domain wall oscillations with $2 \times 10^{-4} < \gamma < 8 \times 10^{-3}$. The continuous (dotted) curve denotes the region for which the domain (uniform-field) solution is stable (unstable). Inset: The normalized field dependence of the carrier velocity used in the calculations

When $\tilde{c} \ll e\,\tilde{n}$ (see (1.8) below and Sect. 1.3), this condition does allow an extremely small charge build-up at the first well and it is compatible with the existence of a stationary solution of the model with the same constant values of the field and the carrier densities for all the QWs [1.13]:

$$\tilde{v}\left(\frac{\tilde{\Phi}}{N\tilde{l}}\right) = \frac{\tilde{J}\,\tilde{r}}{2\,e\,\tilde{\gamma}}\left[\sqrt{\tilde{N}_D^2 + \frac{4\tilde{\gamma}}{\tilde{r}}} - \tilde{N}_D\right], \quad \tilde{E}_j = \frac{\tilde{\Phi}}{N\tilde{l}}, \tag{1.7}$$

for all j. This equation says that the effective electron velocity is (except for constant scale factors) the same as the static *I-V* characteristic curve provided that *the field profile of the SL is static and uniform.* From experiments with undoped SL we know that such electric field profile is observed at low laser power [1.3], so that we can infer the form of $\tilde{v}(\tilde{E})$ from experimental data. Later we shall see when the solution with uniform field profile ceases to be stable at high laser power and domains form. Prengel et al. used a boundary condition related to (1.6) with $c = 0$ for similar reasons (the electron densities at each subband of the first QW are equal to the corresponding values of the zeroth QW) [1.8]. We could have also obtained the velocity curve from microscopic theories of static I-V curves for heterostructures (see [1.8]), but we find our identification with experimental data sufficient.

It is convenient to render the equations dimensionless by adopting as the units of electric field and velocity the values at the first maximum of the velocity curve, $\tilde{v}(\tilde{E})$, $\tilde{E}_M \simeq 10^5$ V/cm and $\tilde{v}_M$. Further we express the carrier and impurity concentrations and bias in terms of

$$\tilde{n} = \frac{\epsilon \tilde{E}_M}{e\tilde{l}} \simeq 10^{17} - 10^{18}\,\text{cm}^{-3}, \tag{1.8}$$

and $\tilde{E}_M\, N\, \tilde{l}$ for $\tilde{l} = 130$ Å. This yields the time and current units, $\tau_p = 1/(\tilde{r}\,\tilde{n})$ and $\tilde{J}_M = e\tilde{v}_M\tilde{n}$; $\tau_p = 10$ ns determines the time scale of the oscillations [1.9]. The dimensionless equations then are:

$$E_j - E_{j-1} = n_j - p_j - \nu, \tag{1.9}$$

$$\beta\,\frac{dE_j}{dt} + v(E_j)\,n_j = J, \tag{1.10}$$

$$\frac{dp_j}{dt} = \gamma - n_j\,p_j, \tag{1.11}$$

$$\frac{1}{N}\sum_{j=1}^{N} E_j = \phi, \tag{1.12}$$

$$E_1 - E_0 = \nu\, c; \qquad c \ll 1. \tag{1.13}$$

Here $\beta = \tilde{r}\tilde{l}\tilde{n}/(\tilde{v}(\tilde{E}_M))$ and $\nu = \tilde{N}_D/\tilde{n}$ go from 0.01 to 1, while ϕ and $\gamma = \tilde{\gamma}/(\tilde{r}\tilde{n}^2)$ are the dimensionless control parameters. These equations are to be solved with initial conditions for the fields, $E_j(0)$, and the hole concentrations, $p_j(0)$, compatible with the bias (1.12) and the boundary condition (1.13). The initial conditions for the electron density, $n_j(0)$, then follow from (1.9).

The rest of this paper is as follows. In Sect. 1.2, we explain how to construct exactly the steady states and the corresponding current-bias diagram in the simpler case $c = 0$. In Sect. 1.3, several results of numerical simulations for doped and undoped SLs with or without photoexcitation are explained. The continuum limit of a long SL with small doping is analyzed in Sect. 1.4. In this limit, an asymptotic theory allows us to qualitatively understand the time-dependent oscillations in superlattices: the oscillations are due to domain wall dynamics [1.14] and are similar to those numerically observed in models of the Gunn effect [1.15] with unusual boundary conditions (corresponding to monopole dynamics [1.16]). This connection might justify the name *discrete Gunn effect* for the oscillations of the current in superlattices.

1.2 Steady States

For given values of γ, ν and J there are stationary solutions of (1.9)-(1.11) compatible with the boundary condition (1.13). Using (1.12), we find the corresponding value of ϕ. From this we can obtain the curves $J = J(\phi)$ appearing in

Fig. 1.2. We have studied two different types of stationary solutions, namely uniform solutions and two-domain non-uniform solutions. For simplicity, we report here on steady states without charge build-up on the first QW: $c = 0$, that is, $E_0 = E_1$. Generalizations to $0 < c \ll 1$ are straightforward and will be omitted.

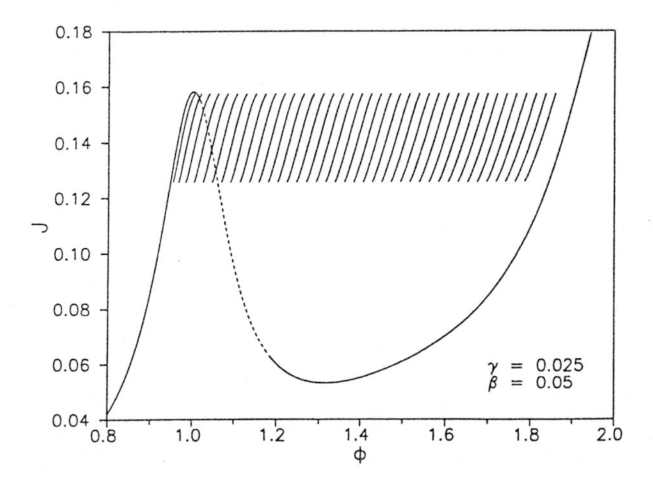

Fig. 1.2. Static characteristic curve: current (J) vs. applied bias (ϕ) for $\nu = 0$, $\gamma = 0.025$ and $\beta = 0.05$. The solid line having the same shape as $v(E)$ corresponds to the stable uniform solution. In the part of the curve with negative slope there is a region where this solution becomes unstable (dashed line). The remaining 39 curves correspond to linearly stable two-domain solutions [1.9]. Important features to stress from this figure are: first, the coexistence, for a given ϕ, not only of several domain solutions but also of domain solutions and the uniform solution (see also Fig. 1.1), something relevant to the appearance of hysteresis and memory effects, and second, the average "flatness" of J in the bias region between the two local maxima of $v(E)$

The stationary versions of (1.10) and (1.11) may be solved for n_j and p_j in terms of the electric field E_j. When the result is inserted in (1.9), we find the following discrete mapping for the stationary electric field profiles $\{E_j,\ j = 1,\ldots,N\}$:

$$E_{j-1} = f(E_j; \nu, \gamma, J) \tag{1.14}$$

with

$$f(E; \nu, \gamma, J) \equiv E + \nu - \frac{J}{v(E)} + \frac{\gamma}{J}\, v(E), \tag{1.15}$$

which should obey the boundary condition $E_0 = E_1$. From these profiles and the bias condition (1.12), we obtain the current–voltage characteristic curve of Fig. 1.2, where only the stable stationary solution branches have been depicted [1.9]. The uniform stationary solution corresponds to the fixed points of the mapping (1.14):

$$E_j = \phi\,, \tag{1.16}$$

$$n_j = \frac{\sqrt{\nu^2+4\gamma}+\nu}{2}, \quad p_j = \frac{\sqrt{\nu^2+4\gamma}-\nu}{2} \qquad (j=1,\ldots,N) \tag{1.17}$$

$$J = \frac{\sqrt{\nu^2+4\gamma}+\nu}{2}\, v(\phi)\,. \tag{1.18}$$

Equation (1.18) yields a static current-voltage characteristic curve which is the same as $v(E)$ except for a scale factor. See Fig. 1.2. It is convenient to separate the analysis of the stationary solutions in three different cases: undoped SL with photoexcitation, dark doped SL, and doped SL with photoexcitation.

1.2.1 Undoped Photoexcited SL, $\nu = 0$, $\gamma > 0$

The mapping (1.14) may have one or three fixed points. The latter is a necessary condition for non-uniform stationary field profiles to exist. For biases ϕ between the two maxima in the inset of Fig. 1.1, we find the linearly stable non-uniform solution branches depicted in Fig. 1.2, with two domains separated by a wall. They correspond to trajectories of the discrete mapping (1.14) that leave the fixed point with lower field on the first branch of $f(E;\gamma,J)$, E_L, after $j=k$ QWs and jump to the third branch of $f(E;\gamma,J)$ for $j>k$. See Fig. 1.3. These profiles exist only for

$$\gamma > \gamma_1 \equiv \frac{1}{4}\left(\min_{1<\phi<E_m}\frac{v(\phi)}{\mid v'(\phi)\mid}\right)^2 \simeq 5.2\times 10^{-3} \tag{1.19}$$

and J on an appropriate interval [1.9]. In general there are several stable stationary solutions for a given bias value, thereby yielding multistability and the possibility of hysteresis, in agreement with experiments [1.2] (see [1.11] for multistability and hysteresis with domain I/domain II stationary branches). There also exist other stationary solution branches which are unstable and have not been depicted in Fig. 1.2. They can be constructed with the help of the mapping (1.14) (they typically correspond to jumps to or from the second branch of $f(E;\gamma,J)$, see Fig. 1.3), and connect the stable branches of Fig. 1.2, [1.9].

1.2.2 Doped SL, $\nu > 0$, $\gamma \geq 0$

These cases may be analyzed with the help of the discrete mapping (1.15). There are non-uniform stationary solutions for $\sqrt{\nu^2+4\,\gamma} > 2\sqrt{\gamma_1} \simeq 0.13$. Unless γ or ν surpass larger values, the only linearly stable stationary solution corresponds to the uniform field profile. See [1.14] for details.

1.3 Phase Diagram and PC Time-Dependent Oscillations

1.3.1 Undoped Photoexcited SL, $\nu = 0$, $\gamma > 0$

With the aim of interpreting the experimental results, we have to delimit the intervals of the parameters γ and ϕ where stable stationary solutions may be

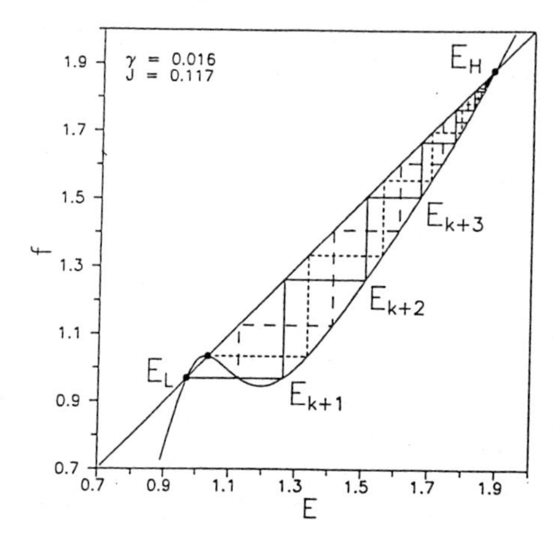

Fig. 1.3. The discrete mapping $f(E;0,\gamma,J)$ and E vs. E for $\gamma = 0.016$ and $J = 0.117$ in the undoped case. Jumps $1 \to 3$ and $2 \to 3$ between the branches of $f(E;0,\gamma,J)$ are allowed. The continuous lines show the discrete mapping process used to construct an $(1 \to 3)$ non-uniform stationary solution with two domains: After remaining at the low field domain value E_L, the field jumps at the $(k+1)$-th well to a different solution of $f(E;0,\gamma,J) = E_L$. In successive iterations the electric field tends to its high-field value E_H. The short-dashed lines show the same construction for a $2 \to 3$ solution, whereas the long-dashed lines correspond to a different (unstable) type of $2 \to 3$ solution explained in [1.9]

found. The $\gamma - \phi$ "phase diagram" of Fig. 1.1 displays the regions where the uniform stationary solution is stable or unstable, the regions where linearly stable solutions with domains exist, and the regions where damped time-dependent oscillations of the photocurrent (PC) may be observed [1.9]. The information on stationary solutions in the phase diagram was obtained with the help of the discrete mapping $f(E;0,\gamma,J)$, whereas that on time-dependent oscillations was found by direct numerical simulation of the model (1.9)-(1.13) with $c = 0$ and an initial condition that differed from the uniform solution in that there was an excess positive charge in the first QW. This corresponds to turning on the laser at $t = 0$ for a fixed value of voltage and laser power [1.9].

The analysis of the steady states in Sect. 1.2 reveals that at low laser power $\gamma < \gamma_1$ ($\simeq 5.2 \times 10^{-3}$ in our rescaled parameters), there is only one steady solution: the uniform one, (1.16) and (1.17), which is linearly stable (to be more precise, the same stands for the whole region outside the full line of the $\gamma - \phi$ phase diagram shown in Fig. 1.1). Inside the full line in Fig. 1.1, multiple stable non-uniform steady states exist. From a dynamical point of view, the simulations show damped time-dependent oscillations of the current in the range $\gamma_0 < \gamma < \gamma_2$ ($\gamma_0 \simeq 2 \times 10^{-4}$, $\gamma_2 \simeq 8 \times 10^{-3}$) for ϕ on a subinterval of $(E_M, E_{M'})$ (the shaded rectangle of Fig. 1.1). In the region of the phase diagram where these oscillations occur we may distinguish two subregions, according to the situation reached after the oscillations stop. For $\gamma_0 < \gamma < \gamma_1$ the oscillation of the PC stops and the electric field profile resembles a two-domain non-uniform stationary solution.

However, no such stationary profile exists as indicated in the previous section. Thus there is a slow drift of the PC and of the field profile that ends at the values corresponding to the uniform stationary solution, after a time much longer than that required for the PC oscillations to stop. We may say that the non-uniform field profile with domains is "metastable" in this case [1.9]. For values of the photogeneration $\gamma_1 < \gamma < \gamma_2$, there are damped oscillations that end with an electric field profile corresponding to the true non-uniform stationary solution with domains. In all the cases, the numerical simulation shows that if the disturbance is large enough, a non-steady step-like electric field profile is formed after a short time. The domain wall of this profile oscillates back and forth in time about a fixed value of the position. As it oscillates, the domain wall changes its width, so that it spreads out over several wells, which makes easier the possibility of experimental detection. The electric field at the two domains in the profile also oscillates in time for a few periods before the oscillations stop (see Fig. 1.4 to visualize this process). Finally, outside the shaded rectangle of Fig. 1.1, the damping is so strong that there are no oscillations of the PC: the step-like electric field profile ($\gamma > \gamma_2$) or the uniform solution ($\gamma < \gamma_1$) are reached in a monotone fashion. These results are visualized in Fig. 1.5 in which the evolution of the current with time is represented for increasing values of γ in a range that covers the three situations just described.

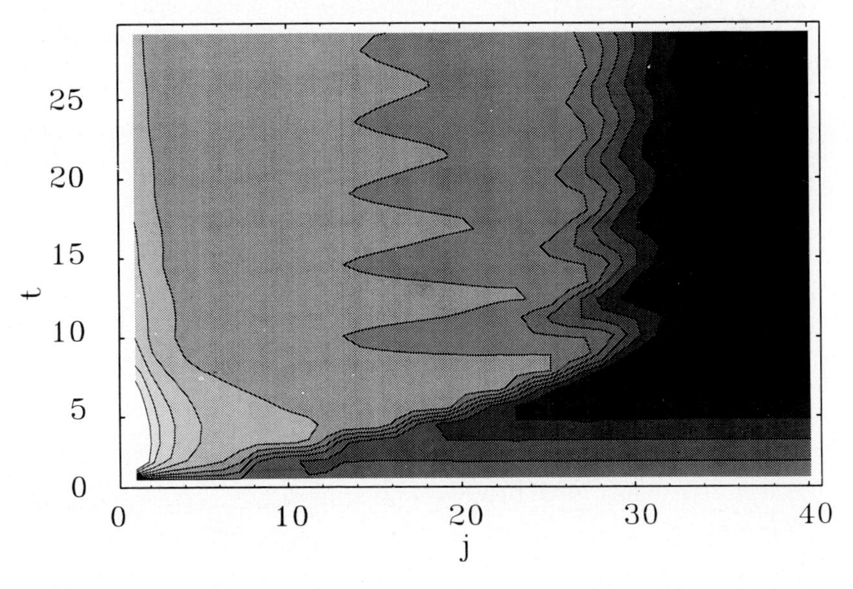

Fig. 1.4. Contour plot of the electric field (with time flowing in the y-axis and the QW number in the x-axis) showing the evolution of the domain wall in time for $\gamma = 2\times10^{-3}$, $\nu = 0$, $\beta = 0.05$ and $\phi = 1.2$. At $t = 0$ the electric field corresponds to the uniform solution except for a small perturbation at the first QW. At $t = 5$ the two domains are already defined (the darker the figure the higher the electric field) and the domain wall begins to oscillate. Most of the wells have an electric field centered around the first maximum of $v(E)$ (low field domain)

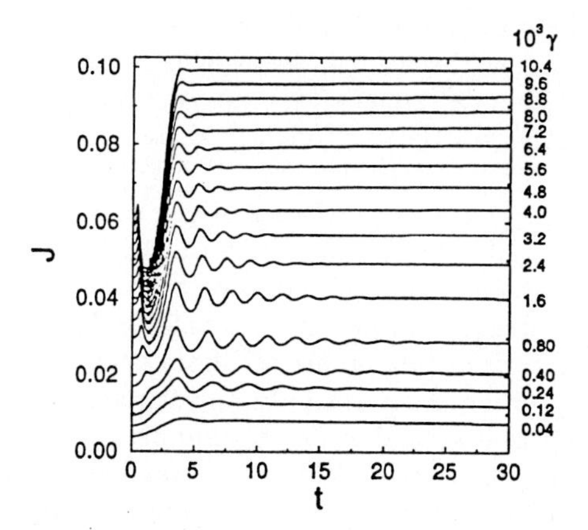

Fig. 1.5. Evolution of the PC, J, for different values of the photogeneration γ (in units of 10^{-3}), which increases from bottom to top of the figure, as indicated at the right margin. The voltage ($\phi = 1.1$) is the same for all the curves

These features of the solution of our model agree with experimental observations [1.3]. In Fig. 1.6 we have plotted the profiles of the electric field and of the electron and hole densities corresponding to a maximum and a minimum of the PC during one oscillation. Notice that the electronic charge oscillates from one side of the domain wall to the other during a period of the PC oscillation [1.3]. These results provide a clear picture of domain formation and posterior time evolution accounting for the time-dependent oscillations of the PC.

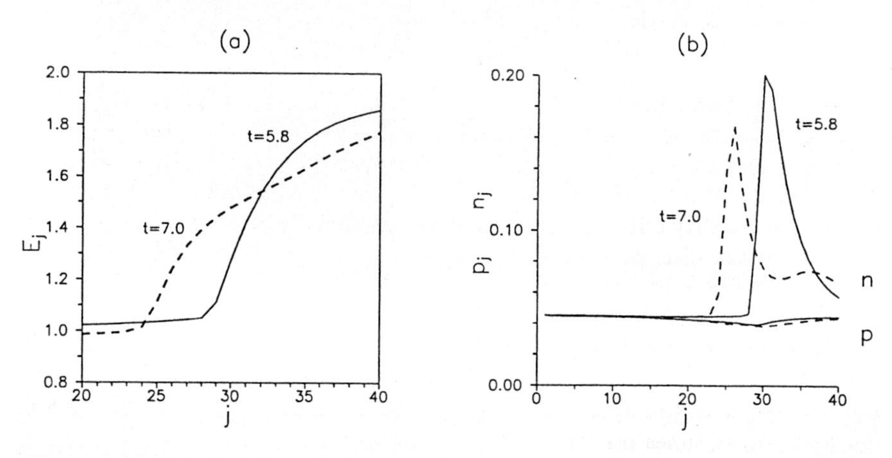

Fig. 1.6. (a) Electric field profile and (b) charge at each well corresponding to a maximum (full line) and a minimum (dashed line) of the PC. Parameter values are as in Fig. 1.4

1.3.2 Doped SL, $\nu > 0$, $\gamma \geq 0$, $0 < c \ll 1$

For a dark n-doped SL which is part of an n^+-n-n^+ diode [1.11], the relevant phase diagram is indicated in the $\nu - \phi$ plane of Fig. 1.7. The appropriate boundary condition is (1.13) with $0 < c \ll 1$. It is important to observe that no time-dependent oscillations of the current exist for $c \leq 0$ [1.14]. Non-uniform stationary solutions with electric field domains are stable only inside the rectangle in Fig. 1.7.

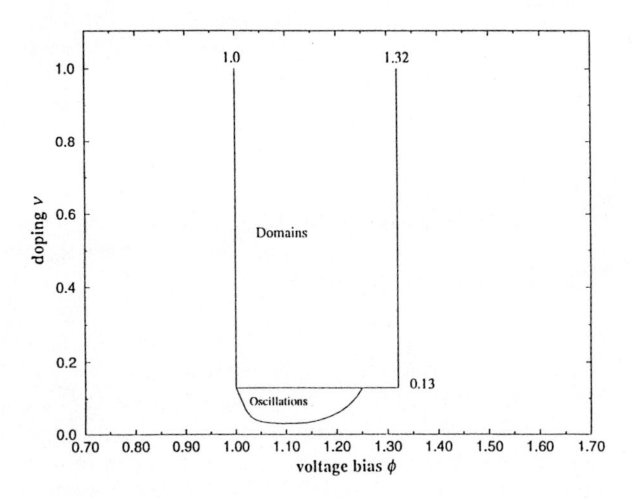

Fig. 1.7. Phase diagram of the doped model: Doping (ν) vs. applied voltage bias (ϕ) in units of, respectively, $\tilde{n}$ and $\tilde{E}_M N \tilde{l}$; see text. Notice the region where time-dependent undamped current oscillations due to domain wall dynamics exist. Outside this region and outside the rectangle, the uniform field solution is stable

While the upper limit of the region where time-dependent current oscillations exist is set by the existence of non-uniform stationary solutions with domains, numerical simulations show that the lower limit of this region is roughly given by $\nu N \simeq 1.0$, for $N > 10$. We conjecture that this relation corresponds to the minimal length stability criterion [1.15] for the continuous limit $N \to \infty$, $\nu \to 0$, $L = N\nu \gg 1$ of the discrete model, given by the system (1.25)-(1.28) of the next section. The values of the doping ν corresponding to the two samples of Kastrup et al.'s experiments [1.11] are $\nu_1 \simeq 0.1$ and $\nu_2 \simeq 3\nu_1$. Time-dependent current oscillations are observed only for the less doped sample [1.12], which agrees with our Fig. 1.7 and supports our theory. Figure 1.8 provides further information on the time-dependent oscillations: the electric field and charge profiles corresponding to different times marked in Fig. 1.8a are depicted in Figs. 1.8b and 1.8c, respectively. Notice that the latter figure shows that two domain walls coexist at the same time for particular subintervals of one period. This fact supports the idea that the time-dependent current oscillations are due to the formation, motion and annihilation of negatively charged domain walls (*monopole dynamics*), which emerges from the asymptotic analysis of the next section.

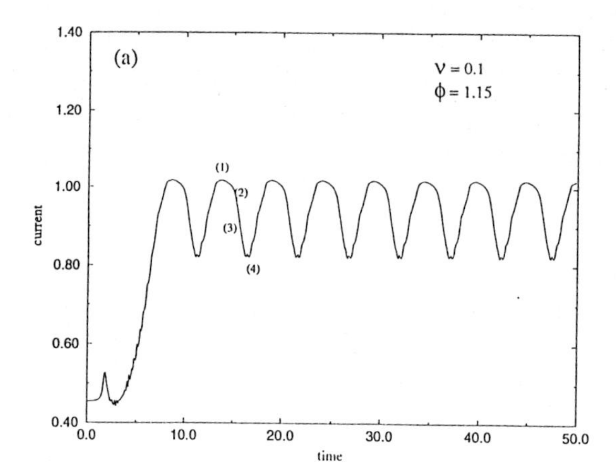

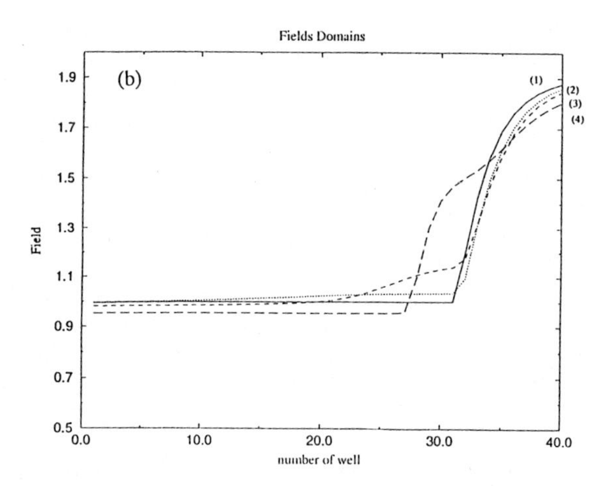

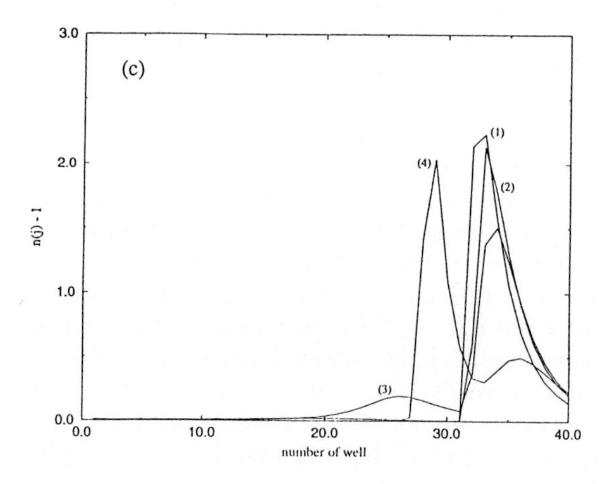

Fig. 1.8. Results obtained by numerically solving (1.20) - (1.23): (a) current vs. time τ, (b) electric field profile and (c) charge at each well for a dark doped SL with $\nu = 0.1$, $\phi = 1.15$, $N = 40$. The profiles in (b) and (c) correspond to the times numbered in (a)

When a doped SL operating in the mode of time-dependent current oscillations is illuminated, numerical simulations of the full model show that their amplitude decreases when the photoexcitation increases. For strong enough photoexcitation, the oscillations become damped and a crossover to the behavior observed for undoped SLs occurs [1.14]. This agrees with experiments [1.12].

1.4 Asymptotics

In this section we will give an asymptotic description of the time-dependent oscillations of the current in the limit $\nu \to 0$, $N \to \infty$ with $L = N\nu \gg 1$. It is convenient to start with the model for a doped SL without photoexcitation,

$$E_j - E_{j-1} = \nu\,(n_j - 1), \tag{1.20}$$

$$\frac{dE_j}{d\tau} + v(E_j)\,n_j = I, \tag{1.21}$$

$$\frac{1}{N}\sum_{j=1}^{N} E_j = \phi, \tag{1.22}$$

$$E_1 - E_0 = \nu c, \quad 0 < c \ll 1. \tag{1.23}$$

Notice that we have set $p_j = 0$ in (1.9): the hole densities approach zero anyway after a transient period because of (1.11) with $\gamma = 0$. We have redefined $n_j = \tilde{n}_j/\tilde{N}_D$ so that $\tau = \nu t/\beta$, $I = J/\nu$ in (1.9) and (1.10). In the continuum limit, $j \to \infty$, $\nu \to 0$ so that

$$x = j\nu \in [0, L], \; L = N\nu \gg 1, \tag{1.24}$$

and the system of equations (1.20) - (1.23) becomes

$$\frac{\partial E}{\partial x} - \frac{\nu}{2}\frac{\partial^2 E}{\partial x^2} = n - 1, \tag{1.25}$$

$$\frac{\partial E}{\partial \tau} + v(E)\,n = I, \tag{1.26}$$

$$\frac{1}{L}\int_0^L E(x,\tau)\,dx = \phi, \tag{1.27}$$

$$\frac{\partial E(0,\tau)}{\partial x} = c, \quad 0 < c \ll 1. \tag{1.28}$$

Up to $O(\nu^2)$ terms, these equations correspond to the classical drift-diffusion model for Gunn oscillations in n-GaAs with a small nonlinear diffusivity $\nu\, v(E)/2$, except that the boundary condition is not the usual one [1.15]. With this boundary condition the time-dependent oscillations of the current $I(\tau)$ are mediated by monopole wavefronts, not by dipole domains (solitary waves) as in the well-known Gunn oscillations. We now adapt our previous asymptotic analysis of monopole oscillations to the present case [1.16].

The keystone of the asymptotic analysis are the *charge monopole wavefronts* which are shock waves joining regions of uniform or almost uniform field profile (the domains!) moving with a speed given by an equal areas rule [1.16]. The periodic motion of these charge monopoles is responsible for the oscillations in $I(\tau)$. To get the monopole wavespeed we go back to the discrete model (1.20)-(1.23). Consider the situation in Fig. 1.9. The shock wave is a field profile that

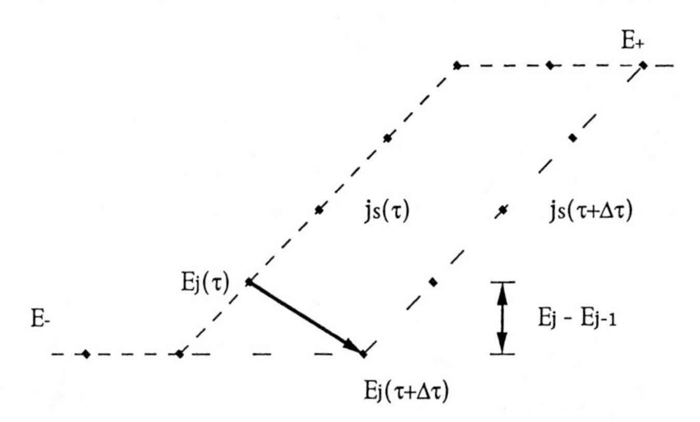

Fig. 1.9. Motion of a shock wave joining regions of uniform fields E_- and E_+

moves rigidly with an average speed $V(E_+,E_-)$, so that $E_j(\tau) = E(j - j_s(\tau))$ with $E(-\infty) = E_-$, $E(+\infty) = E_+$ and $j_s(\tau + \Delta\tau) - j_s(\tau) \sim V(E_+,E_-)\,\Delta\tau/\nu$, for large enough $\Delta\tau$. $j_s(\tau)$ is the QW where the profile $n_j(\tau)$ reaches its maximum value at time τ and $X(\tau) = \nu\, j_s(\tau)$ corresponds to the shock position in the continuum limit. Then $E_j(\tau + \Delta\tau) - E_j(\tau)$ is approximately given by the distance $j_s(\tau + \Delta\tau) - j_s(\tau)$ that the shock has advanced during $\Delta\tau$ times the field difference $-(E_j - E_{j-1})$ at some intermediate time in $(\tau, \tau + \Delta\tau)$. Thus we have

$$\frac{dE_j}{d\tau} \sim -\frac{E_j - E_{j-1}}{\nu}\, V(E_+,E_-). \tag{1.29}$$

If we now sum $(E_j - E_{j-1})$ over the shock region, we obtain $E_+ - E_-$. On the other hand $n_j = O(\nu^{-1}) \gg 1$ and $I = O(1)$ in the shock region, so that (1.20), (1.21) and (1.29) yield

$$\begin{aligned} E_+ - E_- = \sum_{j=-\infty}^{\infty} E_j - E_{j-1} &\sim -\sum_{j=-\infty}^{\infty} \frac{\nu}{v(E_j)}\,\frac{dE_j}{d\tau} \\ &\sim V(E_+,E_-) \sum_{j=-\infty}^{\infty} \frac{E_j - E_{j-1}}{v(E_j)}. \end{aligned} \tag{1.30}$$

In the continuum limit this becomes the equal areas rule

$$\int_{E_-}^{E_+} \left(\frac{1}{v(E)} - \frac{1}{V(E_+,E_-)} \right) dE = 0. \tag{1.31}$$

The same result can be easily derived from the continuum model (1.25) and (1.26) in the limit $\nu \to 0$. To keep $n_j \geq 0$ inside the shock we must add the restrictions $v(E_-) \geq V(E_+, E_-) \geq v(E_+)$ (and hence $V = v(E)$ at only one point x inside the shock). Likewise, no shock with $E_+ < E_-$ arises from realistic initial conditions with $n_j \geq 0$ [1.16]. Let us call $E^{(1)}(I) < E^{(2)}(I) < E^{(3)}(I)$ the fixed point solutions of $v(E) = I$ on the three branches of $v(E)$. There are two possible situations in which a shock joins two regions of uniform or almost uniform field profiles (electric field domains) forming a structure that moves rigidly, the charge monopole: (i) $1/I > 1/V$, and there is a region (the *monopole's tail*) where the field increases monotonically from $E^{(1)}(I)$ to $E_- = E^{(1)}(V)$, with $E_+ = E^{(2)}(I)$; and (ii) $1/I < 1/V$, $E_- = E^{(1)}(I)$ or $E^{(2)}(I)$, the field increases monotonically to the right of the shock from $E_+ = E^{(3)}(V)$ to $E^{(3)}(I)$, and the monopole has a tail to the right of the shock. See Fig. 1.10. The monopole velocity is then

$$U(E_+) = V(E_+, E^{(1)}(V)),$$
$$\text{if} \quad V > I \text{ (monopole with left tail);} \tag{1.32}$$
$$W(E_-) = V(E^{(3)}(V), E_-),$$
$$\text{if} \quad V < I \text{ (monopole with right tail).} \tag{1.33}$$

In either case this velocity depends only on the value of the field at the other side of the tail. The monopoles are archetypes of the field patterns in the semiconductor SL. As there may not be enough room in a finite SL for the monopole tail, these monopoles may never fully come into existence [1.16]. We will now see how to use them to interpret the results of the simulations for a given SL.

Suppose that the initial condition is such that after a short transient, a monopole with shock at $x = X(\tau)$ is moving towards $x = L$. This is the case for an uniform initial profile $E_j = \phi$. The following equations approximately govern the evolution of the monopole:

$$\frac{dX}{d\tau} = V(E_+, E_-), \tag{1.34}$$

$$\frac{dE_+}{d\tau} = I - v(E_+), \quad E_- = E^{(1)}(V), \tag{1.35}$$

$$\frac{X}{L} E^{(1)}(I) + \frac{L - X}{L} E_+ = \phi, \tag{1.36}$$

$$E_+(0) = \phi, \; I(0) = v(E_+(0)). \tag{1.37}$$

Initially all the variables are $O(1)$. The shock has to travel a distance $O(L)$, therefore the transit time is of the same order and $dE_+/d\tau = O(1/L) \ll 1$. Thus $E_+ = E^{(2)}(I)$, and we have soon a monopole with left tail moving with speed $U(E^{(2)}(I))$ given by (1.32) that depends only on the instantaneous value of I. As time grows, $E^{(2)}(I)$ increases and $E^{(1)}(I)$, I and $(L - X)/L$ decrease. After a long time, E_+ grows into the third branch of $v(E)$, and it reaches its maximum value such that $v(E_{+M}) = V(E_{+M}, E^{(1)}(I))$ at a time $\tau = \tau_M$. After this time, the field at the right of the shock can no longer be uniform and a quasistationary tail is formed ahead of the shock. We have now a monopole with a right tail moving with speed $W(E^{(1)}(I))$ given by (1.33) that depends only on

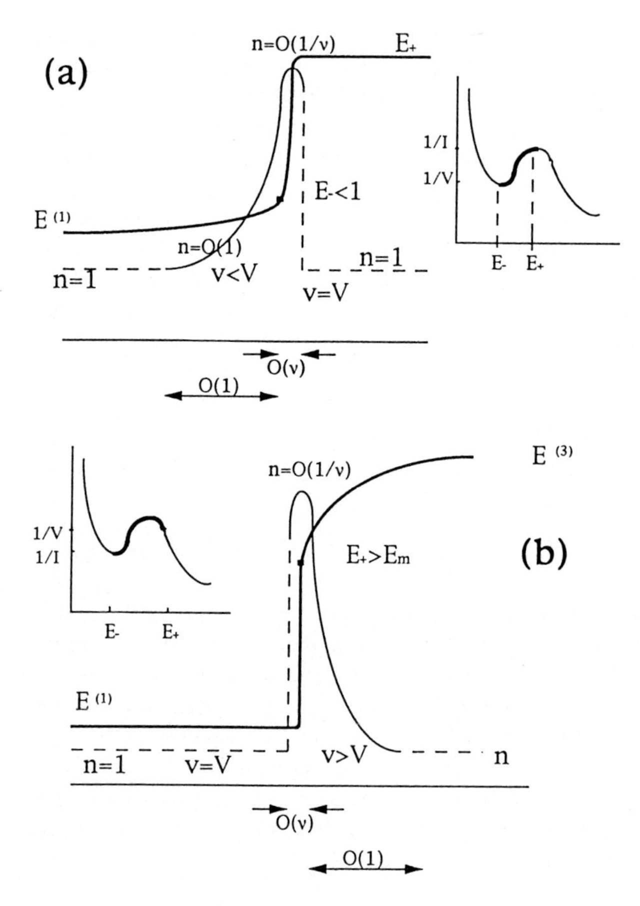

Fig. 1.10. Schematic representation of the electric field (continuous line) and the charge density (broken line) for a monopole wavefront. (a) Tail behind the shock. (b) Tail ahead of the shock

the instantaneous value of I. The equations at this stage are

$$\frac{dX}{d\tau} = V(E_+, E^{(1)}(I)), \tag{1.38}$$

$$v(E_+) = V(E_+, E^{(1)}(I)), \tag{1.39}$$

$$v(E(x;I))\left(\frac{\partial E(x;I)}{\partial x} + 1\right) = I, \qquad E(X;I) = E_+, \tag{1.40}$$

for $X < x < L$, and

$$\frac{X}{L}\, E^{(1)}(I) + \phi_R(\tau) = \phi, \tag{1.41}$$

$$\phi_R(\tau) \equiv \frac{1}{L}\int_X^L E(x;I)\, dx. \tag{1.42}$$

For $\tau > \tau_M$, I increases and a new wave [obeying (1.34) and (1.35) and $\phi - \phi_R(\tau)$ replacing ϕ in (1.36), where $\phi_R(\tau)$ is the average field at the tail region of the monopole near $x = L$] develops near $x = 0$ when $I \sim 1$. A necessary condition to have a time-periodic solution is that the new wave is created while the old one is still present at the sample, near $x = L$. Clearly this is the case if $\phi > 1$. The combination of the dying monopole (1.38) - (1.40) and the new monopole advancing from $x = 0$ and given by (1.34) and (1.35) yields an asymptotic description of the periodic oscillation of the current. Notice that the period is $O(L)$ in the τ time scale, which yields a frequency of $O(\nu/(\beta L)) = O((N\beta)^{-1})$ in the time scale t of (1.9) - (1.13). For $N = 40$ and $\beta = 0.05$, this estimation gives a number of the same order as the frequency in Fig. 1.8.

A more careful consideration of the situation where a new monopole is being born while the old one is dying near $x = L$ yields an upper value of ϕ beyond which no time-periodic monopole dynamics exists. It is easy to show that the disappearance of the old monopole takes place in times which are much smaller than $O(L)$, the transit time for the new monopole. This means that the new monopole does not have time to move away from $x = 0$ while the old one is disappearing and the field is still uniform for $x < X$, where $X(\tau)$ is the position of the shock forming part of the dying monopole. Let us call E to the uniform value of the field at the left of the dying monopole. Then $E = E_-$ for the dying monopole and $E = E_+$ for the newly created monopole. The condition $I = v(E)$ must hold, even when $E > 1$, for otherwise the condition (1.35) for the newly created monopole (with $E_+ = E$) leads to an explosive breakdown in a time of order unity. Under these conditions $\phi_R(\tau)$ given by (1.42) decreases with time both because of the motion of the shock and because the current decays as E grows into the second branch of the curve $v(E)$. We now derive an equation for $(L - X)$ as a function of E by the following procedure. First of all, let us take the time derivative of the balance of area (1.41) with E instead of $E^{(1)}(I)$:

$$\frac{E - E_+}{L}\frac{dX}{d\tau} + \frac{X}{L}\frac{dE}{d\tau} + \frac{1}{L}\frac{dI}{d\tau}\int_X^L \frac{\partial E(x;I)}{\partial I}dx = 0. \tag{1.43}$$

We now eliminate $dI/d\tau = v'(E)\, dE/d\tau$ which follows from the identity $I = v(E)$, and obtain a linear autonomous equation for $X(\tau)$ [1.16]

$$\frac{dX}{d\tau} - \frac{X}{E_+ - E}\frac{dE}{d\tau} = \frac{v'(E)}{E_+ - E}\frac{dE}{d\tau}\int_X^L \frac{\partial E(x;I)}{\partial I}dx. \tag{1.44}$$

Finally, we consider $X = X(E)$, $E_+ = E_+(E)$ (found by solving $v(E_+) = V(E_+, E)$), and change the variable of integration in (1.44) to E by using (1.40). The result is the following equation for $Y \equiv (L - X)/L$ as a function of E:

$$[E_+(E) - E]\frac{dY}{dE} - Y = -1 + \frac{v(E_+(E))^2\,[E(L;I) - E_+(E)]}{v(E)\,v'(E_+(E))\,[E_+(E) - E]\,L}$$
$$- \frac{v'(E)}{L}\int_{E_+(E)}^{E(L;I)} ds \int_{E_+(E)}^{s} \frac{v(r)\,dr}{[v(E) - v(r)]^2}. \tag{1.45}$$

Equation (1.45) should be solved with an initial condition $Y(1) = Y_0$. In it $I = v(E)$. The condition that $Y(E)$ should be a decreasing function of E determines a maximum value of Y_0 compatible with periodic monopole dynamics and this in turn determines the maximum bias [1.16]. Notice that the two last terms in (1.45) are $O(1/L)$ (and if we ignore them, $dY/dE < 0$ because $Y < 1$ and $E_+ > E$) unless the integral in the last term becomes very large, which would require $E(L;I) \to E^{(3)}(I)$. This could occur only if $(L - X)$ is large enough, which, in turn, necessitates values of ϕ close to $E^{(3)}(1)$. Thus it seems that for our velocity curve the monopole branch should exist for almost the entire available bias interval $(1, E^{(3)}(1))$, which is both confirmed by numerical simulations and by a linear stability analysis of the steady states. This large bias interval contrasts with that reported in [1.16] for which the shape of $v(E)$ allowed large values of $E^{(3)}(I)$. This made $O(1)$ the last term in (1.45), thereby rendering possible to achieve a maximum bias on the second branch of $v(E)$.

When the laser illumination is turned on, holes are generated and we have the following three equations replacing (1.20) and (1.21):

$$E_j - E_{j-1} = \nu\,(n_j - p_j - 1), \tag{1.46}$$

$$\frac{dE_j}{d\tau} + v(E_j)\,n_j = I, \tag{1.47}$$

$$\frac{dp_j}{d\tau} = \beta\,(\Gamma - n_j\,p_j), \tag{1.48}$$

to be solved together with (1.22), (1.23) and appropriate initial conditions. Here $\Gamma = \gamma/\nu^2$. The numerical simulations indicate that there is damping and eventual suppression of the current oscillations described above for large enough photoexcitation Γ. This may be qualitatively understood by going back to our derivation of the equal areas rule (1.31). Recall that within the shock, n and $dE/d\tau$ are $O(1/\nu)$. Then (1.48) indicates that $p \to 0$ as the time goes on, unless $\Gamma = O(\nu^{-2})$ [$\gamma = O(1)$ in the previous dimensionless units] for $p = O(\nu^{-1}) \gg 1$. If $\Gamma = O(\nu^{-2})$, we should retain p_j besides n_j when approximating Poisson's law (1.46). Then it becomes possible to get a *stationary* shock with $V = 0$ and $I = O(\nu^{-1})$ [or $J = O(1)$], which corresponds to the non-uniform stationary solutions with domains of Sect. 1.2. Details of the crossover between the undamped current oscillations for $\Gamma = 0$ and the non-uniform stationary states in the continuum limit will be reported elsewhere.

1.5 Final Remarks

We have analyzed a very simple model that satisfactorily explains experimental observations of domain formation (field profiles with domain II/domain III coexistence) and oscillations of the PC in undoped SL [1.3], and also in doped SL with or without photoexcitation [1.12]. Complementary results on formation of stationary field profiles with domain I/domain II coexistence in doped SL were obtained by Prengel et al. [1.8]. They used a more detailed model where electrons belonging to different levels of the QWs were distinguished, and the resonant tunneling current was calculated by perturbation theory in the level splitting due to a small applied electric potential [1.8]. No time-dependent oscillations of the PC were reported in [1.8] nor in the comparison with experiments [1.11]. While the time-dependent oscillations for low enough doping and laser power are undamped [1.14], for large enough photoexcitation a crossover to damped oscillations characteristic of the undoped SL is observed, which agrees with experimental observations by the Grahn group [1.12].

Among interesting open problems related to our work, we outline that of deriving our model from microscopic ones (see [1.8] for a perturbative derivation of a related model for doped SL) and that of the continuum limit of (1.9) - (1.13). In fact, our model resembles known drift-diffusion models in the continuum limit (such as the one in [1.17]) for which the uniqueness of the steady state can be proved easily, at least in the limit of small diffusion coefficient [1.18] (these continuum drift-diffusion models are also known to have Gunn effect oscillations [1.15,16] among their possible solutions [1.19]). As we have explained above, the time-periodic solutions of (1.25) - (1.28) (the continuum limit for dark doped SL) are due to monopole (domain wall) dynamics. They are different from the well-known Gunn effect oscillations due to dipole dynamics [1.15, 16]. As usual, *the boundary condition selects the type of wave responsible for the current oscillations*: if the boundary condition leads to charge accumulation near $x = 0$, we have monopole dynamics whereas dipole dynamics is to be expected when there is charge depletion near $x = 0$ [1.16]. In conclusion, it seems that the discreteness of our model is a crucial ingredient in reproducing both important static and dynamical features of transport in a SL made out of weakly coupled QWs. Many qualitative features of the observed phenomena in finite SL may be understood by analyzing the continuum limit of a long SL [1.14].

I thank the organizers of the minisymposium on Semiconductors within ICPF94 for their invitation. I thank O. M. Bulashenko, J. A. Cuesta, J. Galán, M. Kindelán, S.-H. Kwok, F. C. Martínez, R. Merlin, J. M. Molera, M. Moscoso, G. Platero and S. W. Teitsworth for fruitful discussions and collaboration on related topics and H. T. Grahn and J. Kastrup for showing me their laboratory and experimental results on doped superlattices. This work has been supported by the DGICYT grant PB92-0248, and by the EC Human Capital and Mobility Programme contract ERBCHRXCT930413.

References

[1.1] L. Esaki, L.L. Chang: Phys. Rev. Lett. **33**, 495 (1974)
[1.2] H.T. Grahn, H. Schneider, K. von Klitzing: Phys. Rev. B **41**, 2890 (1990)
[1.3] S.-H. Kwok: Novel Electric Field Effects in GaAs-(Al,Ga)As Superlattices. Ph.D. Thesis, University of Michigan, Ann Arbor (1994); see also S.-H. Kwok, T.C. Norris, L.L. Bonilla, J. Galán, J.A. Cuesta, F.C. Martínez, J.M. Molera, H.T. Grahn, K. Ploog, R. Merlin: unpublished (1994)
[1.4] See for instance the special issue of Physics Today, June (1993)
[1.5] R.F. Kazarinov, R.A. Suris: Sov. Phys.–Semicond. **6**, 120 (1972)
[1.6] B. Laikhtman: Phys. Rev. B **44**, 11260 (1991)
[1.7] B. Laikhtman, D. Miller: Phys. Rev. B **48**, 5395 (1993) and preprint (1994)

[1.8] F. Prengel, A. Wacker, E. Schöll: Phys. Rev. B **50**, 1705 (1994)
[1.9] L.L. Bonilla, J. Galán, J. Cuesta, F.C. Martínez, J.M. Molera: Phys. Rev. B **50**, 8644 (1994)
[1.10] H.T. Grahn, W. Müller, K. von Klitzing, K. Ploog: Surface Sci. **267**, 579 (1992)
[1.11] J. Kastrup, H.T. Grahn, K. Ploog, F. Prengel, A. Wacker, E. Schöll: Appl. Phys. Lett. **65**, 1808 (1994)
[1.12] H.T. Grahn: private communication (June 1994)
[1.13] This boundary condition generalizes that used in [1.9] and in [1.11], $\tilde{c} = 0$. A small charge build-up on the first QW due to the excess doping before the SL is allowed. Then there is a stationary solution with an almost uniform field profile that differs from (1.7) only on the first few QWs.
[1.14] L.L. Bonilla, J. Galán, M. Kindelán, M. Moscoso: unpublished (1994)
[1.15] M.P. Shaw, H.L. Grubin, P.R. Solomon: *The Gunn-Hilsum Effect* (Academic Press, New York 1979); M.P. Shaw, V.V. Mitin, E. Schöll, H.L. Grubin: *The Physics of Instabilities in Solid State Electron Devices* (Plenum Press, New York 1992)
[1.16] F.J. Higuera, L.L. Bonilla: Physica D **57**, 161 (1992)
[1.17] S.W. Teitsworth: Appl. Phys. A **48**, 127 (1989); L.L. Bonilla, S.W. Teitsworth: Physica D **50**, 545 (1991)
[1.18] L.L. Bonilla: Phys. Rev. B **45**, 11642 (1992)
[1.19] I.R. Cantalapiedra, L.L. Bonilla, M.J. Bergmann, S.W. Teitsworth: Phys. Rev. B **48**, 12278 (1993); L.L. Bonilla, I.R. Cantalapiedra, M.J. Bergmann, S.W. Teitsworth: Semicond. Sci. Technol. **9**, 599 (1994)

2 Oscillatory Transport Instabilities and Complex Spatio-Temporal Dynamics in Semiconductors

E. Schöll and A. Wacker

Institut für Theoretische Physik, Technische Universität Berlin,
Hardenbergstr. 36, 10623 Berlin, Germany

Abstract. Instabilities associated with nonlinear electrical transport of hot electrons are widespread in semiconductor devices driven far from thermodynamic equilibrium. They often involve switching behaviour, self-generated current or voltage oscillations, current filamentation, field domain formation and solid-state turbulence. In this paper the theory of such instabilities is reviewed with a special emphasis on recent progress in the description of current filaments and field domains. The occurring instabilities can be considered as nonequilibrium phase transitions. The theoretical tools of nonlinear dynamic systems are applied to describe the emergence of self-organized dynamic spatio-temporal structures: The nucleation and growth of current filaments, their interaction via global couplings through the external circuit, periodic and chaotic oscillatory instabilities in the form of breathing or spiking filaments, filaments travelling laterally in a regular or intermittent way, and the formation of static or oscillating domains giving rise to multistable current-voltage characteristics. The following specific model systems are treated in detail: (i) The dynamic Hall instability in crossed electric and magnetic fields in the regime of low temperature impurity impact ionization. (ii) Vertical electrical transport in layered semiconductor structures like the heterostructure hot electron diode and in superlattices.

2.1 Introduction

Semiconductors represent dissipative nonlinear dynamic systems driven far from thermodynamic equilibrium by application of electric or magnetic fields, or optical excitation. Under such conditions strong nonlinearities are inherent in the charge transport processes and the generation-recombination kinetics of the carriers [2.1]. Since semiconductors are spatially extended systems, the nonlinear coupling of their spatio-temporal degrees of freedom provides a great wealth of complex dynamic phenomena of self-organized pattern formation.

A particular feature, which distinguishes semiconductors from other self-organized nonequilibrium systems like fluids, lasers, or chemical reaction systems [2.2, 3], is the ubiquitous presence of space charges. This, according to Poisson's equation, induces a nonlinear feedback between the charge carrier distribution and the electric potential distribution governing the transport. This mutual interdependence is particularly pronounced in case of semiconductor heterostructures and low-dimensional structures where abrupt junctions between

Springer Proceedings in Physics, Vol. 79 **Nonlinear Dynamics and Pattern Formation in Semiconductors and Devices**
Editor: F.-J. Niedernostheide

different materials on an atomic length scale cause conduction band discontinuities resulting in potential barriers and wells. The local charge accumulation in these potential wells, together with nonlinear transport processes across the barriers have recently been found to provide a number of important mechanisms for nonlinear dynamics and pattern formation in semiconductors [2.4].

Although the microscopic mechanism inducing such behaviour may be very different, the observed macroscopic nonlinear phenomena like negative differential conductivity (NDC), periodic and chaotic current oscillations [2.5,6], and the self-organized formation of spatial patterns such as current filaments and field domains is often quite similar, and reflects universal features of nonlinear dynamics. Depending upon the external circuit to which the semiconductor is connected, in particular the slope of the load line [2.7], the spatially homogeneous states of negative differential conductivity are often unstable against spatio-temporal fluctuations. These may give rise to uniform oscillations, or spatially inhomogeneous distributions of the current density and the electric field.

The most prominent examples of spatial patterns are current filaments, where the current becomes inhomogeneous over the cross-section of the sample with a high-current channel embedded in a low-current state, and field domains, where the electric field distribution breaks up along the direction of the current flow into coexisting regions of high and low field. Stable filaments and domains have been well known as an object of experimental and theoretical investigations for several decades, and they have mostly been associated with S- and N-shaped negative differential conductivity (SNDC and NNDC), respectively [2.8].

It is only recently that research has focussed upon the secondary bifurcations of these elementary spatial structures. Coherent spatio-temporal elementary structures [2.9] like breathing [2.10,11] or travelling or rocking filaments [2.12,13], spatio-temporal spiking [2.14–16], or oscillating domains [2.17], have been identified experimentally and verified theoretically from computer simulations.

In this review we shall present a theoretical description of such complex spatio-temporal dynamics in semiconductor transport. We treat two specific model systems in detail: First, current filamentation in the SNDC regime of low-temperature impurity breakdown is studied. In crossed electric and magnetic fields a dynamic Hall instability occurs [2.18]. Dielectric relaxation of both the applied electric field and the induced Hall field combined with the generation-recombination kinetics, in particular impact ionization of impurities, causes current filaments to travel laterally in a regular or intermittent way [2.19]. Second, vertical electrical transport in layered semiconductor structures is analysed. In the simplest system, thermionic emission and tunnelling across a single heterojunction leads to bistability and SNDC in the Heterostructure Hot-Electron Diode (HHED). Homogeneous relaxation oscillations [2.20], stationary current filaments and periodic or chaotic spatio-temporal spiking is found in different regimes of the parameter space [2.21]. In a superlattice structure, nonresonant and resonant tunnelling across the barriers gives rise to NNDC and field domain formation. Complicated current-voltage characteristics with multiple branches and oscillatory instabilities of the domains are generated.

There are many other recent examples of nonlinear dynamics and pattern formation in semiconductors. Parallel transport in modulation-doped GaAs/AlGaAs heterostructures, for instance, is associated with real space transfer of electrons from the high-mobility undoped GaAs layer into the low-mobility, highly doped AlGaAs layer with increasing field. This results in NNDC [2.22, 23], similarly to the k-space transfer in the Gunn effect. Both spatially homogeneous current oscillations [2.24,25] and moving domains [2.26] occur. Theoretical analyses have established that the oscillations are relatively robust even against weak disorder introduced by fluctuations of the well and the barrier widths, the doping concentration and the alloy fraction [2.27]. Another mechanism for current oscillations at parallel transport in heterostructures is caused by the injection of electrons from the contact into the barrier [2.28].

In the field of optical phenomena in semiconductors, self-organized pattern formation is abundant [2.29, 30]. Examples of particular recent interest include spatio-temporal complexity in the dynamic output of twin-stripe [2.31, 32] and broad-area [2.33] semiconductor lasers. In dependence on the injection current and the stripe separation simulations yield complex dynamical coupling, and chaotically blinking optical filaments. In nonlinear passive optical systems like CdS crystals, optical switching fronts have been predicted theoretically and verified experimentally [2.34]. They traverse the crystal upon excitation with a laser pulse, switching it from a state of low transmissivity to a state of high transmissivity.

A number of recent reviews and monographs exist, focussing on experimental and theoretical aspects of nonlinear dynamics and self-organization in semiconductor transport [2.1, 35–42]. They give an impression of the state of art in the field of pattern formation in semiconductors, which has matured over the past decade. In the present paper we shall not attempt to give a comprehensive overview, but rather concentrate on some specific recent theoretical results which we have obtained, and which are of particular current interest.

2.2 Current Filaments in Crossed Electric and Magnetic Fields

As a first example we consider doped bulk semiconductors at liquid helium temperatures. In particular, high purity semiconductors under the simultaneous application of parallel or crossed electric and magnetic fields have been observed to exhibit self-generated chaotic current or voltage oscillations under dc bias in the regime of impurity impact ionization breakdown [2.43–49]. Such oscillatory instabilites are often connected with SNDC, which gives rise to current filamentation in homogeneously as well as in inhomogeneously [2.50,51] doped material. For perpendicular electric and magnetic fields, the occurring filaments show either asymmetric breathing states [2.48], or travel transversally in the direction of the Lorentz force across the sample [2.52]. While for positive differential conductivity a mechanism for spatially homogeneous magnetic field induced oscillations in terms of a dynamic Hall effect was proposed recently [2.18,53], an extension of this model to filamentary conduction is necessary in order to understand these complex spatio-temporal instabilites. Such a model, which is able to account for the spatio-temporal degrees of freedom of unipolar nonlinear filamentary conduction and instabilities in crossed electric and magnetic fields, was presented in [2.19]. It extends previous work on current filaments [2.1, 54] which was confined to the case without magnetic field. We neglect quantum effects leading to the splitting of the conduction band into discrete Landau levels of spacing $\Delta E = \hbar\omega_c$, where $\omega_c = e\mathcal{B}/m^*$ with magnetic induction $\mathcal{B}$, elementary charge $e > 0$, and effective mass m^*. In the impurity breakdown regime the mean energy E per carrier is comparable to the ionization energy E_{th} of the impurities. A classical treatment will thus be applicable in the magnetic field regime where $\Delta E \ll E_{th}$ holds.

The conduction current density can generally be calculated from the first moment of the Boltzmann equation [2.53]. However, in the regime we are interested in, a constant momentum relaxation time τ_m may be assumed. In the presence of a magnetic field the current density $\mathbf{j} = qn\mathbf{v}$ is then given by [2.18]

$$\mathbf{j} = en\mu_{\mathcal{B}}\frac{\mathbf{F}}{q} - qn\mu\mu_{\mathcal{B}}(\mathcal{B} \times \frac{\mathbf{F}}{q}) + n\mu^2\mu_{\mathcal{B}}\mathcal{B}(\mathcal{B}\mathbf{F}) \quad , \tag{2.1}$$

where n is the carrier density, $\mathbf{v}$ is the mean group velocity of the carriers, $q = \pm e$ is their charge, $\mu = \tau_m e/m^*$ is the mobility for $\mathcal{B} = 0$, $\mu_{\mathcal{B}} = \mu/(1 + \mu^2\mathcal{B}^2)$, $\mathbf{F} = q\mathcal{E} - eDn^{-1}\mu_{\mathcal{B}}^{-1}\nabla_{\mathbf{r}}n$, and D is the diffusion constant.

As the relevant dynamic variables we choose, besides the electric field $\mathcal{E}$ and n, the densities of carriers bound at two shallow impurity levels n_1 and n_2 corresponding to ground and excited states, respectively, of donors or acceptors. Transitions between these levels and the band states are possible due to generation-recombination (GR) processes including impact ionization. It is important that both the applied drift field $\mathcal{E}_x$ and the induced Hall field $\mathcal{E}_z$ are treated dynamically reflecting their finite dielectric relaxation time [2.18], while $\mathcal{B}$ (applied in y-direction) is treated as a control parameter. As the dominant spatial inhomogeneity in the SNDC regime occurs perpendicular to the drift field in the form of current filaments [2.1, 38], we assume spatial inhomogeneity only in the z-direction. The dynamic equations for a sufficiently long sample with ideal planar contacts connected in series with a voltage source U_0 and a load resistor R_L and in parallel with a capacitance C are the continuity equations

$$\frac{\partial n}{\partial t} + \frac{\partial(nv_z)}{\partial z} = \phi(n, \underline{n}_t, \mathcal{E}) \quad , \quad \frac{\partial \underline{n}_t}{\partial t} = \underline{\phi}_t(n, \underline{n}_t, \mathcal{E}) \quad , \tag{2.2}$$

and the dielectric relaxation equations

$$c_r\frac{\partial \mathcal{E}_x}{\partial t} = j_0 - (\mu_{\mathcal{B}}\langle n\rangle + \sigma_l)\mathcal{E}_x - \mu_{\mathcal{B}}\mu\mathcal{B}\left[\frac{\Delta n}{W} - \langle n\mathcal{E}_z\rangle\right] \quad , \tag{2.3}$$

$$\frac{\partial \mathcal{E}_z}{\partial t} = -nv_z \quad , \tag{2.4}$$

supplemented by Maxwell's equations

$$\frac{\partial \mathcal{E}_x}{\partial z} = 0 \quad , \quad \frac{\partial \mathcal{E}_z}{\partial z} = \rho \equiv n + \sum_i^M n_i - 1 \quad , \tag{2.5}$$

where $c_r = 1 + LC/(\epsilon A)$, $A = bW$ is the lateral cross section of the sample, L the length in x-direction, W the width in z-direction and b the thickness in y-direction, ϵ the permittivity, $j_0 = U_0/(R_L A e \mu_l N_A^* \mathcal{E}_0)$, $\sigma_l = \tau_M L/(R_L A \epsilon)$, $v_z = \mu_B(\mathcal{E}_z - (1/n)\partial n/\partial z + \mu \mathcal{B} \mathcal{E}_x)$ is the mean velocity in z-direction, ρ is the local charge density, ϕ, $\underline{\phi}_t$ are the GR rates (We have introduced 2-dimensional vector fields $\underline{n}_t = (n_1, n_2)$ and $\underline{\phi}_t = (\phi_1, \phi_2)$). The brackets $< >$ denote the spatial mean value $\int_0^W dz/W$, and $\Delta n \equiv n(W) - n(0)$. Note that these equations hold for holes ($q = e$), but can easily be adapted to electrons by inverting the sign of $\mathcal{E}_z$. All quantities are given in dimensionless units, i.e., $\mu, t, z, \mathcal{E}$, and $n, \underline{n}_t$ are scaled by the low-field mobility μ_l, the dielectric relaxation time $\tau_M = \epsilon/(e\mu_l N_A^*)$, the Debye length $L_D = (k_B T_L \epsilon/(e^2 N_A^*))^{1/2}$, the thermal field $\mathcal{E}_0 = k_B T_L/(e L_D)$ and the effective acceptor concentration $N_A^* = N_A - N_D$, respectively. The classical static Hall effect is reproduced by the homogeneous steady states (denoted by an asterisk *): $\mathcal{E}_z^* = -\mu \mathcal{B} \mathcal{E}_x^*$, $j_0 = n^* \mu \mathcal{E}_x^*$.

We now consider the case that the static $j_0(\mathcal{E}_x^*)$ characteristic exhibits SNDC. The stability of the homogeneous stationary states $\Phi^* = (n^*, \underline{n}_t^*, \mathcal{E}_x^*, \mathcal{E}_z^*)$ with respect to fluctuations $\delta\Phi(z,t) \propto e^{ikz} e^{\lambda t}$ can be investigated by linearization of (2.2) - (2.5). Due to (2.5) only homogeneous modes ($k = 0$) allow for $\delta \mathcal{E}_x \neq 0$. This motivates a separate treatment for homogeneous and inhomogeneous modes.

The homogeneous modes yield oscillatory instabilites previously found for the dynamic Hall effect [2.18, 53]. For the inhomogeneous case we obtain a new spatio-temporal instability: $\lambda(k) = ivk - \tilde{D}k^2$ with

$$v = (\mu \mu_B)^{\frac{1}{2}} \mathcal{B} \frac{(\partial \rho/\partial \mathcal{E}) \left[(\partial \phi_1/\partial n_1)(\partial \phi_2/\partial n_2) - (\partial \phi_2/\partial n_1)(\partial \phi_1/\partial n_2)\right]}{n^* \left[-(\partial \phi_1/\partial n_1) - (\partial \phi_2/\partial n_2) - (\partial \phi/\partial n)\right]} \tag{2.6}$$

and an effective diffusion coefficient $\tilde{D}$. It can be shown under general conditions, that v has opposite sign as $\mathcal{B}$. Thus the fluctuation moves *transversally in the direction of the Lorentz force.* The velocity is proportional to $\mathcal{B}$ for small $\mathcal{B}$ in agreement with experiment [2.52], until μ_B differs significantly from μ.

To get a physical idea of this remarkable transverse motion of the fluctuations, we consider a carrier density fluctuation around the unstable homogeneous state with and without a magnetic field. This fluctuation will grow in both cases due to the GR-instability. Thus a fluctuation of the transverse electric field will be built up according to (2.4), so that the drift and diffusion currents will tend to cancel each other: $n\,\delta\mathcal{E}_z = \partial\,\delta n/\partial z$. In the case $\mathcal{B} = 0$ this will lead to a *symmetric* change of the absolute value of the electric field, since $\mathcal{E}_z^* = 0$. This local increase of the electric field results in a symmetric increase of the fluctuation due to the GR-processes. In the case $\mathcal{B} \neq 0$ the fluctuation $\delta\mathcal{E}_z$ will *reduce* the absolute value of the electric field on one side, while it will *increase* it on the opposite site. Due to the GR-processes the side of the fluctuation with lower electric field will decrease, the opposite side will increase. As a result, the fluctuation will move in the direction of the Lorentz force. Note that in this picture only the shape of the fluctuation moves transversally while the carriers themselves remain at their transverse positions, performing free to bound transitions and vice versa. This corrects the former view of travelling filaments being due to individual carriers moving in the direction of the Lorentz force [2.46]. Rather, the situation is analogous to the motion of water molecules in a water wave.

We have simulated the system numerically for p-Ge at 4 K with $c_r = 1$, $\sigma_l = 0$ (current control) in the impurity breakdown regime with GR-rates $\phi = X_1^S n_2 - T_1^S n(1 + c - n_1 - n_2) + X_1 n n_1 + X_1^* n n_2$, $\phi_1 = T^* n_2 - X^* n_1 - X_1 n n_1$, $\phi_2 = -\phi - \phi_1$, where $c = N_D/N_A^*$ is the compensation. The GR coefficients have been obtained from a spatially homogeneous Monte Carlo simulation with $\mathcal{B} = 0$ [2.55] and scaling the electric field $\mathcal{E}$ by a factor $(1 + \mu^2\mathcal{B}^2)^{-1/2}$. This is possible because we have assumed constant mobility. The resulting $j_0(\mathcal{E}_x^*)$ characteristic shows SNDC (Fig. 2.1). The Monte Carlo data have been fitted

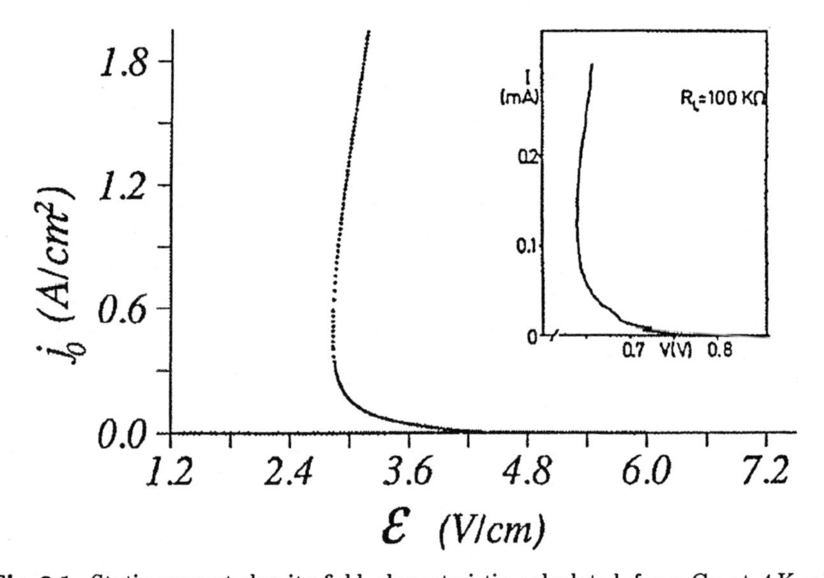

Fig. 2.1. Static current density-field characteristic calculated for p-Ge at 4 K with $N_A = 10^{14}$ cm^{-3}, $N_D = 5 \cdot 10^{12}$ cm^{-3}, $\epsilon = 16\epsilon_0$, $\mu = \mu_l = 10^5$ cm^2/(Vs), $\tau_M = 10^{-12}$ s, $L_D = 5.6 \cdot 10^{-6}$ cm, $\mathcal{E}_0 = 60.8$ V/cm, $X^* = 10^{-15}$, $T^* = 7.21 \cdot 10^{-5}$, $X_1^S = 1.4 \cdot 10^{-6}$, $\alpha = 60.8(1 + \mu^2\mathcal{B}^2)^{-1/2}$, $X_1(\mathcal{E}) = 7.85 \cdot 10^{-4} \exp\left[-11.3(\alpha\mathcal{E})^{-0.745}\right]$, $X_1^*(\mathcal{E}) = 4.18 \cdot 10^{-2} \exp\left[-3.72(\alpha\mathcal{E})^{-0.66}\right]$, and $T_1^S(\mathcal{E}) = -1.2 \cdot 10^{-3} \exp\left[-0.2(-0.254 + \alpha\mathcal{E})^2\right] + 1.73 \cdot 10^{-3}(0.421 + \alpha\mathcal{E})^{-0.887}$.
The inset shows a current-voltage characteristic measured in p-Ge at 4.18 K with a 100 kΩ load resistance [2.11]

by smooth functions and substituted into the differential equations (2.2)-(2.5) which have been solved with the aid of the method of particles [2.56] using cyclic transverse boundary conditions. Our nonlinear spatio-temporal analysis predicts the formation of current filaments on the NDC-branch of the current density-field characteristic. These filaments move in the direction of the Lorentz force (Fig. 2.2). After some transients the filament travels with a constant velocity v. Fig. 2.3 shows v as a function of the applied magnetic field. Additionally, the velocity obtained by the linear stability analysis is plotted. The latter is larger

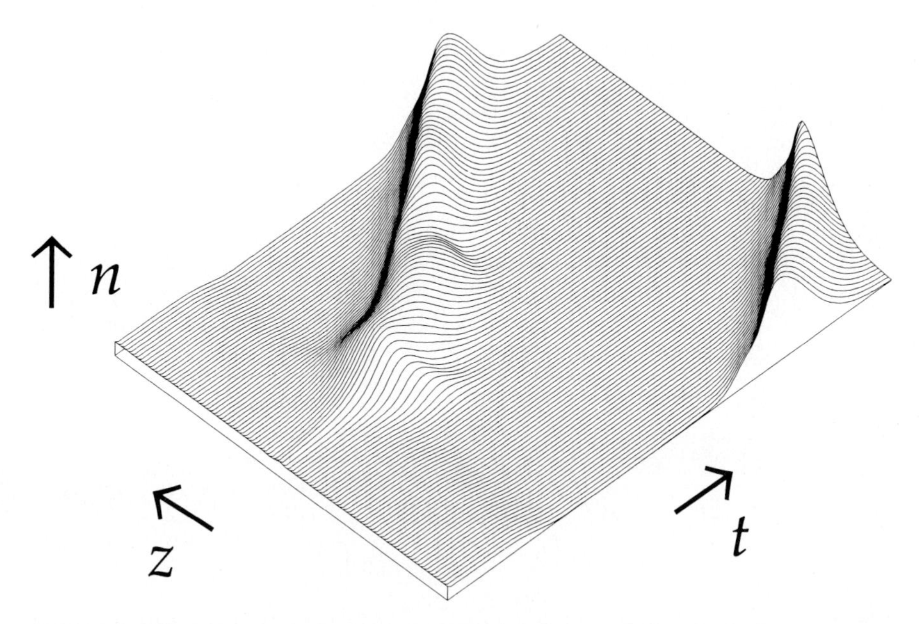

Fig. 2.2. Spatio-temporal dynamics of the carrier density $n(z,t)$ in the presence of an externally applied magnetic field under cyclic boundary conditions for $j_0 = 34$ mA/cm^2, $\mathcal{B} = 10$ mT with planar contacts. The total time is $8.1 \cdot 10^{-6}$ s, the sample width is $W = 90$ μm

because it describes only *small* fluctuations from the *NDC state*. Note that the velocity ($\approx 10^2 - 10^3$ cm/s) is much smaller than the drift velocity of the carriers ($\approx 10^5 - 10^6$ cm/s), in agreement with experiment [2.52].

For a finite sample with Dirichlet boundary conditions [2.54] $\mathcal{E}_z(0,t) = \mathcal{E}_z(W,t) = 0$ the filament travels to the boundary and is pinned there. This behaviour changes if other contact geometries are used. We demonstrate this for triangular contacts (Fig. 2.4) which may also serve as a model for *non-ideal* planar Ohmic contacts [2.52]. In this case the length of the sample $L = L_0 + 2L' \left|z - W/2\right| / W$ is a function of the transverse coordinate. Due to our assumption of homogeneity in the x-direction we restrict ourselves to small contact angles, i. e., L'/W should be small against logarithmic derivatives like $(\partial \ln n / \partial z)$. Qualitatively, these contacts yield the following scenario: The free carriers inside the moving filament will be spread over an increasing effective sample length $L(z)$, which reduces the maximum carrier density in the filament. Additionally, the drift field and therefore the generation rate decrease. Both effects tend to destroy the filament. As a result, the current through the sample decreases, which increases the voltage of the sample and thus enhances the filament due to increased generation rates.

The timescales of these processes are given by the time the filament needs to travel across the sample, the GR-time, and the dielectric relaxation time of the electric field.

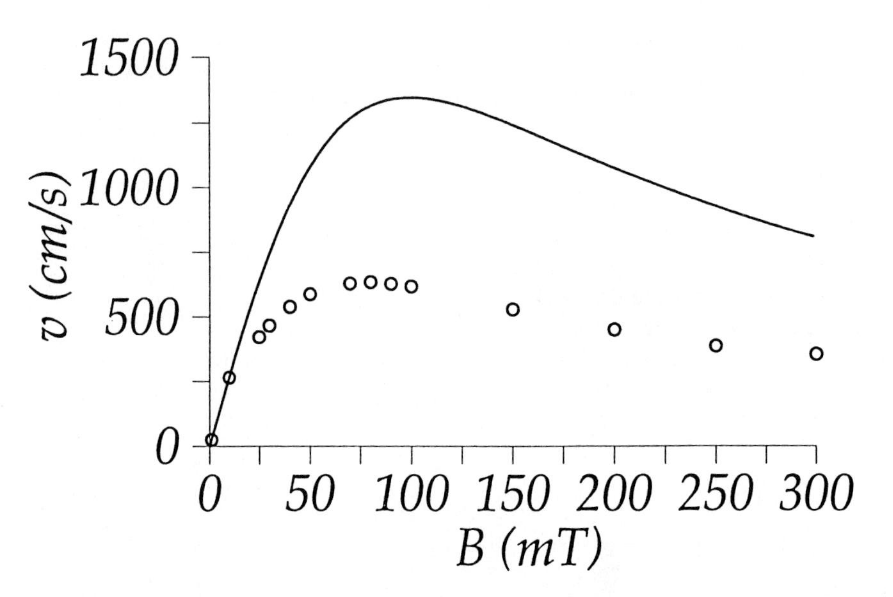

Fig. 2.3. Transverse velocity of the travelling filaments as a function of the magnetic field for $j_0 = 93.76$ mA/cm^2, $W = 120$ μm from the nonlinear simulation (circles). The full line shows the velocity obtained by the linear stability analysis for the same control parameters

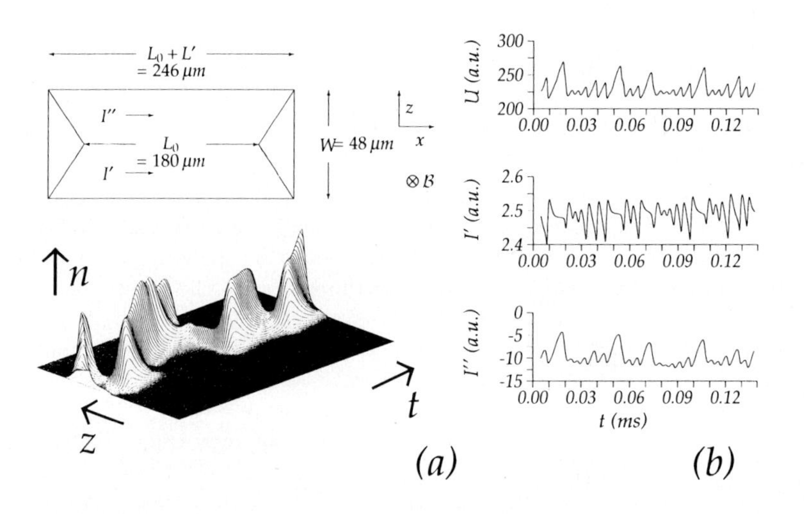

Fig. 2.4. Travelling filaments for $j_0 = 0.85$ mA/cm^2, $B = 20$ mT with triangular contacts. The inset shows the sample geometry. (a): Spatio-temporal dynamics of $n(z,t)$. The total time is $2 \cdot 10^{-5}$ s. (b): Time series of the sample voltage $U(t)$ and currents I', I'' (see inset), all in arbitrary units

Depending upon the ratios of these timescales and the geometry of the contacts, different states of the filament are possible. For small magnetic fields and comparable timescales the filament rests in the middle of the sample and shows asymmetrically breathing filament boundaries as observed in n-GaAs [2.48, 57].

A more complex dynamic behaviour is found if the dielectric relaxation and the destruction of the filament through the GR-processes are slower than the transverse motion of the filament. Such a situation is shown in Fig. 2.4a. The filament travels in the direction of the Lorentz force. Due to the inertia of the dielectric relaxation the drift field increases slowly. This supports destruction of the filament, so that with increasing voltage a new filament in the middle of the sample can be generated. This successive nucleation, travelling, and destruction of filaments leads to slow chaotic voltage and current oscillations (Fig. 2.4b) and has indeed been found in recent experiments on p-Ge [2.52].

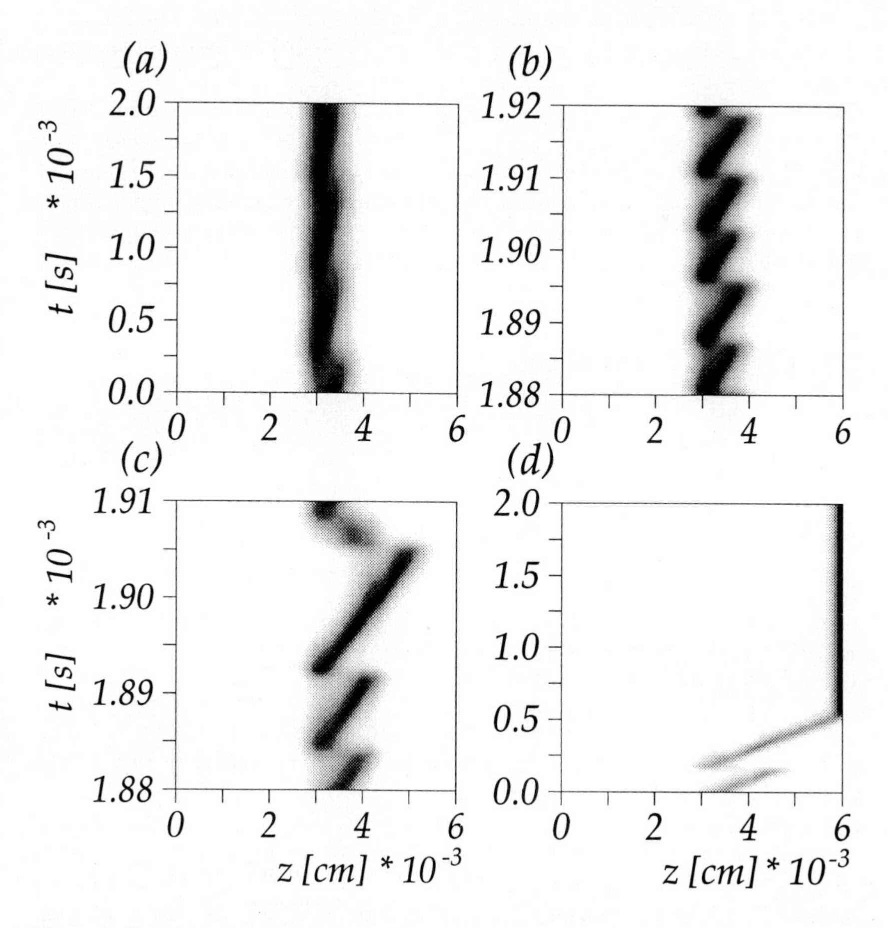

Fig. 2.5. Density plots of the carrier density $n(z,t)$ for different values of the magnetic field. (a) $\mathcal{B} = 11$ mT, (b) $\mathcal{B} = 15$ mT, (c) $\mathcal{B} = 20$ mT, (d) $\mathcal{B} = 25$ mT. Numerical parameters as in Fig. 2.4

As a function of magnetic field, above a minimum threshold $\mathcal{B}$, we find successively asymmetric breathing oscillations (Fig. 2.5a); small regular oscillations due to periodic nascence and destruction of travelling filaments (Fig. 2.5b); chaotic oscillations due to intermittent nascence and destruction (Fig. 2.5c); pinning of the filaments at the transverse boundary after some transient travelling sequences (Fig. 2.5d).

2.3 Spiking in Layered Semiconductor Structures

In layered semiconductors the charge transport is hindered by conduction band discontinuities which leads to charge accumulation in the potential wells. This charge distribution determines the electric field distribution which in turn strongly influences the charge transport. Therefore a condition for selfconsistency exists, which has often more than one solution for fixed applied bias. This indicates an important mechanism for bistability at vertical transport in semiconductor heterostructures.

The simplest example is the heterostructure hot-electron diode (HHED). This diode essentially consists of a *GaAs* and an adjacent $Al_xGa_{1-x}As$-layer with widths of about 100 nm. Both layers are not or only slightly n-doped, so that strong electric fields can be achieved. The layers are surrounded by highly doped contact layers. As shown in Fig. 2.6, in a certain range of voltage U across the sample two states with different current are possible [2.58,59]. In the low current

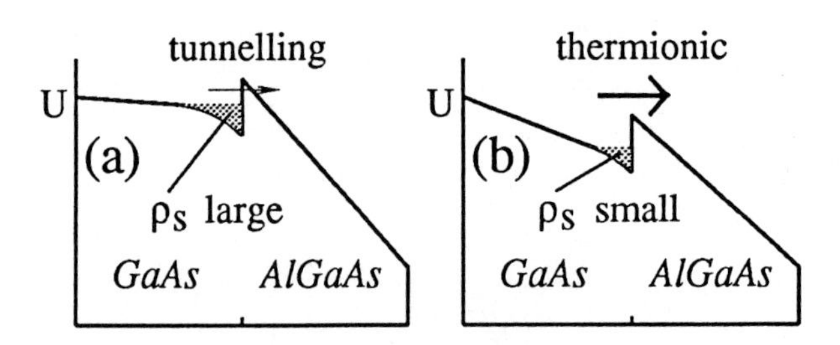

Fig. 2.6. Bistability in the HHED for fixed sample voltage U. The low and the high current state are characterized by a large and a small surface charge density ρ_s between the layers, respectively

state almost the whole voltage drops across the *AlGaAs*-barrier. Due to the small electric field only a low current flows in the first layer. As the electrons do not gain sufficient energy there is only a small tunnelling current through the barrier and the current in the *AlGaAs*-barrier is low. Thus, both currents can be made equal as a necessary condition for stationarity. If the field in the *GaAs*-layer is larger the current increases in this layer and the electrons gain sufficient kinetic energy to cross the barrier thermionically. Therefore, a second stationary state with equal currents in both layers is possible. Both states differ in the distribution of the total bias on the two layers which can be parametrized by the surface charge ρ_s between the layers. Thus, ρ_s is the crucial parameter indicating in which state the diode is operated.

2.3.1 Modelling the Dynamic Behaviour

Let the voltage be applied in the z-direction. Then the interface between the layers extends in the (x, y)-plane. The dynamical behaviour of $\rho_s(x, y, t)$ is given by the continuity equation

$$\frac{\partial \rho_s(x,y,t)}{\partial t} = j_1 - j_2 - \left(\frac{\partial j_x^{\|}}{\partial x} + \frac{\partial j_y^{\|}}{\partial y} \right) \quad , \tag{2.7}$$

where j_1, j_2 denote the current densities in the *GaAs* and the *AlGaAs*-layer, respectively. $j^{\|}$ is the surface current (in units A/cm) in the 2-dimensional electron gas (2DEG) ρ_s.

Now we consider the case that the device is operated under current controlled conditions with a capacitance C_p in parallel. Then the voltage U across the sample is not fixed but must be treated as an additional variable. Its dynamics can be determined as follows [2.60]. Ampère's law gives $\mathrm{div}(\epsilon\dot{\mathbf{F}} + \mathbf{j}) = 0$, where F is the electric field and ϵ is the dielectric contant of the material. From this we obtain:

$$\oint_{\partial V} d\mathbf{A} \cdot (\epsilon\dot{\mathbf{F}} + \mathbf{j}) = 0 \quad . \tag{2.8}$$

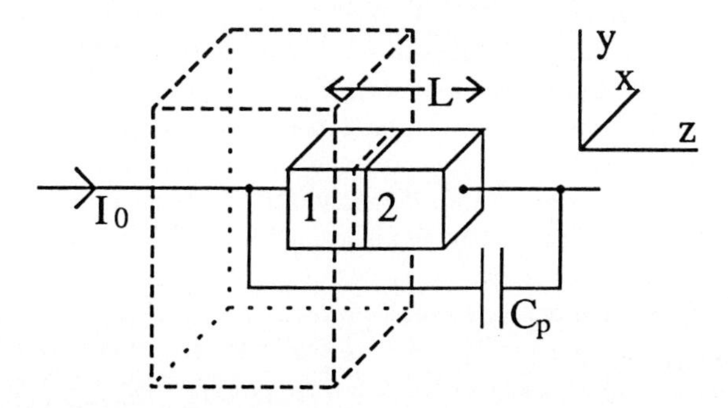

Fig. 2.7. Surface of integration (dotted) used for determining the dynamics of U

Now we integrate over the surface ∂V shown in Fig. 2.7 and find

$$\frac{C_p}{\epsilon(z)} \frac{dU}{dt} + \int_A dxdy \left(\frac{dF_z}{dt} + \frac{j_z}{\epsilon(z)} \right) - \frac{I_0}{\epsilon(z)} = 0 \, , \tag{2.9}$$

where I_0 is the total current and z is the position of the intersection of the surface with the sample of cross-section A. Note that the dielectric constant is only z-dependent due to the geometry of the structure. If we now integrate over the sample length L in z-direction we obtain

$$\frac{dU}{dt}\left(A+C_p\int_0^L dz\frac{1}{\epsilon(z)}\right)+\int_0^L dz\int_A dxdy\frac{j_z}{\epsilon(z)}-\int_0^L dz\frac{I_0}{\epsilon(z)}=0 \quad (2.10)$$

as $\int_0^L dzF_z = U$ is not dependent on (x,y), if we assume ideal metallic planar contacts at $z=0$ and $z=L$ and neglect time varying magnetic fields. Defining the intrinsic capacitance C by $C^{-1}=\int_0^L dz(A\epsilon(z))^{-1}$ we obtain:

$$\frac{dU}{dt}=\frac{1}{C+C_p}\left(I_0-\frac{C}{A}\int_A dxdy\int_0^L dz\frac{j_z}{\epsilon(z)}\right) \quad . \quad (2.11)$$

Assuming that there is no charge accumulation within the two layers except at the interface located at $z=L_1$, the integrals $\int_A dxdyj_z$ do not depend on z within the intervals $[0,L_1)$ and $(L_1,L]$. As $\epsilon(z)$ is constant within these intervals, we find:

$$\frac{dU}{dt}=\frac{1}{C+C_p}\left\{I_0-\int_A dxdy\left(\frac{C}{C_1}j_1+\frac{C}{C_2}j_2\right)\right\} \quad , \quad (2.12)$$

where j_1 and j_2 are the z-components of the current density at some fixed position in the first or the second interval, respectively, and $C_1^{-1}=\int_0^{L_1} dz(A\epsilon(z))^{-1}$ and $C_2^{-1}=\int_{L_1}^L dz(A\epsilon(z))^{-1}$.

As the dynamics of $\rho_s(t)$ and $U(t)$ is slow compared to intrinsic processes like electron heating, e. g., we may assume that all currents can be written as explicit local functions of $U(t)$ and $\rho_s(x,y,t)$. Therefore, (2.7) and (2.12) constitute a closed dynamical system.

In [2.16] and [2.21,61] two different models for the current densities $j_{1/2}(\rho_s,U)$ have been used. Both yield an S-shaped current-voltage characteristic in a certain range of sample parameters. Besides homogeneous stable steady states, homogeneous relaxation oscillations and the formation of current filaments, for both models a new spatio-temporal behaviour has been found where a current filament forms and subsequently vanishes. Due to the significant shape of the voltage-time signal we call this behaviour *spiking*. For a better understanding of this complex phenomenon we have simplified the specific equations $j_{1/2}(\rho_s,U)$ in order to get a very simple "generic" model showing this very effect.

Using dimensionless units (as defined in [2.62]) with $a\sim\rho_s$, $u\sim U$, and $j_0\sim I_0$ the generic model is given by

$$\frac{\partial a(x,y,t)}{\partial t}=\frac{u-a}{(u-a)^2+1}-Ta+\Delta a \quad , \quad (2.13)$$

$$\frac{du(t)}{dt}=\alpha(j_0-\langle u-a\rangle) \quad , \quad (2.14)$$

where $\langle u-a\rangle=\int dxdy(u-a)/A$ is the spatially averaged current through the sample, $\alpha\sim 1/(C+C_p)$ and T is a normalized tunnelling coefficient from the 2DEG into the $AlGaAs$-region.

Note that such global couplings as $< u-a >$ are often introduced in semiconductor charge transport though the external circuit which imposes a restriction on the total current. They have been shown to give rise to a competition between different current filaments similar to Ostwald ripening in equilibrium thermodynamics, i.e., if several filaments are coupled via this nonlocal circuit interaction, the initially largest filament will eventually survive [2.63,64].

2.3.2 Results

The $j_0(u)$ characteristic is obtained by setting the temporal derivatives equal to zero. For $T < 1/8$ we find an S-shaped homogeneous characteristic as shown in Fig. 2.8. In addition to the homogeneous stationary solutions we obtain inhomogeneous solutions. In the following we neglect the y-dependence (i.e. assuming small extension L_y) and use $< u - a >= \int dx(u-a)/L_x$. Furthermore we use $T = 0.05$ and $L_x = 40$ and Neumann boundary conditions. The inhomogeneous solution for these parameters is shown in the inset of Fig. 2.8.

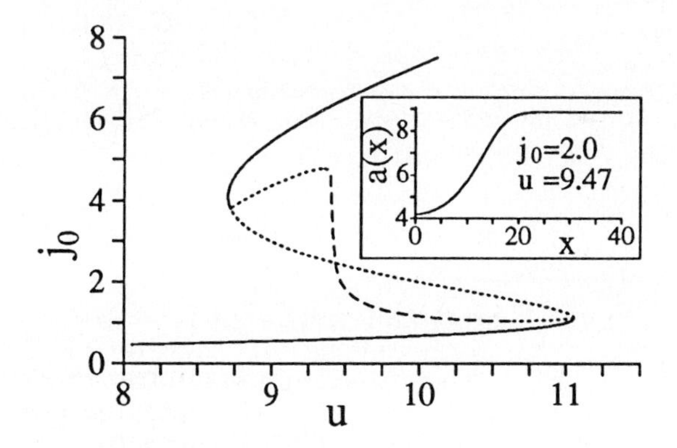

Fig. 2.8. Current-voltage characteristic for the generic model for $L_x = 40$, $T = 0.05$. The full line depicts homogeneous stable states. Inhomogeneous stable states of the shape given in the inset are marked by the dashed line. The dotted lines refer to states which are unstable. The stability is determined for $\alpha = 2$

Depending on the parameter α, i.e., the magnitude of C_p, on the external current j_0, and on initial conditions various spatio-temporal patterns are found in the time-dependent simulation as shown in Fig. 2.9. Next to homogeneous stable states we find homogeneous oscillations (a), stable filaments (b) and spiking which may be periodic (c) or chaotic (d).

A phase diagram of the (j_0, α)-plane indicating the regions where a particular behaviour dominates is provided in Fig. 2.10. It can be clearly seen that the spiking behaviour sets in at $j_0 = 1.138$ where the homogeneous branch becomes unstable against fluctuations of the shape $\delta a(x) \sim \cos(\pi x/L_x)$. In the given parameter range first an inhomogeneous distribution forms within a short time. This distribution breaks down immediately, a nearly homogeneous state is

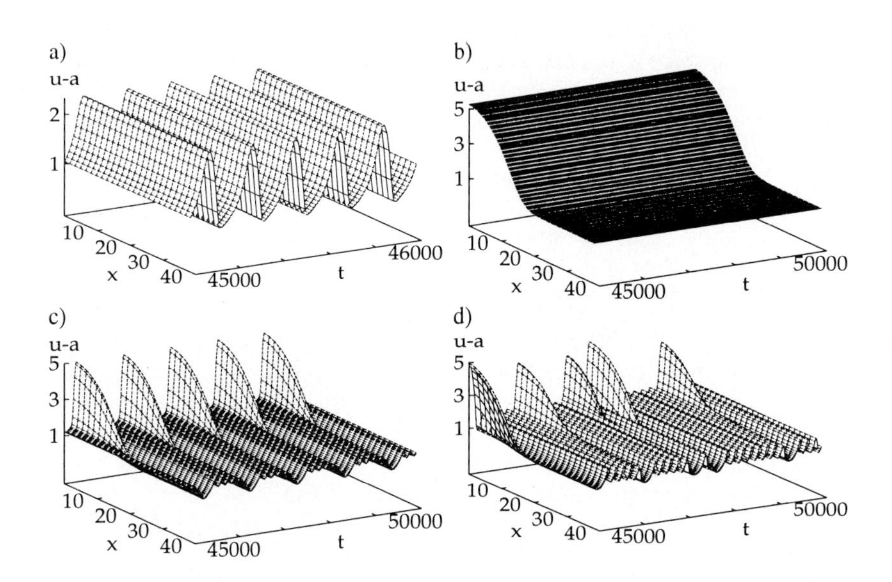

Fig. 2.9. Spatio-temporal patterns of the generic model for $\alpha = 0.035$. **(a)** homogeneous oscillations ($j_0 = 1.40$). **(b)** stationary filament ($j_0 = 2.0$). **(c)** periodic spiking ($j_0 = 1.24$). **(d)** chaotic spiking ($j_0 = 1.31$)

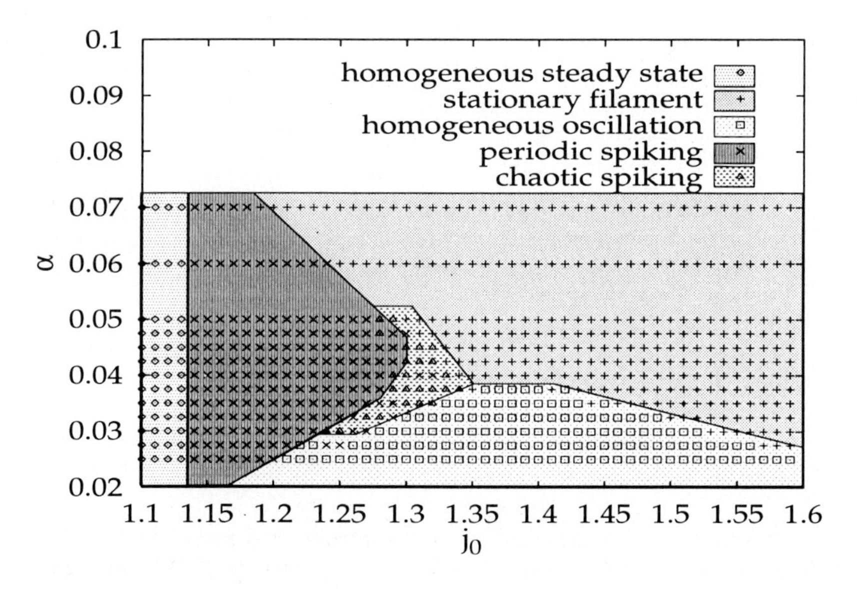

Fig. 2.10. Phase diagram of different asymptotic states in the (j_0, α) control-parameter space

reached again and the same cycle is repeated. The period of the oscillations is mainly determined by the growth of the inhomogeneity which is very sensitive to noise [2.61]. Upon increasing j_0 the spiking disappears via two different scenarios depending on the value of α.

For larger α the spiking mode vanishes via a subcritical Hopf bifurcation and the system switches to the stable filament [2.65]. This is shown in Fig. 2.11 where the minimum inhomogeneity $\delta a = |a(0,t) - a(L_x,t)|$ in time is plotted versus the control parameter. There is bistability between the spiking (s) and the filamentary branch (f) in the voltage interval $1.24 < j_0 < 1.295$ for $\alpha = 0.05$. Upon decreasing j_0 starting on the filamentary branch a subcritical Hopf bifurcation at $j_0 = 1.24$ is found. The unstable limit cycle which bifurcates from there is denoted by (u). It can be visualized for a short time by starting with appropriate initial conditions (see Fig. 2.12) and has the form of a breathing filament.

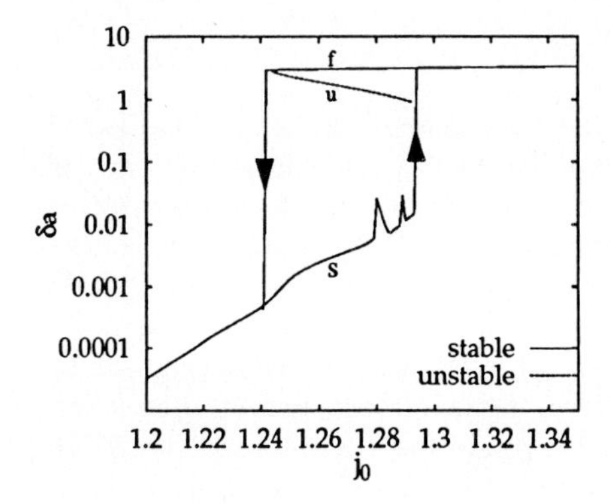

Fig. 2.11. Hysteresis behaviour between spiking (s) and the stable filament (f) with an unstable limit cycle (u) as separatrix. The different cycles are identified by the minimum δa of $|a(0,t) - a(L_x,t)|$ in time ($\alpha = 0.05$)

For lower values of α the intermediate homogeneous state, corresponding to the NDC branch exhibits an oscillatory instability (homogeneous relaxation oscillations). After reaching a sufficient amplitude the Floquet exponent of this limit cycle becomes negative and the homogeneous oscillation becomes an attractor. For values of α which are between these extremes chaotic behaviour occurs which seems to be caused by the interplay of these effects.

The transitions from periodic to chaotic behaviour occurs either via period doubling or by an intermittent scenario. The chaotic nature of the oscillations can be proved by calculating the Lyapunov spectrum which yields one positive exponent [2.65]. The smallness of the Kaplan-Yorke dimension ($d < 2.1$) indicates that the chaos is determined by only a few degrees of freedom. Thus, we do not have spatio-temporal chaos in this model.

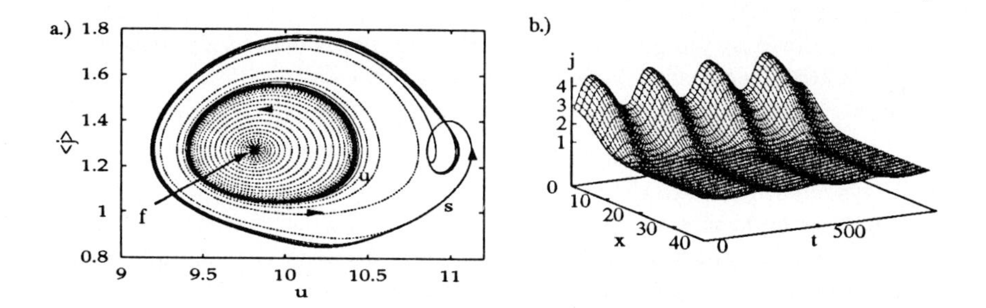

Fig. 2.12. The unstable limit cycle ($\alpha = 0.05, j_0 = 1.27$): (a) gives the phase diagram showing two different trajectories starting near the unstable limit cycle (u) and tending towards the stable filament (f) and the spiking cycle (s), respectively. (b) shows the spatio-temporal plot of the unstable limit cycle

2.4 Field Domains in Superlattices

A further aspect of the perpendicular transport in semiconductor heterostructures lies in the occurrence of resonances in the charge transport from one well to the next. This leads to pronounced maxima in the current-voltage characteristics associated with NNDC. Examples are the Double-Barrier Resonant-Tunnelling Diode (DBRTD) [2.66] and semiconductor superlattices [2.67]. An important aspect of the superlattices is the formation of electric field domains due to NNDC [2.68], which is not possible in the DBRTD as the mechanism for NDC is localized in the transport direction. As mentioned in the last section, bistability is likely to occur in such structures and has been observed [2.69, 70] in both elements. In the following we restrict ourselves to the superlattices as they exhibit a larger wealth of spatio-temporal behaviour due to the domain formation. While this system has been widely studied experimentally [2.71–74] theoretical progress has only been made recently [2.75–78]. Starting with a model for the transport we want to show how multistability appears and to discuss the occurrence of spatio-temporal oscillations.

2.4.1 The Model

We consider a *GaAs/AlAs* superlattice with a period $d = l + b$, where l denotes the width of the GaAs well and b that of the AlAs barrier. For simplicity we focus on the first and second subbands only, although an extension to higher subbands is straightforward. Denoting each well by a superscript (k), $k = 1, ..., N$, we have to find steady-state self-consistent solutions of the rate equations of the 2D electron densities $n_i^{(k)}$ for the first subband

$$\begin{aligned} \dot{n}_1^{(k)} &= A^{(k)} + n_1^{(k-1)} R_1(F^{(k)}) + n_2^{(k-1)} Y(-F^{(k)}) + n_2^{(k+1)} Y(F^{(k+1)}) \\ &\quad - n_1^{(k)} R_1(F^{(k+1)}) - n_1^{(k)} \left(X(-F^{(k)}) + X(F^{(k+1)}) \right), \end{aligned} \tag{2.15}$$

the second subband

$$\dot{n}_2^{(k)} = -A^{(k)} + n_2^{(k-1)} R_2(F^{(k)}) + n_1^{(k-1)} X(F^{(k)}) + n_1^{(k+1)} X(-F^{(k+1)})$$
$$- n_2^{(k)} R_2(F^{(k+1)}) - n_2^{(k)} \left(Y(-F^{(k)}) + Y(F^{(k+1)}) \right) \tag{2.16}$$

(where the dot denotes the temporal derivative), and Poisson's equation

$$\epsilon(F^{(k+1)} - F^{(k)}) = el(n_1^{(k)} + n_2^{(k)} - N_D) . \tag{2.17}$$

Here $F^{(k)}$ is the average field between the wells with index $(k-1)$ and (k), and N_D is the effective electron density provided by the doping in the wells. The different transition rates are sketched in Fig. 2.13.

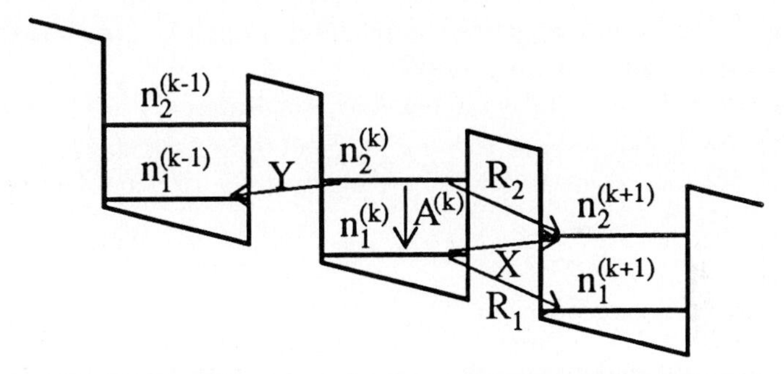

Fig. 2.13. Conduction band vs. position of the superlattice with an applied electric field and various transition rates (schematic)

R_i is the rate of electrons crossing the barrier between two equivalent subbands with index i. We use the expression for transport in a miniband with the dispersion relation $E_i(q) = E_i - \Delta_i \cos(qd)$

$$R_i = \frac{\Delta_i}{\hbar} \sqrt{\frac{\tau_m}{\tau_e}} \frac{F/\tilde{F}}{1 + (F/\tilde{F})^2} \tag{2.18}$$

with $\tilde{F} = \hbar/(ed\sqrt{\tau_m \tau_e})$. This expression has been derived in [2.79] where the influence of the different scattering times τ_m and τ_e for momentum and energy relaxation, respectively, has been considered.

X and Y are the rate coefficients for resonant tunnelling $e_1 \to e_2$ and $e_2 \to e_1$, respectively. They are calculated by perturbation theory [2.77] in the following way: At first we calculate the lowest two eigenstates $|1\rangle$ and $|2\rangle$ of the single quantum wells. By combining the functions $|1\rangle$ and $|2\rangle$ of two neighbouring wells and including the potential drop due to the electric field we obtain the perturbed energy levels $E_\pm$. Starting with an electron in the state $|1\rangle$ at $t = 0$ its wave function $|\Psi(t)\rangle$ will oscillate in time with the frequency $\nu = (E_+ - E_-)/2\pi\hbar$. Let W_{12} be the maximum of $|\langle 2|\Psi(t)\rangle|^2$ in time. If we now assume that the phase is rapidly destroyed once the electron reaches the level $|2\rangle$ then the transition rate from $|1\rangle$ to $|2\rangle$ can be estimated by

$$X(F) = W_{12}\frac{E_+ - E_-}{2\pi\hbar}\,. \tag{2.19}$$

Y is calculated analogously. The rates X and Y become small off resonance as W vanishes rapidly, but do not vanish due to the quantum mechanical energy uncertainty. Note that the inverse tunnelling processes with $X(-F)$ and $Y(-F)$ are very small in general. In the low field case they compensate the processes $X(F)$ and $Y(F)$.

For the intersubband transition rate $A^{(k)}$ we use

$$A^{(k)} = n_2^{(k)}/\tau_{21} \tag{2.20}$$

with $\tau_{21} = 1$ ps, which is justified as the level spacing is larger than the optical phonon energy in the case we consider.

We model the contacts by the boundary conditions $n_i^{(0)} = n_i^{(1)}$ and $n_i^{(N+1)} = n_i^{(N)}$. The fields have to satisfy $U = d\sum_{k=1}^{N+1} F^{(k)}$ for fixed bias U. Thus, we have a closed dynamical system for the evolution of the electron densities for fixed bias U.

2.4.2 Static Characteristic

We consider a superlattice with 40 periods of $l = 90$ Å and $b = 15$ Å. We use the parameters $\epsilon = 13.18\epsilon_0$, $\tau_e = 2$ ps and $\tau_m = 0.005$ ps. Then the subband parameters are $E_1 = 44.6$ meV, $\Delta_1 = 1.4$ meV, $E_2 = 180.2$ meV and $\Delta_2 = 6.1$ meV. In the weakly doped case of $N_D = 1.34 \cdot 10^{14}$ cm^{-3} we obtain a homogeneous field distribution. The characteristic given in Fig. 2.14a shows a first maximum at low electric fields due to miniband transport and a pronounced peak at the field where the first and the second level are in resonance.

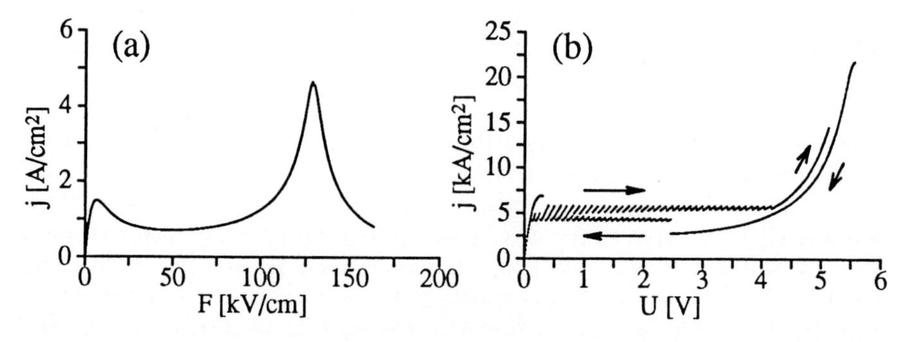

Fig. 2.14. (a) Current-field characteristic for $N_D = 1.34 \cdot 10^{14}$ cm^{-3} where the field distribution is homogeneous along the sample. (b) Current-voltage characteristic for $N_D = 6.7 \cdot 10^{17}$ cm^{-3} for sweep-up and sweep-down of the voltage

For highly doped samples ($N_D = 6.7 \cdot 10^{17}$ cm^{-3}) sufficient charge can be accumulated in the wells so that an electric field domain boundary may form. This leads to several jumps in the characteristic upon sweep-up or sweep-down of the voltage (Fig. 2.14b). In Fig. 2.15 the potential distribution along the sample is shown for the sweep-up. One can clearly see two different regions with distinct

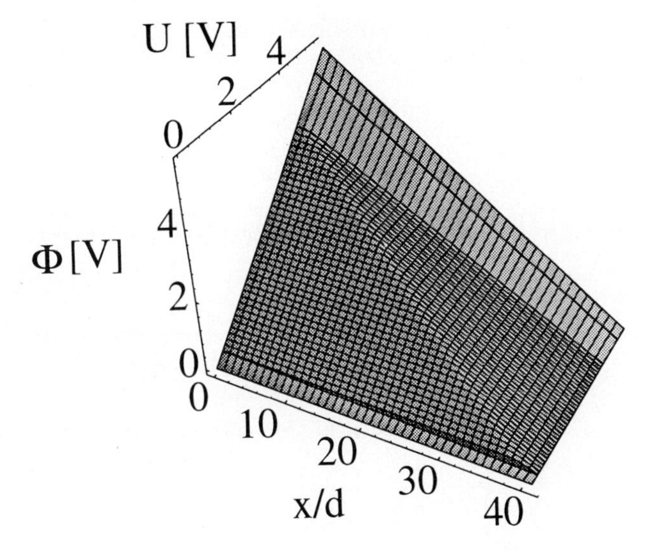

Fig. 2.15. Potential distribution for sweep-up of the voltage as shown in Fig. 2.14b

potential drop per length indicating the region of the two domains. Upon increasing the voltage the length of the high-field domain increases nearly linearly.

The field distributions for different voltages are explicitly shown in Fig. 2.16. The transition between the high and the low field occurs on the length of about two wells. In contrast to the description in [2.78] there are small transition regions between the boundary and both domains. We believe that the difference is caused by the occurrence of currents in both directions in our model, while the transport is modelled uni-directionally in [2.78].

The sweep-up and the sweep-down results in two different curves, which are unconnected in most part of the characteristic. In fact, however, they belong to distinct branches which connect both curves. These branches can be reached by reversing the voltage sweep starting from one point, which is reached by the sweep-up or sweep-down curve. The resulting characteristic is shown in Fig. 2.17a for a small range of voltage. For a fixed bias U up to four stable states with different values of the current may coexist (The same behaviour has been found experimentally [2.70]). These states are distinguished by the position of the domain boundary, which is located at neighbouring wells as shown in Fig. 2.17b. Note that both the high-field and the low-field domain exist at field values which are lower than the value of the resonances in the current (see also [2.80]).

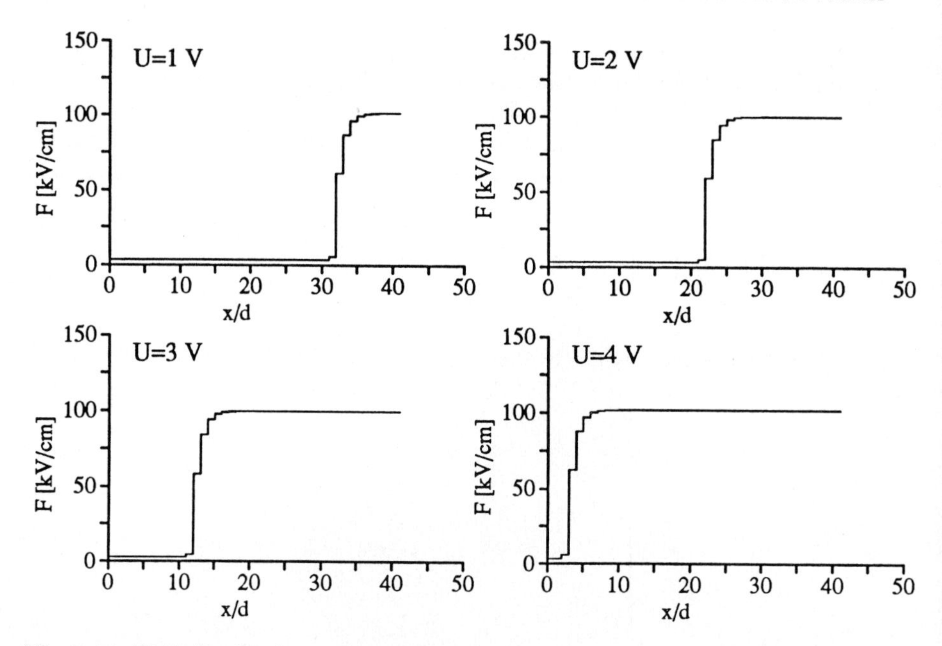

Fig. 2.16. Field distribution at four different voltages for the sweep-up in the characteristic shown in Fig. 2.14b

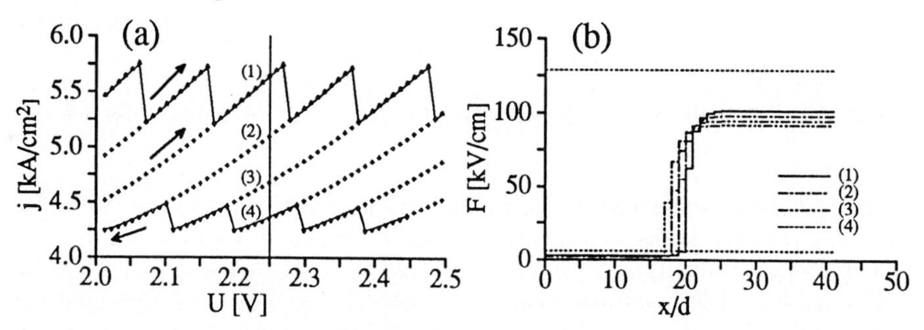

Fig. 2.17. (a) Full characteristic for 2.0 V< U < 2.5 V, where all stable states are shown. The full lines give the curves for sweep-up and sweep-down from Fig. 2.14b. (b) Field profiles for the 4 stable states indicated in (a). The dotted lines mark the position of the maxima in the current-field characteristic from Fig. 2.14a

2.4.3 Spatio-Temporal Oscillations

For lower doping we find oscillatory behaviour as shown in Fig. 2.18a [2.81]. For $N_D = 1.34 \cdot 10^{16}$ cm^{-3} the oscillatory behaviour is transient and a stationary domain state forms.

For higher doping $N_D = 4.03 \cdot 10^{16}$ cm^{-3} we find persistent self-sustained oscillations. The field profile from Fig. 2.18b shows that a domain state forms at first. Then, the domain boundary moves to the anode and vanishes there. Subsequently the cycle is repeated.

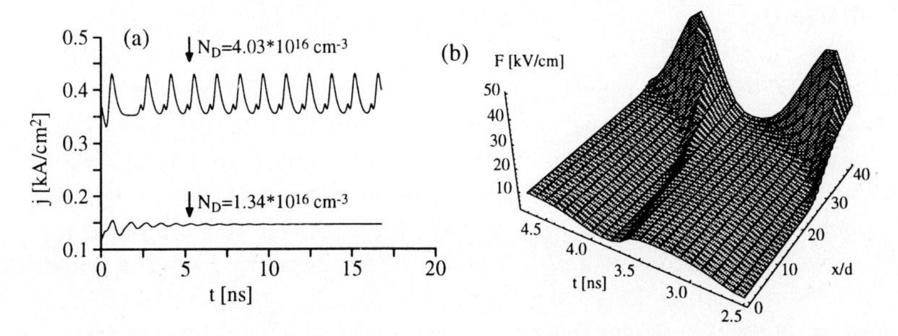

Fig. 2.18. (a) Current oscillations for two different doping densities when the bias is switched from 0.01 V to 0.5 V at $t = 0$. **(b)** Time dependence of the electric field distribution for $N_D = 4.03 \cdot 10^{16}$ cm^{-3}

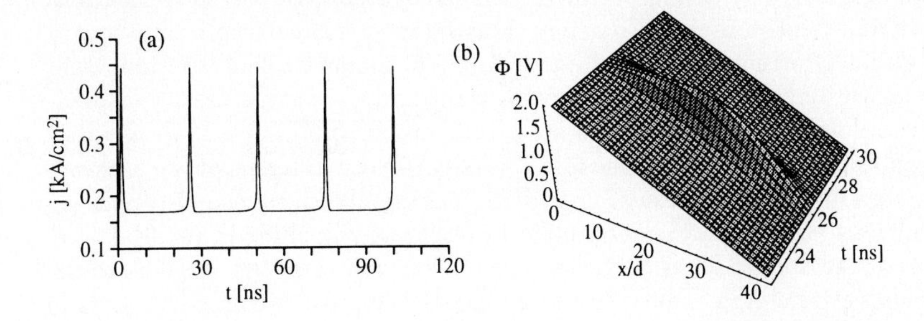

Fig. 2.19. (a) Current oscillation for $N_D = 4.03 \cdot 10^{16}$ cm^{-3} when the bias is switched from 0.01 V to 2.0 V at $t = 0$. **(b)** Time dependence of the potential distribution

If we increase the voltage we find longer intermediate intervals where the field distribution is nearly homogeneous as shown in Fig. 2.19. This behaviour resembles the spiking behaviour dicussed in Sect. 2.3 though the system is completely different (domain formation instead of filaments). Note that the growth of the inhomogeneity might be strongly influenced by noise. This would lead to an irregular spacing of the spikes. We find a variety of such oscillatory behaviour with the following trends [2.82]:

- For fixed bias voltage U the frequency of the self-sustained oscillations decreases sublinearly with increasing doping concentration N_D and more higher-order harmonics appear in the Fourier spectrum. Above some critical value the oscillations vanish and a stable domain solution is reached.
- For fixed N_D and increasing U the form of the oscillations evolves from more regular behaviour shown in Fig. 2.18 to the spiking of Fig. 2.19.
- For fixed U/N and N_D there exists a certain minimum number of periods N for the occurrence of undamped oscillations. With increasing N the oscillation frequency first decreases to a minimum, after which it increases again.

Work geared towards identifying the individual bifurcation types is still in progress.

2.5 Conclusions

We have demonstrated that semiconductors represent convenient model systems for the study of complex nonlinear spatio-temporal dynamics and instabilities far from thermodynamic equilibrium. Progress in the materials engineering of semiconductor microstructures and in the measurement techniques has enabled the direct observation of these phenomena with high spatial and temporal resolution. Moreover, the phenomena of self-organization presented in this review may offer interesting applications for future semiconductor devices like GHz oscillators, switching devices or, for instance, in case of a multistable superlattice, voltage pulse discriminators or memory elements that can store two or more bits "vertically" and thereby increase the lateral memory integration density.

From the viewpoint of nonlinear dynamics, the examples presented involve secondary bifurcations of the primary coherent spatial structures, viz. current filaments and field domains. Filaments breathing and travelling laterally in a regular or intermittent way in crossed electric and magnetic fields, periodic and chaotic spatio-temporal spiking in layered semiconductor structures, and oscillating field domains in superlattices are prominent examples. Those elementary self-organized patterns appear due to a diversity of very different physical mechanisms, e. g., impact ionization of impurities, thermionic emission and tunnelling across a heterojunction, and nonresonant or resonant tunnelling in superlattices. Future research should focus on these coherent structures in the attempt to describe complex and even chaotic spatio-temporal dynamics in semiconductors in terms of simple generic models.

We are grateful to S. Bose, G. Hüpper, R. Kunz, F. Prengel, K. Pyragas, W. Quade, and G. Schwarz for their contributions to the work presented, and to M. Asche, L. Bonilla, H. Grahn, J. Kastrup, H. Kostial, T. Kuhn, S. H. Kwok, F.-J. Niedernostheide, V. Novak, W. Prettl, and J. Wolter for valuable discussions.

References

[2.1] E. Schöll: *Nonequilibrium Phase Transitions in Semiconductors* (Springer, Berlin 1987)
[2.2] H. Haken: *Advanced Synergetics*, 2nd ed. (Springer, Berlin 1987)
[2.3] A.S. Mikhailov: *Foundations of Synergetics 1*, 2nd ed. (Springer, Berlin 1994)
[2.4] E. Schöll: Theory of oscillatory instabilities in parallel and perpendicular transport in heterostructures, in *Negative Differential Resistance and Instabilities in 2-dimensional Semiconductors*, ed. by N. Balkan, B.K. Ridley, A.J. Vickers (Plenum Press, New York 1993) p. 37
[2.5] K. Aoki, K. Yamamoto: Phys. Lett. A **98**, 72 (1983)
[2.6] S. W. Teitsworth, R. M. Westervelt, E. E. Haller: Phys. Rev. Lett. **51**, 825 (1983)
[2.7] A. Wacker, E. Schöll: General conditions for stability in bistable electrical devices with S- or Z-shaped current-voltage-characteristics, in *Quantum Transport in Ultrasmall Devices*, ed. by D.K. Ferry (Plenum Press, New York 1994), in print

[2.8] M.P. Shaw, V.V. Mitin, E. Schöll, H.L. Grubin: *The Physics of Instabilities in Solid State Electron Devices* (Plenum Press, New York 1992)

[2.9] B.S. Kerner, V.V. Osipov: Sov. Phys. Usp. **157**, 101 (1989)

[2.10] E. Schöll, D. Drasdo: Z. Phys. B **81**, 183 (1990)

[2.11] U. Rau, W. Clauss, A. Kittel, M. Lehr, M. Bayerbach, J. Parisi, J. Peinke, R.P. Huebener: Phys. Rev. B **43**, 2255 (1991)

[2.12] C. Radehaus, R. Dohmen, H. Willebrand, F.-J. Niedernostheide: Phys. Rev. A **42**, 7426 (1990)

[2.13] F.-J. Niedernostheide, M. Arps, R. Dohmen, H. Willebrand H.-G. Purwins: phys. status solidi (b) **172**, 249 (1992)

[2.14] R. Symanczyk, S. Gaelings, D. Jäger: Phys. Lett. A **160**, 397 (1991)

[2.15] A. Brandl, W. Prettl: Phys. Rev. Lett. **66**, 3044 (1991)

[2.16] A. Wacker, E. Schöll: Semicond. Sci. Technol. **7**, 1456 (1992)

[2.17] L.L. Bonilla, I.R. Cantalapiedra, M.J. Bergmann, S.W. Teitsworth: Semicond. Sci. Technol. **9**, 599 (1994)

[2.18] G. Hüpper, E. Schöll: Phys. Rev. Lett. **66**, 2372 (1991)

[2.19] G. Hüpper, K. Pyragas, E. Schöll: Phys. Rev. B **48**, 17633 (1993)

[2.20] A. Wacker, E. Schöll: Appl. Phys. Lett. **59**, 1702 (1991)

[2.21] A. Wacker, E. Schöll: Semicond. Sci. Technol. **9**, 592 (1994)

[2.22] Z.S. Gribnikov, M.I. Mel'nikov: Sov. Phys. Solid State **7**, 2364 (1973)

[2.23] K. Hess, H. Morkoc, H. Shichijo, B.G. Streetman: Appl. Phys. Lett. **35**, 469 (1979)

[2.24] E. Schöll, K. Aoki: Appl. Phys. Lett. **58**, 1277 (1991)

[2.25] R. Döttling, E. Schöll: Phys. Rev. B **45**, 1935 (1992)

[2.26] R. Döttling, E. Schöll: Physica D **67**, 418 (1993)

[2.27] O. Rudzick, E. Schöll: Semicond. Sci. Technol., submitted

[2.28] P. Hendriks, E.A.E. Zwaal, J.G.A. Dubois, F.A.P. Blom, J.H. Wolter: J. Appl. Phys. **69**, 302 (1991)

[2.29] C.O. Weiss, R. Vilaseca: *Dynamics of Lasers* (VCH, Weinheim 1991)

[2.30] H. Haug (ed.): *Optical Nonlinearities and Instabilities in Semiconductors* (Academic Press, New York 1988)

[2.31] O. Hess, E. Schöll: Phys. Rev. A **50**, 787 (1994)

[2.32] O. Hess, D. Merbach, H.-P. Herzel, E. Schöll: Phys. Lett. A **194** (1994), in print

[2.33] O. Hess: *Spatio-Temporal Dynamics of Semiconductor Lasers*. Dissertation, TU-Berlin (WuT Verlag, Berlin 1993)

[2.34] R. Schmolke, E. Schöll, N. Nägele, J. Gutowski: J. Crystal Growth **138**, 213 (1994)

[2.35] Y. Abe (ed.): *Special issue on Nonlinear and Chaotic Transport Phenomena in Semiconductors*, Appl. Phys. **A 48**, 93-191 (1989)

[2.36] E. Schöll: Physica Scripta **T29**, 152 (1989)

[2.37] H. Thomas (ed.): *Nonlinear Dynamics in Solids* (Springer, Berlin 1992)

[2.38] J. Peinke, J. Parisi, O.E. Rössler, R. Stoop: *Encounter with Chaos* (Springer, Berlin, Heidelberg 1992)

[2.39] E. Schöll: Nonlinear Dynamics, Phase Transitions and Chaos in Semiconductors, in *Handbook on Semiconductors*, ed. by P.T. Landsberg, Vol. 1, 2nd ed. (North Holland, Amsterdam 1992) pp. 419 - 447

[2.40] N. Balkan, B.K. Ridley, A.J. Vickers (eds.): *Negative Differential Resistance and Instabilities in 2-dimensional Semiconductors* (Plenum Press, New York 1993)

[2.41] B.S. Kerner, V.V. Osipov: *Autosolitons* (Kluwer Academic Publishers, Dordrecht 1994)

[2.42] F.-J. Niedernostheide, M. Ardes, M. Or-Guil, H.-G. Purwins: Phys. Rev. B **49**, 7370 (1994)
[2.43] X.N. Song, D.G. Seiler, M.R. Loloee: Appl. Phys. A **48**, 137 (1989)
[2.44] J. Spinnewyn, H. Strauven, O.B. Verbeke: Z. Phys. B **75**, 159 (1989)
[2.45] K. Aoki, Y. Kawase, K. Yamamoto, N. Mugibayashi: J. Phys. Soc. Jpn. **59**, 20 (1990)
[2.46] W. Clauss, U. Rau, J. Peinke, J. Parisi, A. Kittel, M. Bayerbach, R.P. Huebener: J. Appl. Phys. **70**, 232 (1991)
[2.47] J. Peinke, U. Rau, W. Clauss, R. Richter, J. Parisi: Europhys. Lett. **9**, 743 (1989)
[2.48] A. Brandl, W. Kröninger, W. Prettl, G. Obermair: Phys. Rev. Lett. **64**, 212 (1990)
[2.49] J. Spangler, B. Finger, C. Wimmer, W. Eberle, W. Prettl: Semicond. Sci. Technol. **9**, 373 (1994)
[2.50] H. Kostial, M. Asche, R. Hey, K. Ploog, F. Koch: Jap. J. Appl. Phys **32**, 491 (1993)
[2.51] B. Kehrer, W. Quade, E. Schöll: Monte Carlo Simulation of Low Temperature Impurity Breakdown and Current Filamentation in δ-doped GaAs, in *Proc. 22nd Int. Conf. Phys. Semicond., Vancouver, 1994*, ed. by D. J. Lockwood (World Scientific, Singapore)
[2.52] M. Hirsch, A. Kittel, G. Flätgen, R.P. Huebener, J. Parisi: Phys. Lett. A **186**, 157 (1994)
[2.53] G. Hüpper, E. Schöll, A. Rein: Mod. Phys. Lett. B **6**, 1001 (1992)
[2.54] G. Hüpper, K. Pyragas, E. Schöll: Phys. Rev. B **47**, 15515 (1993)
[2.55] W. Quade, G. Hüpper, E. Schöll, T. Kuhn: Phys. Rev. B **49**, 13408 (1994)
[2.56] R.W. Hockney, J.W. Eastwood: *Computer Simulation Using Particles* (McGraw-Hill, New York, 1981)
[2.57] A. Brandl, M. Völcker, W. Prettl: Appl. Phys. Lett. **55**, 238 (1989)
[2.58] K. Hess, T.K. Higman, M.A. Emanuel, J.J. Coleman: J. Appl. Phys. **60**, 3775 (1986)
[2.59] A.M. Belyantsev, A.A. Ignatov, V.I. Piskarev, M.A. Sinitsyn, V.I. Shashkin, B.S. Yavich, M.L. Yakovlev: JETP Lett. **43**, 437 (1986)
[2.60] S. Bose: Studienarbeit, Technische Universität Berlin (1993)
[2.61] A. Wacker: *Nichtlineare Dynamik bei senkrechtem Ladungstransport in einer Halbleiter-Heterostruktur*. Dissertation, Technische Universität Berlin (Verlag Köster, Berlin 1993)
[2.62] A. Wacker, E. Schöll: Z. Phys. B **93**, 431 (1994)
[2.63] L. Schimansky-Geier, Ch. Zülicke, E. Schöll: Z. Phys. B **84**, 433 (1991)
[2.64] R.E. Kunz, E. Schöll: Z. Phys. B **89**, 289 (1992)
[2.65] S. Bose, A. Wacker, E. Schöll: Phys. Lett. A (1994), in print
[2.66] I.L. Chang, L. Esaki, R. Tsu: Appl. Phys. Lett. **24**, 593 (1974)
[2.67] R.F. Kazarinov, R.A. Suris: Sov. Phys. Semicond. **6**, 120 (1972)
[2.68] L. Esaki, L.L. Chang: Phys. Rev. Lett. **33**, 495 (1974)
[2.69] V.J. Goldman, D.C. Tsui, L.E. Cunningham: Phys. Rev. Lett. **58**, 1256 (1987)
[2.70] J. Kastrup, H.T. Grahn, K. Ploog, F. Prengel, A. Wacker, E. Schöll: Appl. Phys. Lett. **65**, 1808 (1994)
[2.71] Y. Kawamura, K. Wakita, H. Asahi, K. Kurumada: Jap. J. Appl. Phys **25**, L928 (1986)
[2.72] K.K. Choi, B.F. Levine, R.J. Malik, J. Walker, C.G. Bethea: Phys. Rev. B **35**, 4172 (1987)
[2.73] H.T. Grahn, R.J. Haug, W. Müller, K. Ploog: Phys. Rev. Lett. **67**, 1618 (1991)

[2.74] S.H. Kwok, R. Merlin, H.T. Grahn, K. Ploog: Phys. Rev. B **50**, 2007 (1994)
[2.75] A.N. Korotkov, D.V. Averin, K.K. Likharev: Appl. Phys. Lett. **62**, 3282 (1993)
[2.76] B. Laikhtman, D. Miller: Phys. Rev. B **48**, 5395 (1993)
[2.77] F. Prengel, A. Wacker, E. Schöll: Phys. Rev. B **50**, 1705 (1994)
[2.78] L.L. Bonilla, J. Galán, J.A. Cuesta, F.C. Martínez, J.M. Molera: Phys. Rev. B **50**, 8644 (1994)
[2.79] A.A. Ignatov, E.P. Dodin, V.I. Shashkin: Mod. Phys. Lett. B **5**, 1087 (1991)
[2.80] S.H. Kwok, H.T. Grahn, M. Ramsteiner, K. Ploog, F. Prengel, A. Wacker, E. Schöll, S. Murugkar, R. Merlin: Phys. Rev. B, submitted
[2.81] A. Wacker, F. Prengel, E. Schöll: Theory of Multistability and Domain Formation in a Semiconductor Superlattice, in *Proc. 22nd Int. Conf. Phys. Semicond., Vancouver, 1994*, ed. by D.J. Lockwood (World Scientific, Singapore), in print
[2.82] F. Prengel: *Nichtlinearer Ladungstransport in Halbleiter-Übergittern*. Master's thesis, Technische Universität Berlin (1994)

3 Space Charge Instabilities and Nonlinear Waves in Extrinsic Semiconductors

S.W. Teitsworth[1], *M.J. Bergmann*[1], *L.L. Bonilla*[2]

[1] Duke University, Department of Physics and Center for Nonlinear and Complex Systems, Durham NC 27708-0305, USA
[2] Universidad Carlos III de Madrid, Escuela Politécnica Superior, Butarque 15, 28911 Leganés, Madrid, Spain

Abstract. We review the role of nonlinear space charge waves in the electrical conduction properties of extrinsic semiconductors, particularly ultrapure p-type Ge at low temperature. Recent experimental results can be interpreted using a drift-diffusion rate equation model which includes nonlinear electric field dependencies of the drift velocity and impurity capture and impact ionization rates. Under time-independent (dc) current bias, phase plane analysis of a reduced dynamical system indicates the presence of moving solitary waves which are *dynamically unstable*. Under dc voltage bias, numerical simulations for finite-length samples and Ohmic boundary conditions reveal *dynamically stable* solitary waves that travel periodically across the sample. Near the onset of time-dependent behavior we find a Hopf bifurcation to fast small-amplitude current oscillations which are associated with periodic motion of traveling waves that decay before reaching the end of the sample. Numerical estimates of wave speed, wave size and onset behavior are in excellent agreement with available experimental data. Finally, we discuss the sensitivity of dynamical behavior to material parameters such as impurity concentration.

3.1 Introduction

Most semiconductor devices exhibit nonlinear electrical conduction for sufficiently large electric fields or current densities. This behavior is typically manifested in dc current-voltage (I–V) characteristics which are linear (i.e., Ohmic) for small biases, and then deviate from linearity at higher biases as additional microscopic conduction processes are activated due to carrier heating [3.1, 2]. In addition to static nonlinearities, semiconductors have a variety of characteristic time scales including the transit time for a typical charge carrier to cross the sample, the dielectric relaxation time, and trapping times associated with the impurity levels. These time scales describe the return to or departure from steady state behavior.

Neglecting diffusion, the field-dependent current density in a bulk semiconductor can generally be written as

$$j(\mathcal{E}) = ep(\mathcal{E})v(\mathcal{E}), \tag{3.1}$$

where $\mathcal{E}$ is the electric field, p is the (field-dependent) concentration of charge carriers, v is the (field-dependent) drift velocity and e is the electric charge per

Springer Proceedings in Physics, Vol. 79 **Nonlinear Dynamics and Pattern Formation in Semiconductors and Devices**
Editor: F.-J. Niedernostheide

carrier. Ohmic behavior is generally observed for small fields where the mobility approximation (i.e., $v \propto \mathcal{E}$) is valid and the free carrier concentration is field-independent (i.e., $dp/d\mathcal{E} \approx 0$). At larger fields nonlinear I–V behavior occurs because the mobility approximation breaks down, or the carrier concentration becomes field-dependent, or both. The differential conductivity is:

$$\frac{dj}{d\mathcal{E}} = e\left\{p\frac{dv}{d\mathcal{E}} + v\frac{dp}{d\mathcal{E}}\right\} \tag{3.2}$$

and, if either v or p is sufficiently nonlinear, the differential conductivity may become negative giving rise to (voltage-controlled or N-type) negative differential resistance (NDR). We note here that substantial work has also been carried out for current-controlled (also S-type) negative differential conductivity and related instabilities [3.3], but we do not discuss this case further. Using general arguments, Ridley [3.4] showed that voltage-controlled NDR in a bulk material implies that the spatially uniform state is unstable to the formation of high field domains. Thus, NDR is frequently accompanied by the onset of some type of spatio-temporal instability.

Semiconductor instabilities associated with NDR have been studied extensively – in both experiment and theory – since the 1960's. One of the most important examples is the Gunn effect [3.5]: the onset of spontaneous microwave current oscillations in GaAs at room temperature and below due to negative differential mobility which, in turn, is associated with the onset of intervalley scattering of electrons [3.2]. Characteristic frequencies of various space charge instabilities range from a few Hz in extrinsic materials at low temperature [3.6,7], to several GHz in state-of-the-art IMPATT and Gunn diodes [3.2]. Lower frequency instabilities tend be caused by NDR associated with negative differential carrier concentration (i.e., $dp/d\mathcal{E} < 0$ in (3.2)) with oscillation times on the order of impurity trapping times, and these low-frequency oscillations have been observed in Ge [3.6–9], GaAs [3.10,11], InSb [3.12] and other materials. On the other hand, high frequency instabilities are generally associated with negative differential mobility and the oscillation period is typically determined by the carrier transit time.

We focus here on trap-controlled instabilities which occur in extrinsic semiconductors, particularly ultrapure p-Ge at low temperature. We will review the status of the theory of nonlinear space charge waves in the p-Ge system and discuss the extent of agreement with experiments carried out during the last decade. A key element of our approach is the use of singular perturbation techniques and bifurcation methods [3.13–15]. Singular perturbation techniques allow the simplification of the set of equations which comprise the full drift-diffusion model to a "reduced equation" model which appears to capture most of the experimentally observed nonlinear behavior. Bifurcation methods [3.16] are used to study the transitions which occur in the reduced model between time-independent and nonstationary behavior.

A brief chronology of the p-Ge experiments is now given. In 1983, period doubling and temporal chaos associated with rather noisy spontaneous current oscillations were reported in a regime of applied dc voltage just below the threshold for impurity breakdown [3.7]. Experiments were later performed for conditions of dc + ac voltage bias in the pre-breakdown region and also revealed period doubling and temporal chaos as the ac amplitude of voltage was varied [3.17]. These pre-breakdown results were mostly explained in terms of a low-dimensional dynamical model which includes the nonlinear field dependencies of the impurity capture and ionization coefficients but does not retain spatial degrees of freedom [3.18]. Later experiments shifted to the regime of applied voltage above the impurity breakdown threshold and revealed NDR and large amplitude spontaneous current oscillations [3.19]. This post-breakdown instability was then exploited in high precision tests of scaling predictions for the frequency-locking route to chaos [3.20]. Most recently, experiments have capacitively probed the spatial structure of electric fields which accompany the post-breakdown current oscillations and these measurements clearly indicate the presence of solitary waves – in the form of high field domains – which move periodically across the sample [3.21–23].

The organization of this paper is as follows: in Sect. 3.2 we discuss the drift-diffusion model appropriate for the p-Ge system, including a discussion of the relevant transport coefficients and boundary conditions. In Sect. 3.3 we introduce non-dimensional variables and in the process identify important small parameters for the dynamical problem. Taking these small parameters to zero allows us to derive the "reduced equation" model. We first consider the behavior of the reduced model under current bias conditions in a one-dimensional sample of infinite length (i.e., no boundaries) in Sect. 3.4. We treat spatially-uniform stationary states as well as non-trivial stationary states in a co-moving frame, and the latter are seen to correspond to nonlinear space charge waves. Finally, we briefly review theoretical arguments which allow questions of dynamical stability of the nonlinear waves (under current bias) to be addressed. In Sect. 3.5 we summarize results for the case of dc voltage bias and Ohmic boundary conditions. In particular, we construct spatially nonuniform steady states and examine the linear stability thereof. A numerical method is described that produce solitary waves which move periodically and stably across the sample for appropriate voltage levels. Using bifurcation theory supported by numerical simulation we discuss the transition to time-dependent behavior. We conclude in Sect. 3.6 with brief mention of as yet unexplained experimental results and ongoing theoretical and numerical work.

3.2 Full Rate Equation Model

In the drift-diffusion approximation the dynamical equations describing electrical conduction in an extrinsic semiconductor such as p-Ge can be written in the form [3.7,24]:

$$\frac{\partial a_*}{\partial T} = \gamma a_o + p\left[k(\mathcal{E})a_o - r(\mathcal{E})a_*\right], \tag{3.3}$$

$$\epsilon\frac{\partial \mathcal{E}}{\partial T} = j(T) - e\left[pv(\mathcal{E}) - D\frac{\partial p}{\partial X}\right], \tag{3.4}$$

$$\epsilon\frac{\partial \mathcal{E}}{\partial X} = e(p + d - a_*). \tag{3.5}$$

Here $\mathcal{E}$ denotes the electric field; $a_o = a - a_*$, a_* and p represent the neutral acceptor, ionized acceptor and free hole concentrations, respectively. The total acceptor concentration a is a constant. Note that X and T are the dimensional space and time independent variables. Equation (3.3) is a rate equation which describes the local dynamics of trapping and de-trapping of free holes. The term γa_o describes the generation of ionized acceptors from neutral acceptors through absorption of thermal and far-infrared radiation; e.g., γ is proportional to the total photon flux. The rate of impact ionization of neutral acceptors is pka_o, and the rate of hole recombination onto ionized acceptors is pra_*. These processes are illustrated schematically in a conduction band diagram in Fig. 3.1. Equation (3.4) is simply Ampere's law with the drift-diffusion approximation for particle current; v represents the field-dependent drift velocity of free holes and D is the hole diffusivity assumed to be field-independent here. The variable $j(T)$ represents the total current which flows through a hypothetical external circuit. Finally, (3.5) is Poisson's law for the electric field in which d denotes the compensating donor concentration – (all donors are positively charged at temperatures of interest) – and ϵ denotes the semiconductor permittivity.

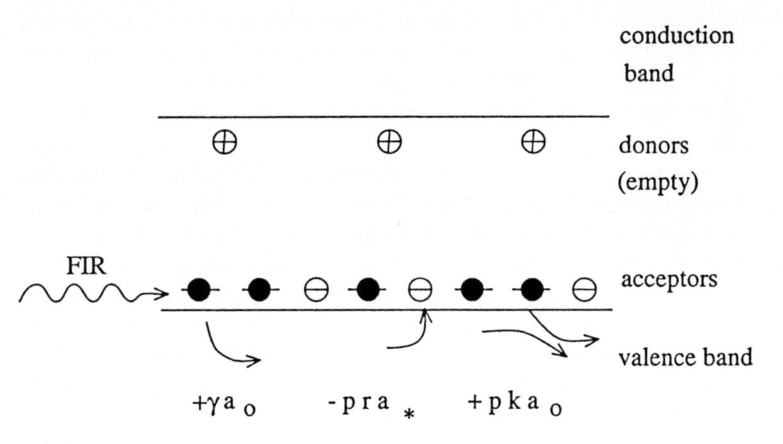

Fig. 3.1. Band diagram of closely compensated p-type germanium, illustrating capture and ionization processes

The detailed physics of these equations is largely determined by the coefficient functions, $v(E)$, $r(E)$, and $k(E)$. We have used the following functional forms in our simulations:

$$v(E) = v_s \left\{ 0.9(E + 0.33) + \frac{1.8}{\pi}(E - 0.5)tan^{-1}[5.0 - 10.0E] \right. \tag{3.6}$$

$$\left. + \frac{0.09}{\pi} ln \left[1.0 + (5.0 - 10.0E)^2\right] \right\},$$

$$r(E) = r_o \left\{ 0.05 + \left[1.04 + 100.0(E)^2\right]^{-1.5} \right\}, \tag{3.7}$$

$$k(E) = k_o \left\{ \left(1.0 + exp\left[\frac{0.55 - E}{0.015}\right] \right)^{-1} \left(0.25 + 2.0 exp\left[\frac{-E}{0.34}\right] \right) + 0.1 \left(\frac{E}{1.15} \right)^4 \right\}. \quad (3.8)$$

These equations are stated with the non-dimensional electric field, E, which is described in Sect. 3.3. Equations (3.6) - (3.8) are based on available data [3.25] and heuristic considerations. The values of r_o, k_o, and v_s are given in Table 3.1. When the electric field is below the breakdown value, on the order of 5 V/cm, the drift velocity increases linearly according to the standard low field mobility approximation, whereas at larger fields the velocity approaches its saturation value v_s due to the onset of various hot electron scattering processes, especially optical phonon emission [3.25]. This behavior is reflected in (3.6). The capture coefficient, which is given in (3.7), is approximately constant for very small fields and then drops off as E^{-3} at larger values of E, reflecting a decrease in the capture cross section as the free carrier kinetic energy increases. The shape of the impact ionization coefficient, (3.8), can be understood as follows: for fields below the breakdown field the average kinetic energy of free holes is insufficient to ionize neutral acceptors and the coefficient k is therefore exponentially close to zero. As the average carrier energy reaches the impurity binding energy, k increases abruptly. Beyond the binding energy threshold k may decrease with field due to a decrease in the cross section for impact ionization [3.24]. Eventually the k coefficient must again increase with field due to the onset of various secondary ionization processes, e. g., optical phonon-induced ionization. Negative differential behavior in k is a key element for NDR and instabilities in our model. The actual coefficient functions used in our numerical simulations are plotted in Fig. 3.2, which also includes the differential mobility, $\mu(\mathcal{E})$.

In the case of a finite sample of length l we must append boundary conditions to the model. In general, the boundary conditions will depend on specific properties of the metal-semiconductor contacts employed. Reasonably realistic macroscopic boundary conditions are [3.26]:

$$\epsilon \frac{\partial \mathcal{E}}{\partial T} + j_{con}(\mathcal{E}) = j(T) \text{ at } X = 0 \text{ and } X = l. \quad (3.9)$$

Here $j_{con}(\mathcal{E})$ is a particular function which models the contact. For contacts with low effective resistivity it is generally acceptable to use the linear (i. e., Ohmic) approximation for the contact characteristic resulting in:

$$j_{con}(\mathcal{E}) = \frac{\mathcal{E}}{\rho_{con}}, \quad (3.10)$$

where ρ_{con} denotes the effective contact resistivity.

The total external current density $j(T)$ is determined by the type of bias condition. In the case of current bias, $j(T)$ denotes a specified function of time which can be inserted into the above dynamical model (i. e., (3.3) - (3.10)). However, experiments commonly use voltage bias and, in this case, we must append

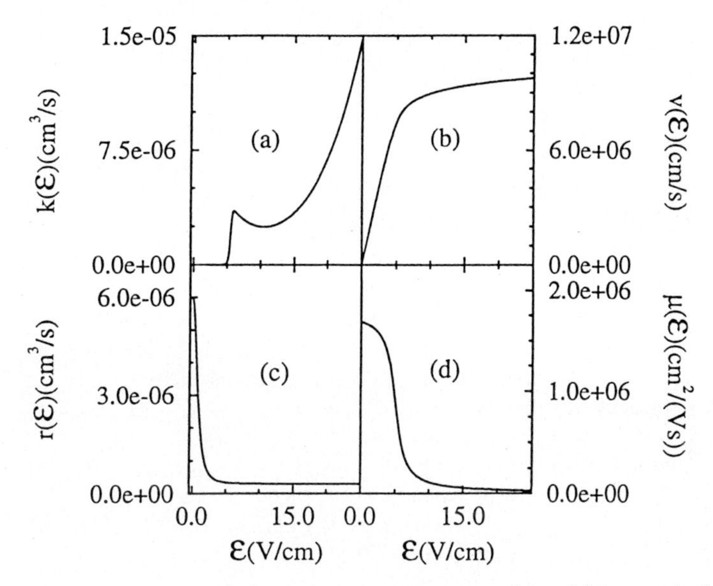

Fig. 3.2. Electric field dependent transport quantities: (a) impact ionization coefficient, (b) drift velocity, (c) capture coefficient, and (d) mobility

to the model the following integral constraint on electric field:

$$\int_{o}^{l} \mathcal{E}(X,T)dX = \Phi, \tag{3.11}$$

where Φ is the total applied voltage. Solving the dynamical equations plus initial and boundary conditions and one type of bias condition now constitutes a well-posed problem. In the above discussion we treat spatial dynamics parallel to current flow and have implicitly neglected transverse dynamics. For the p-Ge system this can be justified because dielectric relaxation occurs rapidly and stably in transverse directions [3.27].

In the table below we indicate various parameter values used in our simulations. They are typical of high purity p-Ge samples used in the experiments of Kahn et al. [3.21,22,24]. At the bottom of this table we have also introduced the two important time scales in the p-Ge system: the impact ionization time T_2, which is the characteristic time for trapping and de-trapping of the free holes relative to acceptors, and the dielectric relaxation time T_1, which is the characteristic time for free holes to move in response to abrupt perturbations in electric field.

3.3 Non-Dimensionalization and the Reduced Model

In order to cast the dynamical model into non-dimensional form we must first identify convenient scales with which to measure time, position, electric field and the various concentrations. By convenient, we mean that typical variations

Table 3.1. Physical parameters and time scales used in the analysis and numerical simulation of the Ge system

Quantity	Value
donor concentration	$d = 8 \times 10^{10}$ cm^{-3}
shallow acceptor concentration	$a = 9.68 \times 10^{10}$ cm^{-3}
compensation ratio	$\alpha = a/d = 1.21$
low field hole mobility	$\mu_o = 1.0 \times 10^6$ cm^2/(Vs)
saturation velocity	$v_s = 1.0 \times 10^7$ cm/s
permittivity	$\epsilon = 16\epsilon_o$
injecting contact resistivity	$\rho_{con} = 585$ Ωcm
lattice temperature	$\mathcal{T} = 4.2$ K
impact ionization coefficient	$k_o = 6.0 \times 10^{-6}$ cm^3/s
infrared generation coefficient	$\gamma = 1.0 \times 10^{-4}$ s^{-1}
recombination ionization coefficient	$r_o = 6.0 \times 10^{-6}$ cm^3/s
diffusivity	$D = \mu_o k_b \mathcal{T}/e = 3.6$ cm/s
impact ionization time	$T_2 = 1/(k_o d) = 2.1 \times 10^{-6}$ s
dielectric relaxation time	$T_1 = \epsilon/(ed\mu_o) = 1.1 \times 10^{-10}$ s
sample length	$l = 1.45$ cm
sample cross sectional area	$A = 0.16$ cm^2

of these quantities across the p-Ge sample will be of order 1 in non-dimensional units. We define the following non-dimensional variables:

$$\begin{gathered} E = \mathcal{E}\mu_o/v_s,\ P = p/d,\ A = a_*/d - 1,\ \tau = T/T_2, \\ J(\tau) = j(T)/(edv_s),\ x = X\mu_o ed/\epsilon v_s,\ V(E) = v(E)/v_s, \\ K(E) = k(E)/k_o,\ R(E) = r(E)/k_o,\ \rho_o = \rho_{con} e\mu_o d. \end{gathered} \tag{3.12}$$

In the above relations the electric field is scaled in terms of the approximate value where velocity saturation sets in and the various concentrations are all measured in terms of the compensating donor concentration d. The characteristic length scale used to define the non-dimensional position variable x is estimated from Poisson's law and represents the typical length over which the electric field changes due to a net charge density on the order of d. We choose to measure time in terms of the relatively slow impact ionization time T_2 because many of the dynamical effects we are seeking to understand occur on this time scale. Current density is measured in terms of the saturation current density and – knowing scales for current density and electric field – the appropriate scaling for contact resistivity follows immediately. Finally, the field-dependent velocity, impact ionization coefficient and capture coefficients are measured in terms of the saturation velocity v_s and the constant k_o.

We may now write the dynamical equations, (3.3) - (3.5), from the preceding section in non-dimensional form:

$$\frac{\partial A}{\partial \tau} = \frac{\Gamma(\alpha - 1 - A)}{\beta} + P[(\alpha - 1)K(E) - R - (K + R)A], \tag{3.13}$$

$$\beta \frac{\partial E}{\partial \tau} = J(\tau) - V(E)P + \delta \frac{\partial P}{\partial x}, \tag{3.14}$$

$$\frac{\partial E}{\partial x} = P - A, \tag{3.15}$$

where we have introduced the following dimensionless small parameters:

$$\begin{aligned} \beta &= T_1/T_2 = 5.2 \times 10^{-5}, \\ \delta &= \mu_o e d D/(\epsilon v_s^2) = 0.04, \\ \Gamma &= \epsilon\gamma/(m u_o e d) = 1 \times 10^{-12}. \end{aligned} \tag{3.16}$$

In the spirit of singular perturbation theory we set the small parameters β, δ and Γ/β to zero which gives the "outer" approximation to the full set of equations, (3.13) - (3.15). Physically, this corresponds to neglecting the displacement current, the diffusion current and the far-infrared generation term. Upon studying the properties of the "outer" solutions, we may find unphysical behavior such as the development of discontinuities (i.e., shocks). Then we must go back to full model and solve the appropriate "inner" problem in the shock region. After that, we would need to match the inner and outer solutions in order to construct a complete solution. We also note that neglecting the Γ term is only reasonable for field strengths above the impurity breakdown threshold. Below breakdown the generation term is critical in determining the photoconductivity properties that are important in practical p-Ge devices.

Setting β and δ to zero allows us to eliminate A and P in favor of E and $\partial E/\partial x$ using (3.14) and (3.15). If we further take Γ/β to zero, we arrive at the following nonlinear second order partial differential equation of hyperbolic type which we call the reduced equation [3.15, 28]:

$$\frac{\partial^2 E}{\partial x \partial \tau} + c_1(E,J)\frac{\partial E}{\partial \tau} + c_2(E,J)\frac{\partial E}{\partial x} + c_3(E,J) = V(E)^{-1}\frac{dJ}{d\tau} \tag{3.17}$$

with coefficients

$$\begin{aligned} c_1(E,J) &= \frac{J}{V(E)^2}V'(E), \\ c_2(E,J) &= \frac{J[K(E)+R(E)]}{V(E)}, \\ c_3(E,J) &= \left[\left\{\frac{\alpha K(E)}{K(E)+R(E)} - 1\right\} V(E) - J\right]\frac{c_2(E,J)}{V(E)}, \end{aligned} \tag{3.18}$$

where

$$V'(E) \equiv \frac{dV}{dE} \tag{3.19}$$

is the dimensionless differential mobility.

Finally, we must write the boundary conditions and bias constraints in nondimensional form. Since (3.17) is first order in its spatial derivatives we can impose only one boundary condition which we do at the injecting contact:

$$E(0,\tau) = \rho_o J(\tau), \tag{3.20}$$

where ρ_0 is the dimensionless contact resistivity. Of course, a similar boundary condition also holds at the receiving contact for the full model. Bonilla [3.15] has shown that this boundary condition can be recovered by inserting a diffusive boundary layer near the receiving contact, that is, by solving and matching the appropriate inner problem to solutions of the reduced model. For the voltage bias case we must also append the integral constraint on E in non-dimensional form,

$$\int_0^L E(x,\tau)dx = \phi, \tag{3.21}$$

where L is the non-dimensional sample length and ϕ is the non-dimensional applied voltage.

3.4 The Case of Time-Independent (dc) Current Bias

3.4.1 Steady States and the *J*-*E* Curve

We now turn to the case of dc current bias in a sample of infinite length. Because we are looking for traveling wave solutions it is convenient to transform the reduced model equation (3.17) to a co-moving reference frame: we introduce the following change of independent variables, $\xi = x - C\tau$ where C is a wave speed parameter expected to be of order 1. Then (3.17) is rewritten as follows:

$$\begin{aligned} \frac{\partial^2 E}{\partial\tau\partial\xi} - \left(\frac{J}{V}\right)'\frac{\partial E}{\partial\tau} = C\frac{\partial^2 E}{\partial\xi^2} - J\left[\left(\frac{C}{V}\right)' + \frac{K+R}{V}\right]\frac{\partial E}{\partial\xi} \\ + \frac{J}{V^2}(K+R)(J-\rho V), \end{aligned} \tag{3.22}$$

where

$$\rho = \frac{\alpha K(E)}{K(E)+R(E)} - 1.0. \tag{3.23}$$

Note that $' \equiv d/dE$ through out the rest of the chapter. In analyzing the properties of (3.22) we must first examine the spatially uniform steady state solutions which are equivalent to fixed points in the terminology of nonlinear dynamics. This is easily accomplished by setting all derivatives to zero, which leaves the following simple equation:

$$J = \rho(E)V(E). \tag{3.24}$$

A key point to note is that this fixed point equation gives the J-E curve for the sample which is closely related to the experimentally measurable dc I–V curve. Numerical evaluations of (3.24) for several different compensation ratios are shown in Fig. 3.3. For compensation ratio α close to 1 a region of NDR is clearly evident just above the breakdown electric field. Impurity breakdown occurs just above 5 V/cm in the figure and is marked by a sharp increase in current density by several orders of magnitude. We note that the J-E curves are not valid below breakdown precisely because (3.24) neglects the far-infrared generation term discussed in preceding sections.

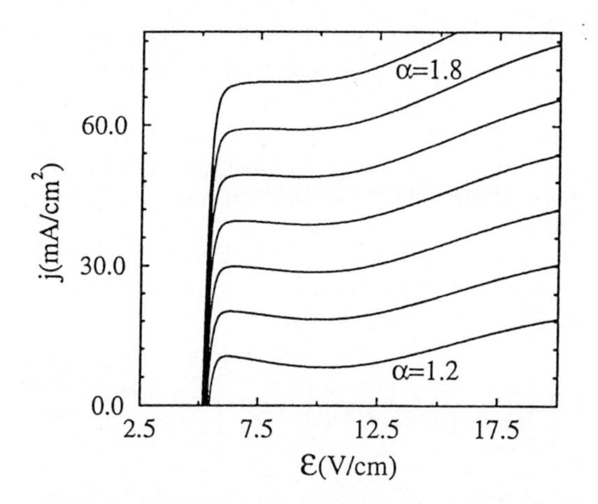

Fig. 3.3. Current density vs. electric field for evenly spaced values of the compensation ratio, α

The presence of NDR in the J-E curves is a direct consequence of the negative differential behavior in the impact ionization coefficient $K(E)$ [3.29]. This is most easily seen by computing the differential conductivity from (3.24) and making the approximations that $R/K \ll 1$ and $R' \approx 0$, both of which are quite reasonable for electric fields above breakdown (refer to Fig. 3.2). A straightforward calculation gives:

$$\frac{dJ}{dE} \approx \alpha \left(\frac{R}{K}\right) \left[\frac{K'}{K} V - V'\right] + (\alpha - 1) V'. \tag{3.25}$$

The first term on the right-hand side is always negative in regions of negative differential impact ionization rate. The second term – which is positive definite – can always be made to have smaller magnitude than the first by choosing the compensation ratio α to be sufficiently close to 1. On the other hand, as α increases away from 1 the second term will eventually dominate the first term (in the limit where R/K is small) and, thus, for larger compensation ratio we predict that NDR will not occur. This behavior is clearly evident in Fig. 3.3 where we see that increasing α from 1.2 to 1.8 leads to a gradual diminution of the NDR region. Using the numerical forms for transport coefficients stated above in (3.6) - (3.8) we find that for α in excess of 1.825 NDR completely disappears. It is interesting to note that all p-Ge experiments showing solitary wave phenomena were performed on closely compensated samples. Some samples were studied which did not show these phenomena and one possible explanation is that their compensation ratio was too large for NDR to occur. Nonetheless, our prediction of this sensitive dependence of NDR on the precise value of α has not been thoroughly tested in experiment.

When the current level is set in the NDR region we see that there are three possible values of electric field consistent with that current. Equivalently, (3.24) has three solutions for a given J. We denote these three solutions in increasing order as E_1, E_2, and E_3. It is interesting to note that both E_1 and E_3 are increasing functions of J while E_2 is a decreasing function of J. Furthermore, it is useful to define J_m and J_M as the respective minimum and maximum current densities for which (3.24) has multiple solutions.

3.4.2 Periodic, Solitary and Monotone Wave Solutions

We now search for nontrivial traveling wave solutions to (3.22). Since rigid traveling waves appear as spatially non-uniform steady states in a co-moving reference frame we drop the time derivatives from (3.22) which gives:

$$C\frac{\partial^2 E}{\partial \xi^2} - J\left[\left(\frac{C}{V}\right)' + \frac{K+R}{V}\right]\frac{\partial E}{\partial \xi} + \frac{J}{V^2}(K+R)(J-\rho V) = 0. \qquad (3.26)$$

This is a two-dimensional nonlinear ordinary differential equation and it is useful to think of C as a bifurcation parameter which can be fine-tuned to give interesting waveforms. This equation can be productively studied using phase plane techniques [3.16] where the phase plane axes are taken as E and $E_\xi \equiv dE/d\xi$. In treating such a dynamical system it is useful to first identify the fixed points and their stability type. Here we are talking about stability in the phase plane which has nothing to do *a priori* with dynamical stability properties of the full time-dependent problem.

Provided the current level J is in the NDR range – i.e., $J_m < J < J_M$ – there are three fixed points. These are obtained from (3.26) by setting all derivatives to zero and are simply E_1, E_2, and E_3 defined in the preceding section. Linearization of (3.26) about these fixed points gives their stability types: E_1 and E_3 are saddle points (i.e., with one stable and one unstable manifold) while E_2 is a stable or unstable spiral point or a node depending on the value of C. Of course, if J is outside the NDR region there is only one fixed point which is of saddle type in the phase plane.

Periodic and solitary wave solutions correspond to closed trajectories (i.e., orbits) in the phase plane. If J is outside the NDR region then there is only one fixed point of saddle type and a closed trajectory can be ruled out on the basis of general topological arguments. Thus, nonlinear wave solutions are only possible for current biases J inside the NDR region, $J_m < J < J_M$. To find the closed trajectories we conceive of C as a bifurcation parameter and find that closed orbits are produced by Hopf bifurcation about the middle fixed point E_2. Performing a linear stability analysis about E_2 one can easily show that the stability eigenvalues form a pure imaginary pair for a unique wave speed $C = C^*$ where [3.13]:

$$C^* = \left.\frac{(K+R)V}{V'}\right|_{E_2} \qquad (3.27)$$

and provided that the following two conditions are satisfied:

$$\rho'|_{E_2} < -\left.\frac{JV'}{V^2}\right|_{E_2} < 0. \qquad (3.28)$$

The Hopf bifurcation can be subcritical, in which case small-amplitude limit cycle orbits occur for wave speeds above the bifurcation value, or it can be supercritical, and in this case the limit cycles occur for wave speeds decreasing from the bifurcation value. We have performed normal form calculations to determine the Hopf bifurcation type and these are reported in [3.13]. In numerical simulations we have found that both types of Hopf bifurcation occur depending on the precise value of current bias: nearer to J_m the bifurcation tends to be subcritical while nearer to J_M it tends to be supercritical.

In both the sub- and super-critical cases as C departs from C^* the amplitude of the limit cycle grows smoothly until it collides with one of the saddle points (i.e., E_1 or E_3) at a unique value of wave speed $\overline{C}$. At this point the limit cycle becomes a homoclinic orbit in the phase plane: it departs from the saddle point along the unstable direction, loops around E_2 once, and then returns to the same saddle point along the stable direction. Plotting these closed orbits in the $E-\xi$ plane we see that a homoclinic orbit emanating from E_1 (E_3) corresponds to a solitary wave in the form of a high (low) field domain. The high field domain case is indicated in Fig. 3.4a where we show the homoclinic orbit in the phase plane on the left-hand side and the corresponding solitary wave on the right-hand side. For a unique value of J the limit cycle will collide simultaneously with both saddle points and then two heteroclinic trajectories are formed. When these are plotted in the $E-\xi$ plane it is easily seen that they correspond to increasing and decreasing monotone wavefronts as shown in Fig. 3.4b. Finally, we note that the limit cycle orbits can also be plotted in the $E-\xi$ plane and they correspond to simple periodic waves with amplitude uniquely determined by the values of C and J.

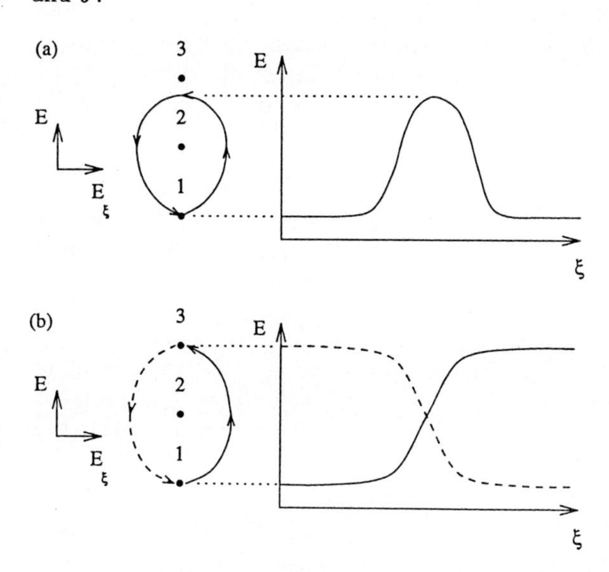

Fig. 3.4. Phase plane (left) and real space (right) illustrations of (a) the solitary wave and (b) the monotone waves. The labeled points, 1, 2, and 3, are the fixed points and the arrows indicate the direction of phase plane trajectories

We conclude this subsection by noting that the analytical criterion for C^* in (3.27) provides a very good estimate for the actual speed $\overline{C}$ of the solitary and monotone waves. We have verified this behavior in several numerical simulations and find that these two speeds are generally no more than 10% apart.

3.4.3 Dynamical Stability of Periodic, Solitary and Monotone Waves

Having demonstrated the existence of various traveling wave solutions to the current bias problem we now turn to the question of the dynamical stability of these waves, with the understanding that – in order to be experimentally observable – they should be stable with respect to small perturbations. To address this point we must perform a linear stability analysis of (3.22) about the traveling wave solution which we denote by $E^\circ(\xi)$. We assume the following separation of variables form for $E(\xi,\tau)$:

$$E(\xi,\tau) = E^\circ(\xi) + \eta\varphi(\xi)e^{\lambda\tau} \ , \ \eta \ll 1. \tag{3.29}$$

After substituting this into (3.22) and retaining only linear terms in the small parameter η we obtain the following eigenvalue problem:

$$-C\varphi_{\xi\xi} + [\lambda + f_1(\xi)]\,\varphi_\xi + [\lambda f_2(\xi) + f_3(\xi)]\,\varphi = 0, \tag{3.30}$$

where the functions f_1, f_2 and f_3 are specified by:

$$f_1(\xi) \equiv J\left[\left(\frac{C}{V}\right)' + \frac{K+R}{V}\right]\Bigg|_{E^\circ(\xi)}, \tag{3.31}$$

$$f_2(\xi) \equiv c_2(E^\circ(\xi), J), \tag{3.32}$$

$$f_3(\xi) \equiv \frac{df_1}{d\xi}\frac{dE^\circ}{d\xi} - \frac{d}{d\xi}\left\{J\frac{K+R}{V^2}[J-\rho V]\right\}\Bigg|_{E^\circ(\xi)},$$

and $\varphi_\xi \equiv d\varphi/d\xi$. Clearly if any eigenvalues λ exist with positive real part, then the traveling wave state is unstable. Bonilla and Teitsworth [3.13] showed that all periodic traveling waves are unstable using arguments based on standard Floquet theory combined with some recent work of Maginu on the Nagumo equation for nerve conduction [3.30].

We turn to the case of solitary and monotone waves. It is relatively easy to show that $\lambda = 0$ is always an eigenvalue of (3.30) and that the corresponding eigenfunction is $dE^\circ/d\xi$ which decays to zero exponentially as ξ approaches plus or minus infinity. Following Bonilla and Vega [3.31] we now introduce the following change-of-variable in order to eliminate first derivative terms:

$$\varphi(\xi) = \psi(\xi)e^F, \ F = \frac{1}{2C}\left\{\lambda\xi + \int f_1(\xi)d\xi\right\}. \tag{3.33}$$

Substituting this into (3.30) we obtain a transformed eigenvalue problem:

$$-\psi_{\xi\xi} + \left\{\Lambda^2 + w(\xi)\Lambda + \Omega(\xi)\right\}\psi = 0 \tag{3.34}$$

with

$$\Lambda \equiv \frac{\lambda}{2C}, \tag{3.35}$$

$$w(\xi) \equiv J \left[\frac{CV' + (K+R)V}{CV^2} \right] \Bigg|_{E^\circ(\xi)} > \Delta > 0, \tag{3.36}$$

and

$$\Omega(\xi) \equiv \left[\frac{f_1(\xi)}{2C} \right]^2 - \frac{3}{2C} \frac{df_1}{d\xi} \frac{dE^\circ}{d\xi} + \frac{1}{C} f_3(\xi), \tag{3.37}$$

where the small positive parameter Δ is used in (3.36) to make clear that w is positive and bounded away from zero. Equation (3.34) is in the form of a Schrödinger equation in which the eigenvalue enters quadratically rather than linearly. Bonilla and Vega [3.31] have used techniques from spectral theory to prove the existence of a one-to-one correspondence between eigenvalues of (3.30) and (3.34). Thus, results obtained for the spectrum of (3.34) are directly applicable to the original stability problem. In particular, if there are eigenvalues such that $Re(\Lambda) > 0$ then the wave is unstable.

We note that a solution to (3.34) with $\Lambda = 0$ is:

$$\psi_o(\xi) = \frac{dE^\circ}{d\xi} exp \left[-\frac{1}{2C} \int f_1(\xi) d\xi \right], \tag{3.38}$$

where E° refers to either the solitary or monotone wave state. If E° refers to a solitary wave then $dE^\circ/d\xi$ and hence ψ_o has one zero. In the context of the one-dimensional Schrödinger equation, such a function corresponds to the first excited state. On the other hand, for a monotone wave $dE^\circ/d\xi$ has no nodes and ψ_o would then correspond to the ground state. Bonilla and Vega [3.31] have been able to establish a rigorous connection between (3.34) and the standard theory of one-dimensional Schrödinger equations. In particular they show that for $Re(\Lambda) > -\Delta/2$, Λ is necessarily real. Then they consider an auxiliary Schrödinger equation

$$-\psi_{\xi\xi} + \left\{ \Lambda^2 + w(\xi)\Lambda + \Omega(\xi) \right\} \psi = \mu(\Lambda)\, \psi, \tag{3.39}$$

and prove that the "effective" energy parameter $\mu(\Lambda)$ is a monotonic increasing function of Λ, with one of the eigensolutions being ψ_o as in (3.38) and corresponding eigenvalue $\mu(0) = 0$. Notice that, in general, those values of Λ for which there are solutions of (3.39) with $\mu(\Lambda) = 0$ correspond to eigenvalues of (3.34). Now, if E° corresponds to a monotone wavefront (resp. solitary wave), $\mu(0) = 0$ is the ground state (resp. first excited state) of (3.39) with $\Lambda = 0$. Since $\mu(\Lambda)$ is increasing, $\mu(\Lambda) > 0$ for monotone wavefronts when $\Lambda > 0$. Therefore, all zeros of $\mu(\Lambda)$ occur for $Re(\Lambda) < 0$ and the wavefronts are linearly stable. For solitary waves there is a ground state of (3.39) with $\mu(0) < 0$. Thus, there will be a positive Λ^* which makes $\mu(\Lambda^*) = 0$, and the solitary wave is dynamically unstable. Because the monotone wave occurs only for a unique value of current bias J it would be difficult to observe in a current bias experiment. The slightest fluctuation in J due, for example, to external noise would convert it to a very wide but unstable solitary wave; in this sense the monotone wave is not structurally stable.

3.5 The Case of dc Voltage Bias

3.5.1 Position-Dependent Steady States

The spatio-temporal dynamics of the electric field in a p-Ge sample of finite length and under voltage bias is described within the reduced model by (3.17)-(3.21). In order to understand the nature of solutions to this problem we first turn to the case of the steady states which satisfy the following equation:

$$c_2(E_{ss}, J)\frac{\partial E_{ss}}{\partial x} + c_3(E_{ss}, J) = 0, \tag{3.40}$$

where $E_{ss}(x)$ denotes the steady state and the coefficients c_2 and c_3 have been specified previously. This equation is obtained from (3.17) by eliminating all time derivatives and is subject to a boundary condition and voltage constraint similar to that for the full problem, that is:

$$E_{ss}(0) = \rho_\circ J \tag{3.41}$$

and

$$\int_0^L E_{ss}(x)dx = \phi. \tag{3.42}$$

An analytical solution to these equations can be obtained by straightforward integration. The exact steady state is thus given in implicit form by:

$$x = -\int_{\rho_\circ J}^{E_{ss}(x)} \frac{c_2(E, J)}{c_3(E, J)} dE, \tag{3.43}$$

with the current J determined by the voltage constraint. Bonilla has carried out an extensive analytical study of these steady states both inside and outside of the NDR region and we summarize some of his results here [3.15]. It can be shown that J is a strictly increasing function of applied voltage ϕ. This immediately implies that there is a unique steady state under dc voltage bias for each value of ϕ. Concerning the shape of the $J - \phi$ curve we note that, for a long sample (i.e., $L \gg 1$), the $J - \phi$ curve follows the $J - E$ curve (see Fig. 3.3) closely outside the NDR region provided the E-axis is multiplied by the sample length to give voltage units. However, inside the NDR region, it has been shown that the $J - \phi$ curve is almost flat with small positive slope, while the $J - E$ curve shows explicit NDR.

Within the NDR region, there are three basic forms of steady state solution which depend critically on the contact resistivity $\rho_\circ$. This is easy to see by superposing the contact characteristic $J = E/\rho_\circ$ onto the same graph as the sample $J - E$ curve. For intermediate values of contact resistivity typical of p-Ge experiments the contact characteristic intersects the sample characteristic in the NDR region, that is, the intersection occurs for a value of current $J = J_c$ such that $J_m < J_c < J_M$. In this case, the contact characteristic also has at least one more intersection for low current $J < J_m$ on the stable low-field branch of the sample characteristic. Examining (3.43) for the steady state solution it is not hard to show that $E_{ss}(x)$ has the following approximate form:

$$E_{ss}(x) \approx \begin{cases} E_2(J), & 0 \le x < \overline{x} - \Delta x \\ E_1(J), & \overline{x} + \Delta x \le L, \end{cases} \tag{3.44}$$

where Δx refers to the approximate width of a transition layer connecting E_1 and E_2 and $\overline{x}$ refers to the position of the transition layer in the sample. The parameter $\overline{x}$ is determined by the voltage bias constraint and one can also show that Δx is much smaller than both $\overline{x}$ and the sample length. Equation (3.44) tells us that the position-dependent steady state electric field starts at the injecting contact with value E_2, remaining approximately constant as one moves into the sample, then undergoes a relatively abrupt transition about $\overline{x}$ to the value E_1, and remains constant at E_1 up to the receiving contact. For the purposes of stability analyses presented later, we frequently assume that the transition width can be taken to zero so that $E_{ss}(x)$ is a piecewise constant function. In this case the position of the transition layer is determined by:

$$\overline{x} = \frac{\phi - E_1 L}{E_2 - E_1} + O(1), \tag{3.45}$$

which tells us that, as the voltage increases the transition layer moves farther into the sample. In Fig. 3.5 we plot the result of numerical integration of (3.40) for an applied (dimensional) voltage of 6.048 V/cm, which confirms and illustrates the above predictions.

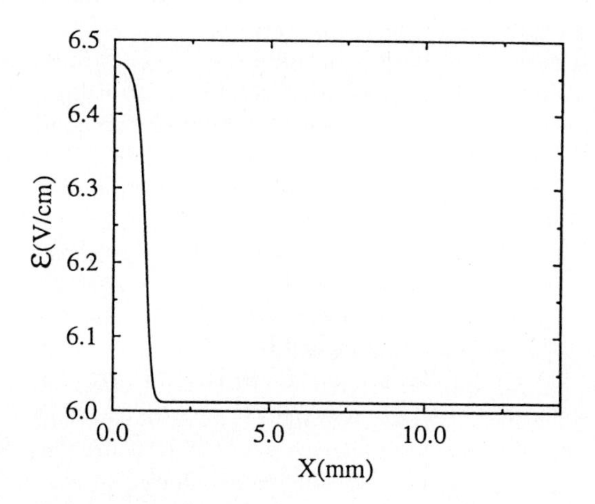

Fig. 3.5. The spatial profile of the electric field for $\Phi/l = 6.048$ V/cm, slightly below the onset value for time-dependent behavior

We turn briefly to the cases of high and low contact resistivity which have been exhaustively treated by Bonilla [3.15]. For very small contact resistivity the contact characteristic rises with a very steep slope and does not intersect the sample characteristic at all. In the NDR region the steady state electric field starts at the injecting contact with a very small value, namely $E = \rho_0 J$ and immediately rises to the field value E_1 as one moves into the sample for $\phi/L < E_M$, where E_M denotes the local maximum of the sample J-E characteristic. Recall that E_1 has been defined in a preceding section as the smallest root of the equation $J = J(E)$. For a long sample, the J-ϕ curve is similar to the case of intermediate contact resistivity described above with J_c pinned at J_M. Furthermore, the steady state electric field profile has a transition layer from $E_1(J_M)$ to $E_3(J_M)$ at a position $\overline{x}$ determined by the voltage constraint. The case of very large contact resistivity is analogous except that electric field starts at the injecting contact with a very large field, then immediately decays to the field value E_3 or E_1 as one moves into the sample, depending on the voltage constraint.

3.5.2 Numerical Studies of Solitary Waves

We now present the results of numerical simulation of the space- and time-dependent reduced model subject to the dc voltage bias constraint, equations (3.17) to (3.21). The numerical method is somewhat complicated due to the non-local nature of the voltage bias constraint. A detailed account will be presented elsewhere [3.32] and we content ourselves here with a brief summary of the main features.

The computer algorithm we use to integrate (3.19) under voltage bias is centered around a large matrix solver. The matrix to be solved arises from an explicit finite difference approximation to the reduced equation, where we are simultaneously solving a set of coupled equations for the time advanced electric field profile. We discretize the spatial derivatives with a second-order upwind finite difference scheme and advance the solution in time with a second-order Runge-Kutta method [3.33]. The hyperbolic nature of the reduced equation forces us to use upwind spatial differencing in order to obtain numerical stability. The pseudolinear nature of the reduced equation allows a simple matrix formulation of the problem. The discretized reduced equation and global voltage boundary condition, along with the injecting contact boundary condition, produces a doubly-bordered band matrix. With some manipulation, we reduce the system to an equivalent system of two tridiagonal matrix equations. We solve the tridiagonal system with a standard LINPACK matrix solver.

The two important types of spatio-temporal electric field behavior observed in numerics are shown in Figs. 3.6 and 3.7. These two figures also indicate the spatio-temporal behavior of the net space charge density ρ/e, the hole concentration p and the net concentration of trapped negative space charge $a_* - d$. The two plots (a) and (b) in Fig. 3.8 indicate the time-dependent currents corresponding to voltage bias conditions used in Figs. 3.6 and 3.7, respectively. We calculate currents assuming a sample cross-sectional area of 0.16 cm^2 which is equivalent to experimental values; we also assume an intermediate value of contact resistivity as defined in the previous subsection.

Figure 3.6 corresponds to an average field $\Phi/l = 6.055$ V/cm just above the threshold field for which the steady state first becomes unstable. The threshold

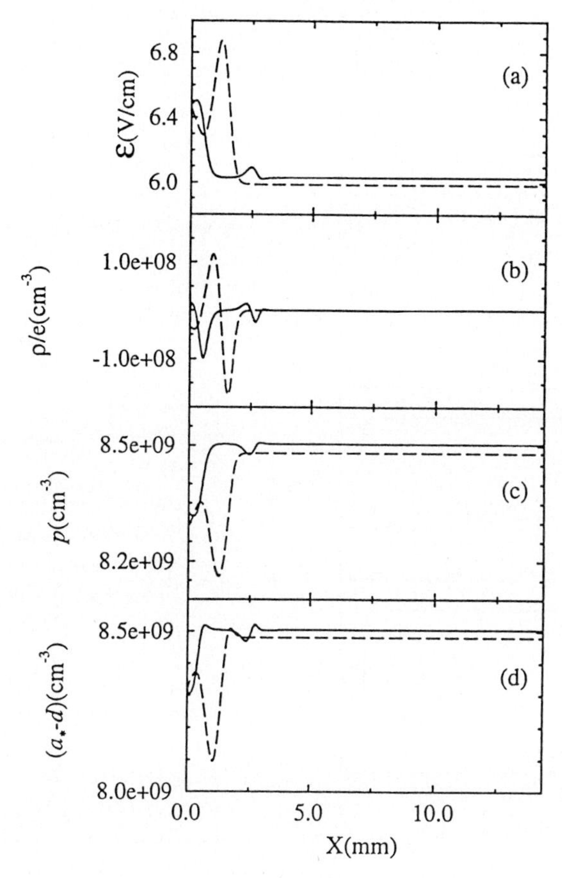

Fig. 3.6. The spatial profiles of (a) the electric field, (b) the net charge concentration, (c) the free hole concentration, and (d) the net negative trapped charge concentration for a small amplitude fast oscillation state. The average electric field, Φ/l, is 6.055 V/cm. The dashed and solid curves are separated by approximately 1/2 of a period of oscillation

field is defined as Φ_α/l and is approximately equal to 6.049 V/cm for parameters used in the simulations. Most of the space- and time-dependent behavior is strongly localized near the injecting contact. A wave of larger electric field is injected into the sample and then decays rapidly, having "died out" completely before reaching the receiving contact. As the wave begins to die a new wave is injected at the contact in order to maintain constant voltage. The solid trace in Fig. 3.6b is a snapshot taken as one wave is dying away in the sample and a new one is just being injected, while the dashed trace shows the wave when it achieves maximum amplitude. Figure 3.6b makes clear that the region of high electric field is associated with a space charge dipole as expected from Poisson's law. This dipole is formed by the net positive and negative space charge con-

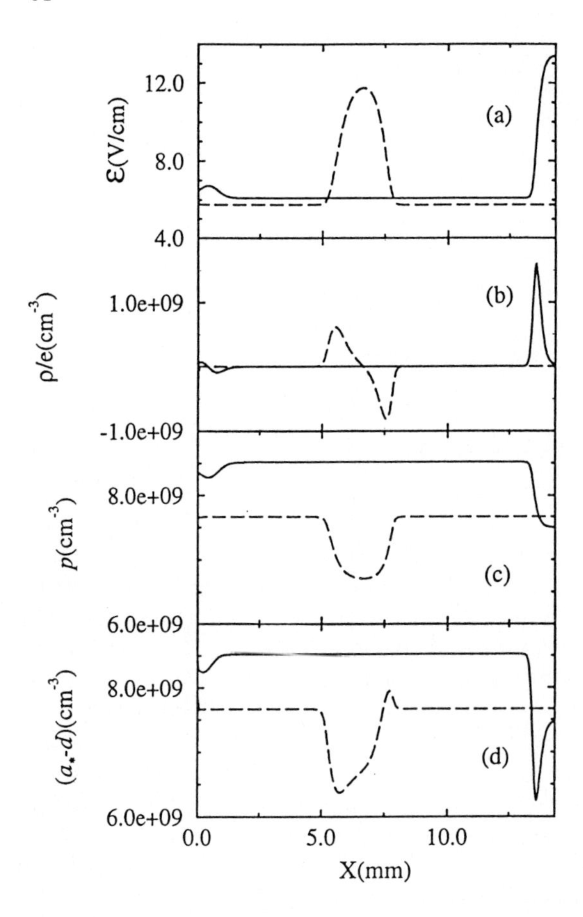

Fig. 3.7. The spatial profiles of (a) the electric field, (b) the net charge concentration, (c) the free hole concentration, and (d) the net negative bound charge concentration for a stable solitary wave state. The average electric field, Φ/l, is 6.5 V/cm. The two curves are separated in time by approximately 1/3 of a period of oscillation

centrations depicted in parts (c) and (d) of the figure. The current corresponding to this behavior is shown in Fig. 3.8a and we clearly detect the presence of small-amplitude oscillations with frequency 2.36 kHz. These oscillations are fairly sinusoidal in shape which suggests that they may be associated with a supercritical Hopf bifurcation of the steady state [3.16], an idea we pursue below. The frequency and small amplitude of these oscillations are in good agreement with noisy current oscillations observed in p-Ge experiments near the onset of time-dependent behavior [3.21].

Figure 3.7 corresponds to an average applied electric field of $\Phi/l = 6.5$ V/cm and in part (a) we clearly see propagating solitary waves in the form of high field domains. As in Fig. 3.6 above we show two snapshots of the field profile: the dashed profile shows the domain moving through the middle of the sample,

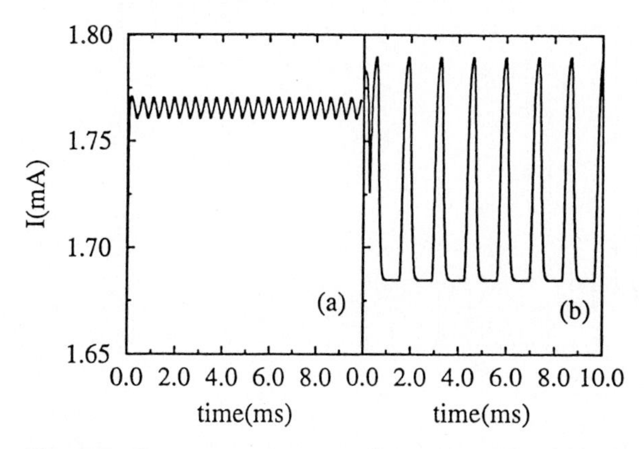

Fig. 3.8. Current vs. time associated with (a) the decaying waves of Fig. 3.6 and (b) the stable solitary waves of Fig. 3.7

while the solid profile shows a mature domain leaving the sample at the right-hand (i.e., receiving) contact with a new domain being injected at the left-hand contact. Figure 3.7b shows that the high-field domain is associated with a space charge dipole, comprised of the free and trapped charge contributions shown in parts (c) and (d), resp. It is apparent that the trapped charge lags behind the free charge, illustrating that the wave velocity is "controlled" by the characteristic time scale for impurity trapping. By way of contrast, for the Gunn effect in n-GaAs the domain velocity is determined by free carrier or drift velocity. Figure 3.8b shows the corresponding time-dependent current which is strongly non-sinusoidal and has relatively large amplitude. The shape of the current trace is easily understood: the flat portion corresponds to those times when the domain is moving in the middle of the sample (i.e., far from the contacts), while the pulses correspond to those times when an old domain is leaving and new domain is entering the sample thus providing an additional contribution to current. The speed and shape of numerically generated solitary waves as well as the calculated form of current traces are all in excellent agreement with p-Ge experiments. We also note that the numerical velocity of solitary waves are closely approximated by (3.27) for the wave velocity under dc current bias.

Our numerical results can be concisely interpreted with the help of the global bifurcation diagram (time-averaged current vs. dc voltage bias) shown in Fig. 3.9a. The time-averaged current appears as the solid dots in this figure while the vertical bars denote the extrema of current oscillations. Figure 3.9b shows the fundamental frequency of the oscillating current vs. dc voltage bias and clearly indicates an abrupt jump at $\Phi_\beta/l = 6.055\,\mathrm{V/cm} > \Phi_\alpha/l$ which corresponds to the transition from the decaying periodic waves of Fig. 3.6 to the fully developed solitary waves of Fig. 3.7. The inset clearly shows that this transition is slightly hysteretic. Finally, we note that the threshold field $\Phi_\omega/l = 12.975\,\mathrm{V/cm}$ indicates the point above which the steady state is again stable. We discuss analytical criteria for both Φ_α and Φ_ω in the subsection below.

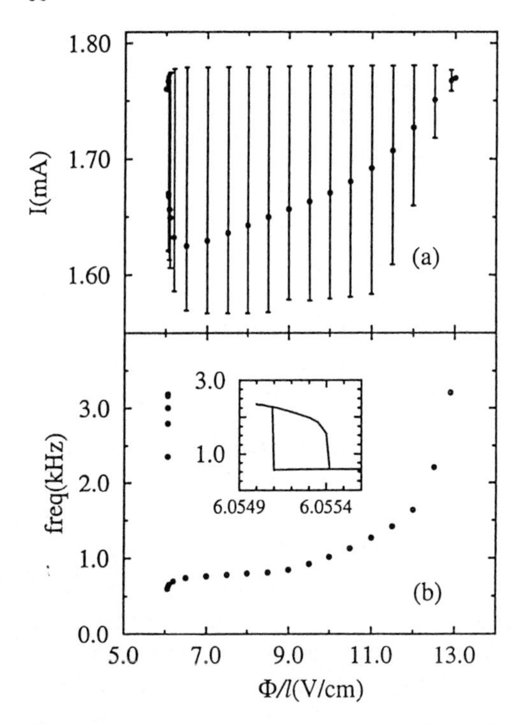

Fig. 3.9. (a) Current vs. average electric field, where the vertical bars represent the extent of the current oscillations and the solid circles are the time-averaged values. (b) The fundamental frequency vs. average electric field. The inset illustrates the hysteretic nature of the transition from the decaying wave branch to the solitary wave branch

3.5.3 Stability Analysis and Hopf Bifurcations from the Steady State

In this subsection we treat the linear stability of the spatially-dependent steady state discussed above. The stability problem for the solitary wave state is very complex and has not been fully addressed for the p-Ge system. Bonilla [3.14] has presented some qualitative arguments which indicate that solitary waves are dynamically stable while they are traveling far from the contacts. On the other hand, the stability problem when solitary waves are in the neighborhood of contacts has not been treated. (However, Bonilla and Venakides [3.34] have been able to obtain rigorous stability results in a simplified model of the Gunn effect.) For these reasons we limit ourselves to the linear stability of the steady state.

We start with a standard separation of variables Ansatz:

$$E(x,\tau) = E_{ss}(x) + \eta \hat{E}(x) e^{\lambda\tau} \tag{3.46}$$

and

$$J(\tau) = J_{ss} + \eta \hat{J} e^{\lambda\tau},\ \eta \ll 1, \tag{3.47}$$

where $E_{ss}(x)$ is the position-dependent steady state described above and J_{ss} denotes the steady state current which is determined by the voltage constraint.

Substituting these two forms into the full reduced model equations (3.17)-(3.21) gives the following eigenvalue problem:

$$\frac{d}{dx}\left[(\lambda + c_2)\hat{E}\right] + \left[\lambda c_1 + \frac{c_2 J'}{V}\right]\hat{E} = \left[\frac{\lambda + c_2}{V} - \frac{\partial c_2}{\partial J}\frac{dE_{ss}}{dx}\right]\hat{J}. \tag{3.48}$$

Equation (3.48) is to be solved subject to the Ohmic boundary condition,

$$\hat{E}(0) = \rho_o \hat{J} \tag{3.49}$$

and the voltage bias constraint can be written in the following form:

$$Z(\lambda) \equiv \int_0^L \frac{\hat{E}}{\hat{J}}dx = 0, \tag{3.50}$$

where $Z(\lambda)$ is recognized as the differential impedance. Thus the eigenvalues of the linear stability problem are simply the zeroes of the differential impedance within the complex λ-plane. If any of these zeroes have positive real part then the steady state is unstable. Evaluating the coefficients appearing in (3.48) about the approximate steady state specified by (3.44) with Δx taken to be zero we arrive at the following condition for the zeroes of Z:

$$\frac{f(\lambda)}{L} exp\left\{\left[1.0 - \frac{|J_2'| + c_{12}V_2}{c_{22}\,|J_2'|}\lambda\right]\frac{|J_2'|\overline{x}}{V_2}\right\} \sim 1, \tag{3.51}$$

where the additional subscripts on c_1, c_2, V, and J' indicate that these functions are evaluated at E_2. Also, the function $f(\lambda)$ is an analytic function of the form:

$$f(\lambda) = f(0) + f'(0)\lambda + O(\lambda^2),\ 0 < f(0) \sim O(1). \tag{3.52}$$

Thus, for λ values of order 1, the exponential term in (3.51) will have to be of order L and we can replace f by 1 for purposes of obtaining approximate eigenvalues. The condition for instability threshold is $Re(\lambda) = 0$, and substituting this into (3.51) we find:

$$\lambda \sim \frac{i(2n+1)\pi c_{22}V_2}{(|J_2'| + c_{12}V_2)\overline{x}_c},\ n = 0, \pm 1, \ldots o(L). \tag{3.53}$$

where

$$\overline{x}_c \sim \frac{V_2}{|J_2'|}log(L) + O(1) \gg 1. \tag{3.54}$$

The $n = 0, -1$ complex eigenvalue pair of (3.53) corresponds to the first mode to go unstable as voltage increases and $\overline{x}_c$ denotes the critical position of the transition layer where the steady state first loses stability through a Hopf bifurcation. Combining the results of (3.53) and (3.54) with (3.45) we can estimate ϕ_α and the frequency of small amplitude current oscillations; the results are in excellent agreement with numerical and experimental data [3.28]. In the limit of very long samples, the higher order modes of (3.53) become unstable for a sequence of ϕ values infinitesimally greater than ϕ_α; we can then argue that the steady state loses stability through a quasi-continuum of Hopf bifurcations. Furthermore, using multiscale analysis at this transition [3.35] (see a detailed description of our methods in [3.36] for the classical Gunn effect in GaAs), we have been able to construct the time-dependent electric field profile associated with the small amplitude oscillations, and the results are in qualitative agreement with numerics (see, e.g., Fig. 3.6).

3.6 Concluding Remarks

In this paper we have reviewed the theory of nonlinear traveling waves in a drift-diffusion model of electrical conduction for extrinsic semiconductors in the post-impurity breakdown regime. A reduced model has been derived using singular perturbation techniques and appears to accurately describe complex experimental behavior relating to NDR and solitary space charge waves under dc current or dc voltage bias conditions. Nonetheless, a number of interesting questions remain unanswered. For example, does the abrupt transition at ϕ_β observed in numerical simulations produce intermittent switching between solitary wave and decaying wave branches when a small noise term (e. g., in the applied voltage) is added to the reduced model? Such behavior has been seen in recent experiments near the onset of the solitary wave regime [3.23]. Similarly, one might wonder about the effect of a spatially-varying compensation ratio on the transitions at both ϕ_α and ϕ_β. When p-Ge crystals are grown using standard Czochralski methods some variation in impurity concentration always occurs due to segregation effects [3.37]. We have seen above how sensitive the dc J-E characteristics are to material composition, and we would expect to see the same level of sensitivity for various dynamical effects. Another open question concerns whether or not temporal and spatio-temporal chaotic behavior observed in p-Ge experiments under dc + ac voltage bias [3.22] can be captured by the reduced model.

Acknowledgments. It is a pleasure to acknowledge beneficial conversations and collaborations with I. R. Cantalapiedra, E. E. Haller, F. J. Higuera, A. M. Kahn, D. J. Mar, J. M. Vega, S. Venakides, R. M. Westervelt, and X. Zhang. We acknowledge financial support from the National Science Foundation through grant DMR-9157539, the Spanish DGICYT through grant PB92-0248 and the NATO Travel Grant Program through grant CRG-900284.

References

[3.1] E.M. Conwell: In *High Field Transport in Semiconductors, Solid State Physics Suppl. 9*, ed. by F. Seitz, D. Turnbull, H. Ehrenreich (Academic Press, New York 1967)
[3.2] S.M. Sze: *Physics of Semiconductor Devices*, 2nd ed. (Wiley, New York 1981)
[3.3] E. Schöll: *Nonequilibrium Phase Transitions in Semiconductors* (Springer, Berlin 1987)
[3.4] B.K. Ridley: Proc. Phys. Soc. **82**, 954 (1963)

[3.5] J.B. Gunn: Solid State Communications **1**, 88 (1963)
[3.6] B.K. Ridley: Proc. Phys. Soc. **86**, 637 (1965)
[3.7] S.W. Teitsworth, R.M. Westervelt, E.E. Haller: Phys. Rev. Lett. **51**, 825 (1983)
[3.8] R.G. Pratt, B.K. Ridley: J. Phys. Chem. Sol. **26**, 11 (1965)
[3.9] J. Peinke, A. Mühlbach, R.P. Huebener, J. Parisi: Phys. Lett. A **108**, 407 (1985)
[3.10] P.W. Dorman: Proc. IEEE **56**, 372 (1968)
[3.11] K. Aoki, K. Miyame, T. Kobayashi, K. Yamamoto: Physica **117 & 118B**, 570 (1983)
[3.12] D.G. Seiler, C.L. Littler, R.J. Justice, P.W. Milonni: Phys. Lett. A **108**, 462 (1985)
[3.13] L.L. Bonilla, S.W. Teitsworth: Physica D **50**, 545 (1991)
[3.14] L.L. Bonilla: Physica D **55**, 182 (1992)
[3.15] L.L. Bonilla: Phys. Rev. B **45**, 11642 (1992)
[3.16] J. Guckenheimer, P. Holmes: *Nonlinear Oscillations, Dynamical Systems and Bifurcations of Vector Fields* (Springer, New York 1983)
[3.17] S.W. Teitsworth, R.M. Westervelt: Phys. Rev. Lett. **56**, 516 (1986)
[3.18] S.W. Teitsworth, R.M. Westervelt: Phys. Rev. Lett. **53**, 2587 (1984)
[3.19] S.W. Teitsworth, R.M. Westervelt: Physica D **23**, 181 (1986)
[3.20] E.G. Gwinn, R.M. Westervelt: Phys. Rev. Lett. **57**, 1060 (1986)
[3.21] A.M. Kahn, D.J. Mar, R.M. Westervelt: Phys. Rev. B **43**, 9740 (1991)
[3.22] A.M. Kahn, D.J. Mar, R.M. Westervelt: Phys. Rev. Lett. **68**, 369 (1992)
[3.23] A.M. Kahn, D.J. Mar, R.M. Westervelt: Phys. Rev. B **46**, 7469 (1992)
[3.24] R.M. Westervelt, S.W. Teitsworth: J. Appl. Phys. **57**, 5457 (1985)
[3.25] L. Reggiani: *Hot-Electron Transport in Semiconductors*, 1st ed. (Springer, New York 1985)
[3.26] M.P. Shaw, H.L. Grubin, P.R. Solomon: *The Gunn-Hilsum Effect* (Academic, New York 1979)
[3.27] S.W. Teitsworth: Ph.D. thesis, Harvard University (1986)
[3.28] I.R. Cantalapiedra, L.L. Bonilla, M.J. Bergmann, S.W. Teitsworth: Phys. Rev. B **48**, 12278 (1993)
[3.29] S.W. Teitsworth: Appl. Phys. A **48**, 127 (1989)
[3.30] K. Maginu: J. Math. Biol. **10**, 133 (1980)
[3.31] L.L. Bonilla, J.M. Vega: Phys. Lett. A **156**, 179 (1991)
[3.32] M.J. Bergmann, S.W. Teitsworth, L.L. Bonilla, I.R. Cantalapiedra: unpublished (1994)
[3.33] D. Kincaid, W. Cheney: *Numerical Analysis: Mathematics of Scientific Computing* (Brooks/Cole, Pacific Grove, California 1991)
[3.34] L.L. Bonilla, S. Venakides: to appear in SIAM J. Appl. Math (1994)
[3.35] L.L. Bonilla, I.R. Cantalapiedra, M.J. Bergmann, S.W. Teitsworth: Semicond. Sci. Technol. **9**, 599 (1994)
[3.36] L.L. Bonilla, F.J. Higuera: to appear in SIAM J. Appl. Math (1995)
[3.37] E.E. Haller, W.L. Hansen, F.S. Goulding: Advances in Physics **30**, 93 (1981)

4 Autosolitons in the Form of Current Filaments and Electric Field Domains in Semiconductors and Devices

B.S. Kerner

SIC GbR, P.O. Box 50 06 02
D-70336 Stuttgart, Fed. Rep. Germany

Abstract. A review of the theoretical, numerical and experimental results of investigations of autosolitons in the form of *current filaments and electric field domains* in semiconductors and semiconductor devices is presented. Main attention is given to a qualitative description and to the physics of phenomena. Properties of current filaments and electric field domains which appear in semiconductors and devices with a *positive* differential conductivity are discussed.

4.1 Introduction

The phenomena of self-organization in semiconductors and semiconductor devices are often linked to spontaneous appearance of current filaments or electric field domains. In many cases such phenomena of self-organization are connected with the *negative differential conductivity* of the sample which results in an S-shaped or N-shaped current-voltage characteristic, respectively, when current filaments or electric field domains appear (e. g., [4.1–22]). The results of theory and experimental observations of current filaments and electric field domains in semiconductors and semiconductor devices with negative differential conductivity can be found in reviews [4.10, 20] and books [4.21, 22]. From the nonlinear theory of current filaments and electric field domains in semiconductors and semiconductor devices with negative differential conductivity it has been found that only *a single* current filament or *a single* electric field domain can be stable in such initially homogeneous media. Besides, a load resistor in an external circuit of a sample should be large enough for a stability of a current filament; an electric field domain can be stable if the load resistor is low enough [4.20–22]. These conclusions of the theory of semiconductors with negative differential conductivity have been confirmed by experimental observations of these phenomena in many semiconductors and semiconductor devices (see, e. g., [4.1, 3–5] and the references in [4.10, 20–22]).

However, in experimental investigations of current filaments and electric field domains in semiconductors and semiconductor devices often a totally different behaviour of these structures has been found:

Springer Proceedings in Physics, Vol. 79 **Nonlinear Dynamics and Pattern Formation in Semiconductors and Devices**
Editor: F.-J. Niedernostheide

1. Current filaments and electric field domains in semiconductors and semiconductor devices appeared spontaneously, although the differential conductivity of a sample was *positive* (e.g., [4.23–27]).
2. An instability in semiconductors led to a formation of *multi*-filament or *multi*-domain stable states (e.g., [4.24–29,43]).
3. A very complex dynamics of current filaments in semiconductors, which has led in a lot of cases to an appearance of chaos, has been observed (e.g., [4.30–41,44]).
4. Stationary or randomly rocking and also pulsating and travelling current filaments have been found [4.42–46].

These nonlinear effects observed in semiconductor and semiconductor devices may find an explanation within the bounds of the nonlinear theory of *autosolitons* - localized dissipative structures of large amplitude which can occur in a wide class of active nonequilibrium systems. Indeed, in active nonequilibrium systems, corresponding to the theory of autosolitons, static, stationary or randomly pulsating, rocking, travelling and more complex types of autosolitons can appear. Autosolitons can show very complex spatiotemporal behaviour. Autosolitons can spontaneously transform into autosolitons of another type. They can spontaneously appear and disappear in different parts of the system. Due to these properties of autosolitons and also due to their sophisticated interactions, diverse scenarios of self-organization and of an appearance of turbulence can be realized in active nonequilibrium systems, in particular, in semiconductors and semiconductor devices. In the latter case autosolitons can have a form of current filaments or electric field domains (strata). The results of the theory of autosolitons and the corresponding references can be found in reviews [4.47,48] and books [4.49,50].

The general theory of autosolitons in active systems predicted a huge number of nonlinear phenomena of self-organization and turbulence connected to the spontaneous appearance, disappearance, interactions and evolution of different types of autosolitons [4.47–50]. Some of these phenomena have already been observed in experimental and numerical investigations of semiconductors (e.g., [4.23–29,38–46,51–71]), gas plasmas, nonlinear optical active media, nonequilibrium superconductors, hydrodynamical systems, chemical and biological reactions, etc. (see references in [4.50]).

In this review article, based on the results of the general theory of autosolitons in active systems [4.47–50], the qualitative consideration of properties of current filaments and electric field domains in semiconductors and devices will be given. For simplicity we will restrict the consideration to autosolitons in one-dimensional systems. Besides, we will leave out the precise description of models of semiconductors and also the discussion of the corresponding equations. Main attention will be given to qualitative descriptions and to the physics of phenomena. Firstly, based on a consideration of some relative simple semiconductor systems, physics of the appearance of current filaments and electric field domains (strata) in semiconductors with a *positive differential conductivity* will be discussed. Secondly, the possible forms and types of autosolitons in semiconductors will be considered. Finally, some general nonlinear properties of autosolitons which may be realized in different semiconductors and semiconductor devices will briefly be reviewed.

4.2 Formation of Current Filaments and Electric Field Domains in Semiconductors with Positive Differential Conductivity

The possibility of instabilities in semiconductors which lead to a formation of current filaments or electric field domains at the ***positive differential conductivity*** of a sample has been found in [4.72], based on the theoretical analysis. It has also been found that a current-voltage characteristic of a semiconductor for a homogeneous distribution of the current density (or electric field) can be even ***single-valued***, but nevertheless current filaments or electric field domains can spontaneously appear in a sample [4.72–75].

In this section four examples, where such an effect can be realized, will be qualitatively considered:

- current filaments in transistor structures,
- multi-filament stable states in reverse-biased *p-n* junctions.
- current filaments in dense electron-hole plasmas.
- electric field domains (strata) in electron-hole plasmas heated by an electric field.

4.2.1 Current Filaments in Transistor Structures

Hot spots, i.e. thermal current filaments, were discovered in the earliest experiments with transistor structures [4.76, 77]. This effect was explained by the thermal instability in the transistor in connection with the thermally activated character of the current of the forward-biased *p-n* junction (emitter junction) of the transistor [4.77, 78]. In [4.72] it was noticed that the thermal instability in a transistor structure can be realized at the positive differential conductivity of a sample. Let us consider the physical reason for this effect.

Fig. 4.1. Schematic representation of the *p-n-p* transistor structure

The injection current j in the emitter junction (lower *p-n* junction in Fig. 4.1) as a function of the temperature T and of the voltage drop V_e across the emitter

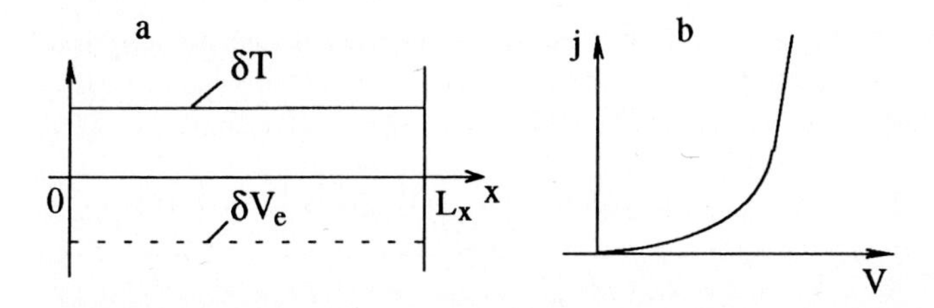

Fig. 4.2. (a) Schematic representation of homogeneous temperature fluctuations δT and the voltage drop δV_e across the emitter junction in the transistor structure; (b) qualitative form of the current-voltage characteristic of the structure for homogeneous current-density distributions

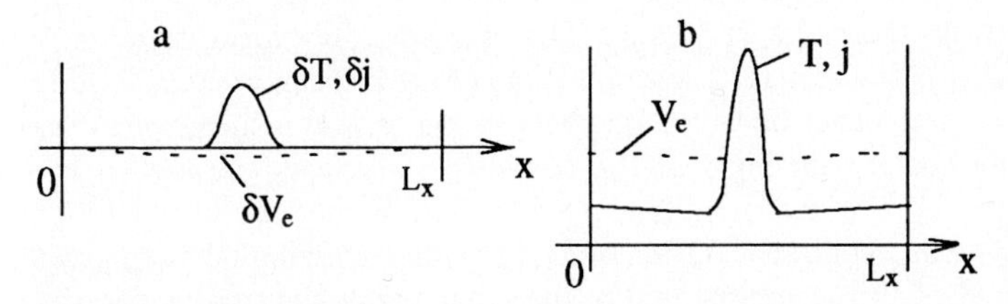

Fig. 4.3. (a) Schematic representation of *local* temperature perturbations δT and the voltage drop δV_e across the emitter junction; (b) typical form of a spike current filament in the transistor

junction can be written in the form (e. g., [4.79]):

$$j = j_0 \exp(-\frac{E_g - eV_e}{kT}). \tag{4.1}$$

In (4.1) E_g is the band gap of a semiconductor; $eV_e < E_g$. Although the injection current j has a thermally activated character in the transistor structure, the current-voltage characteristic of this structure, which corresponds to a homogeneous current-density distribution, is single-valued (Fig. 4.2). This occurs, because a homogeneous increase of the temperature, δT (Fig. 4.2), is suppressed by a corresponding homogeneous decrease of the voltage drop δV_e across the emitter junction (Fig. 4.2). In other words, the value j of the homogeneous current-density distribution practically does not depend on the temperature.

Let us explain this circumstance for some examples in more detail. In the transistor structure shown in Fig. 4.1 the value j of the homogeneous current-density distribution may be determined by a leakage through the emitter or the collector junctions. Taking into account, for example, only the leakage through the collector junction, we get

$$j = \frac{V_c}{\rho_c(1 - \alpha_I M)}. \tag{4.2}$$

In (4.2) V_c is the voltage drop across the collector junction, $V_c = V - V_e$, V is the total voltage drop across the structure, $V_c \gg V_e$; ρ_c is the specific leakage resistance of the collector junction (for simplicity we do not include in (4.2) the thermal current of the reverse-biased collector junction which is supposed to be many orders of magnitude less than the current due to the temperature-independent leakage); α_I is the transfer coefficient of the emitter current; M is the coefficient of avalanche multiplication in the collector junction which is an increasing function of the voltage V_c [4.79]. Because the values α_I and M usually depend only weakly on the temperature, the current-voltage characteristic of the transistor structure, as it follows from (4.2), is single-valued (Fig. 4.2). This explains the mentioned fact that homogeneous fluctuations of the temperature δT are suppressed by the corresponding decrease of the voltage drop δV_e across the emitter junction. In other words, the differential conductivity of the transistor structure is *positive* at any value of the current under consideration.

In spite of this, a current filament can spontaneously appear in this structure. To explain this effect, let us consider the behaviour of a *local* increase of the temperature $\delta T(x)$ in the transistor (Fig. 4.3). The corresponding local decrease of the voltage drop across the emitter junction $\delta V_e(x)$ (Fig. 4.3) is now considerably less than for the homogeneous fluctuation shown in Fig. 4.2. It is linked with the base current along the n base layer of the transistor which appears when the voltage distribution V_e becomes nonhomogeneous: The higher the base current is, the less is the possible spatial variation of $\delta V_e(x)$ which corresponds to a given local increase $\delta T(x)$ of the temperature. On the other hand, the smaller the size of the region of the localization of the perturbation is, the higher is the value of the base current, i.e., the more are the spatial variations of $\delta V_e(x)$ damped. For this reason, an increase of the temperature (of the current density) in a small enough region of the structure occurs while $V_e \approx const.$ In other words, this local increase of the current density cannot be suppressed by an appropriate local change in $V_e(x)$. Therefore, a local increase of the current density can lead to an instability taking place at a positive differential conductivity of a sample. As a result of this instability a current filament is formed (Fig. 4.3). These conclusions have been confirmed both by the results of the experimental investigations and by the nonlinear theory of current filaments which can be found in [4.23].

4.2.2 Multi-Filament Current States in Reverse-Biased *P-N* Structures

Stratification of an initially homogeneous distribution of the avalanche current density and spontaneous formation of multi-filament states at a positive differential conductivity of a sample have been found out experimentally in the investigation of a reverse-biased p-n structure made of α-SiC [4.27].

It has been found that, if an orientation of the crystal is chosen in such a way that the electric field in the reverse-biased p-n junction is directed parallel to the natural superlattice of α-SiC, an initially homogeneous distribution of an avalanche current, beginning at current densities $j > j_c = 10 - 20\,\mathrm{A/cm^2}$, stratifies. The stratification occurs at a positive differential conductivity of the sample. It is manifested as a change with increasing current from a uniform to a non-uniform breakdown pattern. This nonuniform breakdown corresponds to a two-dimensional dissipative structure in the form of numerous and brightly

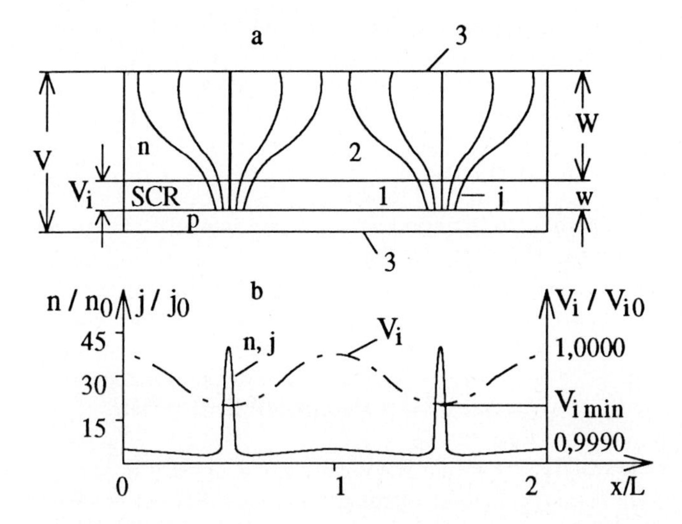

Fig. 4.4. Formation of current filaments in a reverse-biased *p-n* structure: (a) schematic representation of the structure (lines indicate the current lines within two periods of current filaments; space charge region (SCR) of the *p-n* junction (1), resistive layer (2), metallic electrode (3)); (b) numerically calculated distributions of n, j, and V_i [4.63]

luminous avalanche-current filaments of different dimensions which are not periodically distributed over the area of the sample. The result of the theory of this phenomena developed in [4.27] and of the numerical investigations [4.63] are in good agreements with the experiment. Let us consider the physics of this effect [4.27].

A natural superlattice exists along the C crystallographic axis in α-SiC crystals (e. g., [4.80]). The existence of such a superlattice is known to cause a qualitative difference between the heating of electrons in an electric field directed parallel to the superlattice axis and that in an electric field directed perpendicular to this axis. The difference is that the heating of electrons in a field parallel to the axis of the superlattice is hindered by the miniband nature of the electron spectrum. In other words, in the case of a semiconductor with a superlattice the Bragg reflection from the miniband edge hinders the heating of the electron gas by a field parallel to the superlattice axis and, consequently, hinders the impact ionization process [4.81]. When an electron acquires energy, the transition of the electron from one miniband to another is possible only by virtue of a quasielastic scattering of electrons by phonons [4.82]. Such a situation is characteristic for low current densities.

As the current density is increased, the transition of the electron from one miniband to another may be possible also due to electron-electron collisions. This may occur, because the rate of electron-electron collisions is proportional to the electron concentration n, i. e., this rate is an increasing function of the density of the avalanche current. The latter effect may be responsible for a rising dependence of the average ionization rate ν_i in the space charge region of a *p-n* junction on the carrier density n. This rising dependence of ν_i on n can cause a growth of a local perturbation of the current density in the structure. The latter can lead to a spontaneous appearance of current filaments [4.27].

Indeed, the avalanche current density is:

$$j = env_e, \tag{4.3}$$

i.e., it is proportional to the electron concentration n in the space charge region of a p-n junction, because the drift velocity of electrons v_e in this region is practically constant (e.g., [4.79]). The local increase of the current density implies the local increase of the electron concentration inside the space charge region of a p-n junction and, therefore, it implies the increase of the average ionization rate ν_i. The latter causes a further local increase of the previous local perturbation of the current density and by this the further local growth of the average ionization rate ν_i and so on.

Considering this avalanchelike growth of the local current density in the p-n junction, the voltage drop across the quasineutral parts, i.e., the resistive layers of the structure (Fig. 4.4), has not been taken into account. For simplicity, we take into account the specific resistivity ρ of only one of the quasineutral regions (the n region in Fig. 4.4) and get the following equation for the total voltage drop V across the structure:

$$V = j\rho + V_i, \tag{4.4}$$

where V_i is the voltage drop across the p-n junction. Equation (4.4) is valid, if the current-density distribution is *homogeneous*.

If the value of the total voltage drop V is constant, it follows from (4.4) that a *homogeneous* increase of the current density, δj, leads to a decrease of the voltage drop V_i across the p-n junction:

$$\delta V_i = -\delta j \rho. \tag{4.5}$$

On the other hand, it is well-known that the average ionization rate ν_i is a strong increasing function of the voltage drop V_i. Therefore, if the value ρ is not very small, the decrease of V_i (4.5) suppresses the growth of the homogeneous current-density fluctuations induced by the rising dependence of the average rate of ionization ν_i on the carrier density n. In other words, the homogeneous perturbations of the current density cannot grow. For this reason, the current-voltage characteristic of the whole structure is single-valued [4.27].

A different situation is encountered when the growth of the current density is nonhomogeneous. Let us consider the growth of the current density in a small *local* region of size l of the area of the p-n junction (see current lines in the space charge region (SCR) in Fig. 4.4). If the thickness W of the resistive layer of the structure is large enough,

$$W \gg l, \tag{4.6}$$

the current density spreads over the quasineutral n region of the structure, as shown by the lines in Fig. 4.4. Because of that, the change in the voltage drop across the n type region, i.e., the value $V_\rho = V - V_i$, produced by a *local* increase of the avalanche current in the small neighbourhood of size l, will be much less, in proportion to $l/W \ll 1$, than that in the above case of a uniform increase of the avalanche current. In other words, a local increase of the current density causes a very small change of the voltage drop across the resistive layer: $\delta V_\rho \ll \delta j\rho$. As a result, taking into account that $\delta V_i = -\delta V_\rho$, instead of (4.5), we have

$$| \delta V_i | \ll \delta j\rho. \tag{4.7}$$

Corresponding to (4.7), the decrease of the value of the voltage drop δV_i across the p-n junction can be so small that it cannot suppress the effect of the growth of the current density in the local region under consideration, where the average rate of ionization is an increasing function of the current density. Therefore, if the current density increases in the local regions of the very small size l (4.6), the voltage drop δV_ρ is not large enough to prevent the further growth of the current density in these small local regions of the structure. So, local solitary regions of high current densities may arise in a homogeneous p-n structure (Fig. 4.4).

Note that the value l, determing the size of the region in which a local growth of the avalanche current density can be realized, is defined by diffusion of free carriers that spread out during the time $\tau = w/v_e$; τ is the time carriers need to cross the space charge region of the p-n junction. Thus, $l \approx (D\tau)^{1/2}$, where D is the diffusion coefficient of charge carriers. In the case under consideration [4.27] we have $l \sim 10^{-6}$ cm and $W \sim 10^{-1}$ cm. Therefore, the condition (4.6) which is important for the formation of multi-filament current states at a positive differential conductivity of the structure is really fulfilled.

4.2.3 Current Filaments in Dense Electron-Hole Plasmas

Multi-filament current states can also spontaneously be formed at a positive differential conductivity of a sample in an electron-hole plasma heated by an electric field [4.75]. This effect can be realized, for example, in a dense electron-hole plasma in semiconductors and semiconductor devices.

For simplicity, let us consider here the model of a "symmetrical" electron-hole plasma, where the parameters of the electrons and the holes are the same. In a dense enough electron-hole plasma the mobility of carriers is defined by collisions between electrons and holes moving in opposite directions in an electric field. In this case, because of the scattering of electrons by holes, the characteristic time of collisions of electrons and holes τ_{eh} is proportional to $T^{3/2}n^{-1}$, where T is the effective carrier temperature and n is the carrier concentration. Hence, the mobility of the carriers is

$$\mu \propto T^{3/2}n^{-1}. \tag{4.8}$$

Accordingly, the conductivity of the dense electron-hole plasma,

$$\sigma = e\mu n \propto T^{3/2}, \tag{4.9}$$

does not depend on the carrier concentration. The amount of power supplied to the carriers by the electric field E correspondingly to (4.9) is

$$W_j = \sigma E^2 \propto T^{3/2}, \tag{4.10}$$

i. e., it also does not depend on the carrier concentration. The power transmitted from the system of hot carriers to the lattice due to their interactions with lattice phonons is

$$P = 2n\frac{T - T_l}{\tau_\epsilon(T)}, \tag{4.11}$$

where T_l is the lattice temperature; $\tau_\epsilon(T)$ is the characteristic time of relaxation of carriers with phonons.

For a *homogeneous* electron-hole plasma with carrier concentration n we can find the effective carrier temperature T from the equation of energy balance of charge carriers $W_j = P$:

$$\sigma E^2 = 2n\frac{T - T_l}{\tau_\epsilon(T)}, \tag{4.12}$$

where we have taken into account (4.10) and (4.11). From (4.8), (4.9), (4.12) and the equation for the current density $j = \sigma E$ we obtain

$$\sigma_d = \sigma\left[1 + \frac{3}{2}E^2\frac{\sigma\tau_\epsilon}{nT_l}\frac{1}{s + 3/2 - (s + 1/2)(T/T_l)}\right] \tag{4.13}$$

for the differential conductivity $\sigma_d = dj/dE$ of the semiconductor plasma. In this equation we have used the designation

$$s = \frac{\partial \ln \tau_\epsilon(T)}{\partial \ln T}. \tag{4.14}$$

In semiconductors the condition $s \leq -1/2$ is often fulfilled (e. g., [4.83]). In

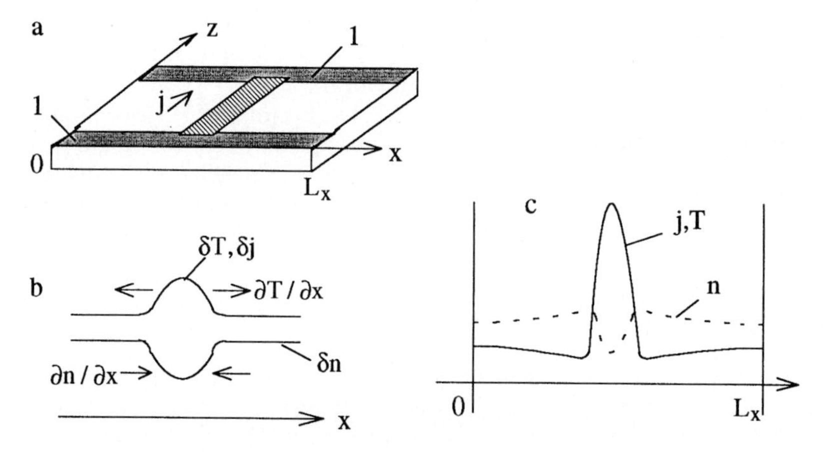

Fig. 4.5. (a) Schematic representation of the current filament (hatched region) in the semiconductor sample (1 - metallic electrodes; the arrow indicates current lines); (b) schematic form of the local perturbations of the temperature δT, the current density δj and the carrier concentration δn; (c) one of the possible forms of a localized current filament

these cases the differential conductivity σ_d of a sample, as it follows from (4.13), is *positive* at any value of the electric field. Nevertheless current filaments can spontaneously appear in the sample (Fig. 4.5) [4.52,53,75].

To understand the physics of this effect [4.75], let us consider the behaviour of a *local* increase of the carrier temperature, δT, in the directions transverse to the direction of electric current (Fig. 4.5). In this case the electric field does not change, i.e., $E = const.$ The local increase of the temperature of the carriers forces the carriers to leave this local region. This is due to the appearance of the thermodiffusion flux which is proportional to the temperature gradient $\partial T/\partial x$. As a result, the carrier concentration reduces in this local region (Fig. 4.5). If the size of the region, where the temperature increases, is considerably less than the diffusion length of the carriers, the decrease of the concentration in this region is [4.75]:

$$\delta n = -\frac{n}{T}\delta T. \tag{4.15}$$

In other words, in such a local region the pressure of the electron-hole plasma, nT, practically does not change in space, i.e., the function

$$\eta = n(x)T(x) \approx const. \tag{4.16}$$

The condition (4.16) is satisfied due to the competition between the thermodiffusion flux out the local region and diffusion flux into this region from the periphery.

Although the concentration decreases in the local region under consideration (Fig. 4.5), the amount of power supplied to the carriers W_j increases in the local region of the higher temperature corresponding to (4.10). Indeed, the power W_j does not depend on the carrier concentration in the case of a dense electron-hole plasma. Besides, the power P transmitted from the hot carriers to the lattice, as it follows from (4.11), is proportional to the carrier concentration n which decreases in the local region where the temperature increases (see (4.15)). Therefore, the decrease of the carrier concentration in the local region, where the carrier temperature is elevated, can lead to a further local increase of T. This implies a stronger thermodiffusion flux of carriers from the local region under consideration, i.e., it leads to a further decrease of the carrier concentration and correspondingly to a further increase of the temperature. Because the current density in a dense plasma does not depend on the carrier concentration and is the higher the higher the carrier temperature is, such an avalanchelike process of the local growth of the temperature causes the formation of local current filaments (Fig. 4.5).

In these current filaments the carrier temperature is higher and the carrier concentration is lower than in the initial homogeneous plasma (Fig. 4.5). The region of low carrier concentration in the centre of a current filament does not spread out, because the diffusion inflow of carriers from the periphery, which is proportional to $\partial n/\partial x$, is practically counterbalanced by the thermodiffusion outflow, i.e., inside the current filament the condition (4.16) is fulfilled.

This competition, as it follows from the theoretical analysis of autosolitons in a dense plasma [4.75], can lead to the formation of current filaments, if the following necessary conditions are fulfilled:

$$s + 3/2 > 0, \quad L \gg l, \tag{4.17}$$

where l is the length of energy relaxation and L is the diffusion length of carriers.

The first of the conditions (4.17) can be obtained by putting the concentration n, taken from (4.16), into (4.12). As a result we get

$$\sigma E^2 = 2\frac{\eta}{T}\frac{T - T_l}{\tau_\epsilon(T)}. \tag{4.18}$$

This equation describes the energy balance of electron-hole plasmas for the *local* region where the temperature increases, if the size of this region is considerably less than the diffusion length of carriers. From (4.18), taking into account (4.9) for the conductivity of the plasma, one can derive the condition for the *local* instability of the plasma. Indeed, the local instability in a semiconductor, which leads to the appearance of a current filament, occurs when the local increase of the power W_j supplied to the carriers becomes higher than the increase of the power P transmitted from the hot carriers to the lattice of the semiconductor, i.e., when $\delta W_j > \delta P$. Taking into account (4.16) and $E = const$, we can write this condition in the form

$$\frac{3}{2}\frac{\sigma E^2}{T} > \frac{\eta}{\tau_\epsilon T}\frac{(s+1)T_l - sT}{T}. \tag{4.19}$$

Using (4.18), it can be shown that the inequality (4.19) is fulfilled when the plasma temperature exceeds the critical value T_c:

$$T > T_c = \frac{s + 5/2}{s + 3/2}T_l. \tag{4.20}$$

From (4.20) we obtain the first of the conditions (4.17) being necessary for local plasma instability.

The conditions (4.17) can be fulfilled in many semiconductors both at low and at room temperature of the lattice T_l [4.24, 52, 53, 58, 75]. This effect may, for example, be realized at room temperature in GaAs. It may explain the occurrence of multi-filament current states which have been found out in thin GaAs film [4.24] and GaAs transistors with Schottky barriers [4.25, 26]. Note that the numerical investigations of current filaments in a dense electron-hole plasma [4.52,53] are in a good agreement with the theoretical results concerning this effect [4.75].

4.2.4 Electric Field Domains in Hot Electron-Hole Plasmas

If the density of an electron-hole plasma is not very large, the mobilities of electrons and holes are determined by the collisions of carriers with lattice phonons. Therefore, the conductivity of the plasma is in this case proportional to the carrier concentration. Then, in contrast to the dense plasma, the electric field domains can spontaneously appear in semiconductors and devices [4.74]. We restrict our considerations to the model of "symmetrical" electron-hole plasmas, as we have already done for the dense plasma. In this case electric field domains formed do not move in the electric field. In a real "nonsymmetrical" electron-hole plasma moving electric field domains can occur [4.60, 75]. Some properties of such electric field domains will be discussed below.

In the case of a "symmetrical" electron-hole plasma the conductivity of the plasma is

$$\sigma = 2en\mu(T), \tag{4.21}$$

where $\mu(T)$ is the carrier mobility which depends on the effective carrier temperature T. If the concentration of the plasma is known, the effective carrier temperature can be found from the energy balance of charge carriers

$$W_j = P. \tag{4.22}$$

In this equation the density of Joulean power supplied to carriers by an electric field is

$$W_j = j^2/\sigma \tag{4.23}$$

and the density of power P transferred from carriers to the lattice is given by (4.11). Using (4.22), (4.11), (4.23), and $j = \sigma E$, we find that the differential conductivity of the plasma is

$$\sigma_d = \sigma \left[1 + 2\alpha E^2 \frac{\sigma \tau_\epsilon}{nT_l} \frac{1}{s + \alpha - (s + \alpha - 1)(T/T_l)} \right], \tag{4.24}$$

where

$$\alpha = \frac{\partial \ln \mu(T)}{\partial \ln T}. \tag{4.25}$$

As it follows from (4.24), the current-voltage characteristic of the plasma will be single-valued if the condition

$$s + \alpha \leq 1 \tag{4.26}$$

is satisfied, where s is given by (4.14). The condition (4.26) is usually fulfilled for semiconductors [4.83].

Although the current-voltage characteristic of an electron-hole plasma under condition (4.26) is single-valued, electric field domains can be formed in the sample [4.74]. To understand the physics of this effect, let us consider the behaviour of a *local* increase of the carrier temperature, δT, in the direction parallel to the direction of electric current (Fig. 4.6). Similar to the above considered case of a dense plasma, the local increase of the carrier temperature forces the carriers to leave this local region. This is due to the appearance of the thermodiffusion flux which is proportional to the temperature gradient $\partial T/\partial z$ of charge carriers. If the size of the region of the local temperature increase is considerably less than the diffusion length L of carriers, the carrier concentration is reduced by

$$\delta n = -n \frac{1 + \alpha}{T} \delta T. \tag{4.27}$$

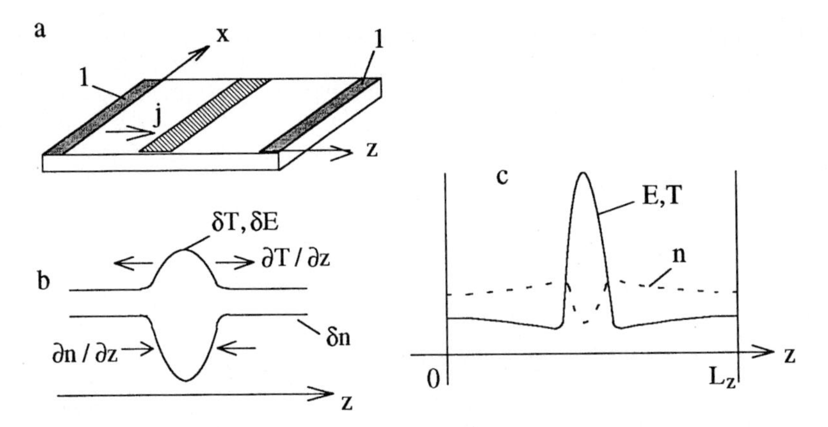

Fig. 4.6. (a) Schematic representation of the electric field domain (hatched region) in the semiconductor sample (1 - metallic electrodes; the arrow indicates current lines); (b) schematic form of the local perturbations of the temperature δT, the electric field δE and the carrier concentration δn; (c) one of the possible forms of a localized electric field domain

In other words, in the local region under consideration the function

$$\eta = n(x)\mu(T(x))T(x) \approx const, \tag{4.28}$$

i.e., η practically does not change in space. The condition (4.28) is satisfied due to the competition between the thermodiffusion flux out the local region and the diffusion flux into this region from the periphery.

As it follows from (4.27), likewise the carrier conductivity (4.21) in this local region is reduced by

$$\delta\sigma = -\frac{\sigma}{T}\delta T. \tag{4.29}$$

On the other hand, the current density does not change along the sample:

$$j = const. \tag{4.30}$$

This means that a decrease of the conductivity (4.29) causes an increase of the electric field $E = j/\sigma$ given by:

$$\delta E = -\frac{E}{\sigma}\delta\sigma. \tag{4.31}$$

Therefore, from (4.29) and (4.31) we obtain that an initial local increase of the temperature causes a local increase of the electric field:

$$\delta E = \frac{E}{T}\delta T. \tag{4.32}$$

Because of $j = const$ (4.30) the amount of power supplied to the carriers, $W_j = jE$, also increases in the local region of the elevated temperature proportionally to the increase of the electric field, i.e., corresponding to (4.32):

$\delta W_j = (jE/T)\delta T$. This increase of W_j can cause a further increase of temperature and a further growth of the electric field. As a result, a localized region of high electric field, where the carrier temperature is high and the carrier concentration is low, can be formed (Fig. 4.6). Such a *thermodiffusion* localized electric field domain exists owing to the competition of diffusion flux of carriers into the domain and thermodiffusion flux of the hot carriers out the domain. This competition can lead to the formation of the localized electric field domain, if the following necessary conditions are fulfilled [4.74]:

$$s + \alpha > -1, \quad L \gg l, \tag{4.33}$$

where the designations s, l, L are the same as in (4.17).

The first of the conditions (4.33) can be obtained by putting the concentration n, taken from (4.28), into (4.22), where W_j and P are given by (4.23) and (4.11), respectively. Taking into account (4.28) and $E = j/\sigma$, we get:

$$\frac{j^2 T}{2e\eta} = \frac{2\eta}{\mu(T)T}\frac{T - T_l}{\tau_\epsilon(T)}. \tag{4.34}$$

This equation describes the energy balance of an electron-hole plasma for the *local* region where the temperature increases, if the size of this region is considerably less than the diffusion length of carriers. The local instability in the semiconductor occurs when the local increase of the power W_j supplied to the carriers becomes higher than the increase of the power P transferred from the hot carriers to the lattice of the semiconductor, i.e., when $\delta W_j > \delta P$. As it follows from (4.34), the latter inequality can be written in the form

$$\frac{j^2}{2e\eta} > \frac{2\eta}{\mu\tau_\epsilon T}\frac{(s + \alpha + 1)T_l - (s + \alpha)T}{T}. \tag{4.35}$$

Using (4.34), we find that the condition (4.35) is fulfilled when the plasma temperature exceeds the critical value T_c:

$$T > T_c = \frac{s + \alpha + 2}{s + \alpha + 1}T_l. \tag{4.36}$$

From (4.36) we obtain the first of the conditions (4.33) being necessary for a local plasma instability.

The conditions (4.33) can be fulfilled in many semiconductors both at low and at room temperature of the lattice T_l [4.51,59,60,74]. The discussed effect of the occurrence of localized electric field domains in a semiconductor plasma [4.74] has been experimentally found in a light-generated plasma in Ge at a lattice temperature $T_l = 77\,\mathrm{K}$ [4.64].

4.3 Local Active and Damping Processes. Concept of "Activator" and "Inhibitor"

In this section a concept of an activator and an inhibitor applying to semiconductors and devices will be discussed. By means of this concept in the next sections possible forms and properties of autosolitons in semiconductors and semiconductor devices will be considered.

4.3.1 Possibility of the Appearance of Current Filaments or Electric Field Domains and Form of the Current-Voltage Characteristic of Semiconductors

An instability in semiconductors which leads to a formation of current filaments or electric field domains is, as a rule, linked to an avalanchelike *local* increase of the current density (or electric field) in a sample at some critical value of a control parameter. In the case under consideration the control parameter is the voltage or the current of a source of energy in the external circuit of the sample. For the control parameter we will use the designation A.

A correlation between the critical point of the *local* instability and the critical point at the current-voltage characteristic, where the differential conductivity of a sample becomes negative, found in many systems with S-shaped or N-shaped current-voltage characteristic (see references in [4.10, 20–22]), can exist for the following reasons:

- The nonlinear properties of both effects - the local instability and the negative differential conductivity of a sample - are essentially determined by the same active processes in a semiconductor.
- The local instability is determined by only these active processes.

In many cases, however, the local instability in semiconductors and semiconductor devices is not linked only to some active but also to some damping local processes. Strictly speaking, this local instability is often connected with a very complex ***space-time competition*** of some active and damping processes. On the other hand, due to these possible damping processes in semiconductors the above-mentioned correlation between the critical point of instability in a semiconductor and the critical point at the current-voltage characteristic, where the differential conductivity of a sample becomes negative, may not be realized. As a result, the local instability in semiconductors, which leads to a formation of current filaments or electric field domains, can spontaneously appear at a ***positive*** differential conductivity [4.72]. A current-voltage characteristic of a semiconductor for a homogeneous case can be even ***single-valued***, but nevertheless current filaments or electric field domains can spontaneously appear there [4.23, 24, 72, 75]. In other words, the form of the current-voltage characteristic may not be correlated at all with the conditions of the appearance of current filaments or electric field domains in semiconductors and semiconductor devices. Examples of such situations have been considered in Sect. 4.2.

Nonlinear properties of current filaments and electric field domains, which appear in semiconductors due to the competition between active and damping processes, can also be qualitatively different from the properties of current filaments and electric field domains in systems with S-shaped or N-shaped current-voltage characteristic mentioned in Sect. 4.1. For example, owing to the competition between local active and damping processes it is possible that

- stable multi-filament or multi-domain states can be realized in semiconductors [4.24–27, 29, 43, 51–53, 55, 58, 60, 61, 74, 75],

- transitions between different multi-filament or multi-domain stable states appear leading to numerous scenarios of self-organization in semiconductors when the control parameter is changed [4.24,27,29,53,58,74],
- pulsating, rocking, travelling and more complicated autosolitons in the form of current filaments or electric field domains occur [4.23,41–47,70,71,84],
- autosolitons transfer to autosolitons of another type, when the control parameter is changed, or due to interactions between them [4.42,47,71,84–86],
- sophisticated scenarios of self-organization and chaos spontaneously occur [4.44,47,48,55–57,70,71,84].

We will briefly discuss these properties of autosolitons below.

It is found that due to the competition between local active and local damping processes, dissipative structures and sophisticated scenarios of self-organization may develop also in a huge number of other physical systems, chemical reactions and biology (see references, e. g., in reviews [4.87,88] and books [4.50,89–92]).

In many nonequilibrium media the processes of self-organization and of spatiotemporal chaos are linked with a spontaneous appearance and subsequent evolution of localized dissipative structures - *autosolitons* [4.47–50]. Current filaments or electric field domains in semiconductors and semiconductor devices are examples of autosolitons. Although the physics of nonequilibrium systems can be entirely different, the nonlinear properties of autosolitons realized in these systems may be qualitatively the same. This occurs if the *nonlinear properties* of local active and local damping processes causing an appearance of autosolitons have qualitatively the same character.

In particular, it can occur that a local active process in a semiconductor is due to the positive feedback in one of its parameters. Such a parameter is called *an activator* (see, e. g., references in books [4.50,89]). We will use the designation θ for the value of the activator. The local damping process which controls the local change of the activator can also be fulfilled by some other parameters (or a function of parameters) of the semiconductor. This parameter is called *an inhibitor*. We will use the designation η for the value of the inhibitor. In different systems entirely different variables can play the roles of the activator and the inhibitor. In the examples considered in Sect. 4.2 we have found the following:

1. In the transistor structure (Sect. 4.2.1) the lattice temperature of the semiconductor plays the role the activator, i. e. $\theta = T$, and the voltage drop across the emitter junction plays the role of the inhibitor, i. e. $\eta = V_e$.
2. In the reverse-biased *p-n* structure (Sect. 4.2.2) the carrier concentration in the space charge region of the *p-n* junction plays the role of the activator, i. e. $\theta = n$, and the voltage drop across the *p-n* junction plays the role of the inhibitor, i. e. $\eta = V_i$.
3. In the dense electron hole plasma (Sect. 4.2.3) the effective carrier temperature plays the role of the activator, i. e. $\theta = T$, and the pressure of the gas of carriers plays the role of the inhibitor, i. e. $\eta = nT$.
4. In the plasma, where electric field domains can appear (Sect. 4.2.4), the effective carrier temperature plays the role of the activator, i. e. $\theta = T$, and a function of the temperature and the carrier concentration plays the role of the inhibitor: $\eta = n\mu(T)T$.

From the examples considered it follows that the spatial distribution of the activator θ qualitatively corresponds to the distribution of the current density in current filaments or to the distribution of the electric field in electric field domains (Figs. 4.3-4.6). This situation is typical of the formation of current filaments and electric field domains in semiconductors. For this reason, as of now we will use (especially in the figures) the designation θ as generalization of the activator distribution for both the current density j in current filaments or the electric field E in electric field domains. Besides, considering the properties of current filaments and electric field domains we will use the term *autosolitons* for both current filaments and electric field domains in semiconductors.

4.3.2 An Activator-Inhibitor Model

If the spatiotemporal behaviour of the activator and the inhibitor satisfies some conditions [4.93] (activator-inhibitor "principles" [4.94]), a *general* nonlinear theory of such autosolitons, inclusive the theory of current filaments and electric field domains in semiconductors and semiconductor devices, can be developed [4.93, 95, 96]. The results of this theory can be found in reviews and books [4.47–50].

From the mathematical point of view this theory is based on the analysis of the activator-inhibitor model, i. e., the system of two nonlinear equations for the activator and for the inhibitor with corresponding boundary conditions (see (9.1), (9.2) in [4.50]). However, as it follows from many experimental and numerical investigations, the qualitative nonlinear properties of autosolitons which follow from this analysis can be applied to many systems described by a set of equations which cannot directly be reduced to the activator-inhibitor model [4.47–50]. This is linked with the fact that the qualitative nonlinear properties of autosolitons essentially depend on the qualitative character of the process of activation and inhibition and their characteristic times and lengths scales. Therefore, for a lot of cases it is enough to distinguish clearly the processes of activation and inhibition to predict the possible forms and properties of autosolitons. For this reason, we will discuss below the properties of different types of autosolitons rather than consider separately the different types of semiconductors and semiconductor devices, where autosolitons can appear.

4.4 Autosolitons in Semiconductors

In this section the form and the basic types of autosolitons, which can be formed in semiconductors and semiconductor devices, will be discussed.

4.4.1 Form of Autosolitons

The form of an autosoliton essentially depends on the range of activator values

$$\theta_0 < \theta < \theta_0', \tag{4.37}$$

(i.e., on the range of the current density $j_0 < j < j_0'$ or the electric field $E_0 < E < E_0'$), where the *local* increase of the activator leads to a positive feedback in the medium and causes a further increase of the local activator value.

If the bottom limit θ_0 and the top limit θ_0' in (4.37) are of the same order of magnitude, i.e., $\theta_0 \sim \theta_0'$, then *wide* autosolitons can usually be formed. If, on the contrary, the condition $\theta_0' \gg \theta_0$ is valid for the medium, or if the top limit does not exist at all, *spike* autosolitons can be formed [4.95,96]. In systems with $\theta_0' \gg \theta_0$ a spike autosoliton can be transformed into a wide autosoliton, when the control parameter A is increased (see references in [4.47–50]).

Wide autosolitons. The form of wide autosolitons has been found in [4.96]. The wide autosoliton consists of two regions of the medium, where the activator slightly changes in space, and two walls, where the activator sharply changes in space (Fig. 4.7). The two regions of the wide autosoliton, separated by the walls of the autosoliton, correspond to the stable states of the medium. In other words, the positive feedback due to the process of activation is realized neither in the centre of the autosoliton nor on its periphery. In contrary, such positive feedback is localized in the walls of the autosoliton. However, due to the competition between this local activating and the inhibiting processes in the autosoliton's walls, the autosoliton can be stabilized in a definite range of the control parameter.

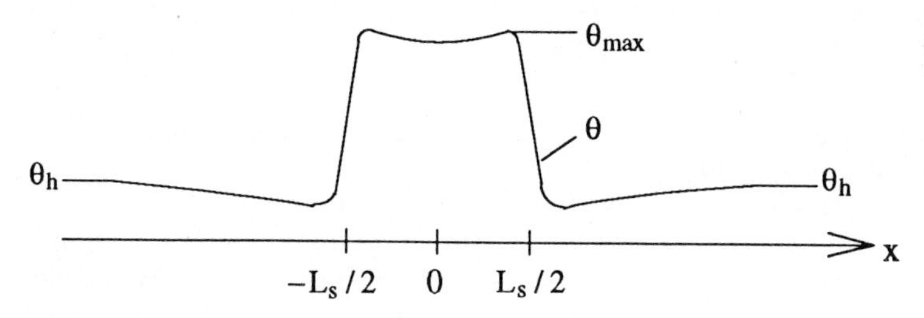

Fig. 4.7. Schematic representation of the activator distribution (current density or electric field) in a wide autosoliton

In the centre of a wide autosoliton the activator value is of the order of magnitude of the value θ_0'. On the periphery of the wide autosoliton the activator tends to its value of the homogeneous stable state θ_h. It can reach this value if the length of the system is large enough. Note that in semiconductors with a single-valued current-voltage characteristic the stable region in the autosoliton's centre does not correspond to a homogeneous state of the medium. In contrary, this stable state is self-formed in the medium owing to the appearance of the autosoliton. When the control parameter is changed, the width L_s of the wide autosoliton is also changed, but its amplitude θ_{max} remains almost constant and equals to some value which is characteristic for the medium. These properties of wide autosolitons [4.93,96] have been confirmed both by numerical investigations of dense electron-hole plasmas [4.52,53] and by experimental investigations of electrical networks representing an active medium [4.97,98]. The theory of wide autosolitons, results of numerical and experimental investigations are reviewed in [4.50].

Spike autosolitons. In contrast to wide autosolitons, the positive feedback caused by the process of activation exists almost in the whole spike of autosolitons, in particular, in the centre of *spike* autosolitons [4.47–50, 95, 99]. The amplitude of the spike θ_{max} remains however finite because of an appropriate local change of the inhibitor which prevents uncontrolled growth of the activator in the spike. The competition between the positive feedback caused by the activation process in the spike of the autosoliton and the mentioned local change of the inhibitor determines the amplitude of the autosoliton. The amplitude of a spike autosoliton θ_{max} considerably rises when the control parameter is increased because of the increasing activation in the centre of the spike autosoliton. The form and these properties of spike autosolitons have been found in [4.95, 99].

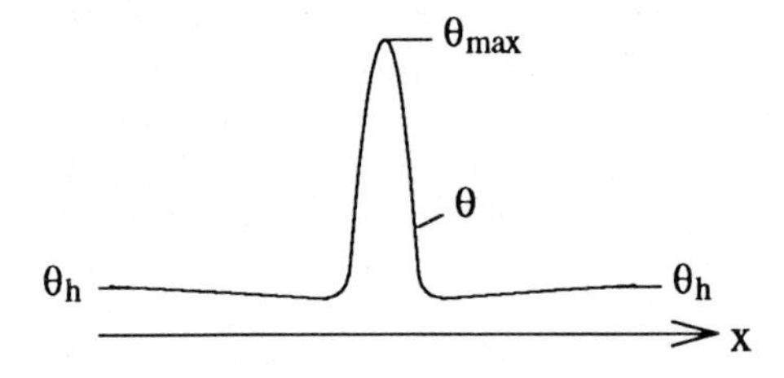

Fig. 4.8. Schematic representation of the activator distribution (current density or electric field) in a spike autosoliton

There are two types of spike autosolitons: *narrow* [4.95, 99] and *broad* spike autosolitons [4.51, 59, 100]. The width of narrow spike autosolitons is of the order of the characteristic length of the change of the activator. The width of broad spike autosolitons is of the order of the characteristic length of the change of the inhibitor. The theory of spike autosolitons has been reviewed in [4.47–50].

"Ball lightning" in semiconductors. Spike autosolitons can be realized in many cases of local instabilities in semiconductors and devices. In particular, narrow spike autosolitons are realized both in the transistor structure (Subsect. 4.2.1) [4.23] and the reverse-biased *p-n* structure (Subsect. 4.2.1) [4.62, 63]. Both narrow and broad spike autosolitons can appear in electron-hole plasmas at a relative high effective carrier temperature (Subsects. 4.2.3 and 4.2.4) [4.51, 55–57, 59]. Narrow spike autosolitons can explain the occurrence of microplasmas even in ideally homogeneous *p-n* junctions [4.62]. A microplasma is a glowing narrow spike autosoliton with a diameter of about 10^{-4} cm. In the centre of microplasmas the current density is 10^6 to 10^9 times higher than outside the autosoliton. Microplasmas are examples of high nonequilibrium regions in a weak nonequilibrium system.

This fact that the amplitude of spike autosolitons θ_{max} may be as high as to exceed the value of the activator at the periphery θ_h by several orders of magnitude is the usual property of spike autosolitons [4.47–50]. On the other hand, there is a certain lower threshold of the control parameter $A = A_b$ below which the spike autosoliton suddenly disappears: it cannot exist when $A < A_b$. The following statement is valid for spike autosolitons in some cases: The smaller the threshold value $A = A_b$ is, the higher is the value of the activator θ_{max} in the centre of the autosoliton [4.101, 102]:

$$\theta_{max}(A_b) \propto A_b^{-a}, \tag{4.38}$$

where $a > 0$. For the activator-inhibitor model (Sect. 4.3.2) equation (4.38) follows from the results of investigations of narrow spike autosolitons. For example, in some activator-inhibitor models for narrow spike autosolitons we have [4.47, 102]:

$$A_b \propto \left(\frac{l}{L}\right)^{1/2}, \quad \theta_{max}(A_b) \propto \left(\frac{L}{l}\right)^{1/2}.$$

Thus, the smaller the characteristic length l of activator changes is (in comparison with the length L for the inhibitor), the lower is the value of the control parameter $A = A_b$ at which it is still possible to excite a narrow spike autosoliton and the higher is the amplitude θ_{max} of the autosoliton. In other words, a spike autosoliton of gigantic amplitude, in some cases resembling a "ball lightning", may occur in semiconductors not far from equilibrium [4.47, 50, 101, 103].

4.4.2 Basic Types of Autosolitons

In semiconductors and semiconductor devices a lot of different types of autosolitons in the form of both wide and spike current filaments or electric field domains, as discussed in Subsect. 4.4.1, can be realized: static, pulsating, rocking, travelling autosolitons, etc. Here we will discuss some typical types of autosolitons for one-dimensional systems.

Static autosolitons. Static wide (Fig. 4.7) and spike autosolitons (Fig. 4.8) are usual for many semiconductors and devices, where the process of inhibition is faster and spatially more extended than the process of activation [4.93, 104]. Such active systems are called K-systems [4.47–50]. Static autosolitons have been investigated both theoretically [4.93, 96], experimentally [4.97, 98], and numerically [4.52, 53, 60]. Static multi-filament or multi-domain states can be stable because of an appropriate local change of the inhibitor which prevents a decrease of the number of autosolitons in the system in some range of the control parameter.

There are semiconductors, where the process of inhibition is slower and spatially more extended than the process of activation. Such systems are called $K\Omega$-systems [4.47–50]. In these media the characteristic relaxation time of the inhibitor τ_η exceeds the characteristic relaxation time of the activator τ_θ:

$$\tau_\eta > \tau_\theta. \tag{4.39}$$

As a result, static autosolitons exist in a narrower range of the control parameter. The reason for this is that in such semiconductors static autosolitons can loose their stability with respect to growing local perturbations of the activator with some frequency $\omega = \omega_c$. This instability arises because under the condition (4.39) the local change of the inhibitor is too slow to damp the critical fluctuation of the activator whose frequency is $\tau_\eta^{-1} < \omega_c < \tau_\theta^{-1}$. Due to such Hopf bifurcations static autosolitons can spontaneously transform into *pulsating* autosolitons.

Pulsating autosolitons. Apparently, the condition for the appearance of a pulsating autosoliton was first established from the stability analysis of spike current filaments in transistors [4.23]. This condition was later generalized for the activator-inhibitor model in [4.84], where also the classification of different kinds of pulsating autosolitons, based on the stability analysis of static autosolitons, was given. The properties of pulsating wide (Fig. 4.9) and pulsating spike

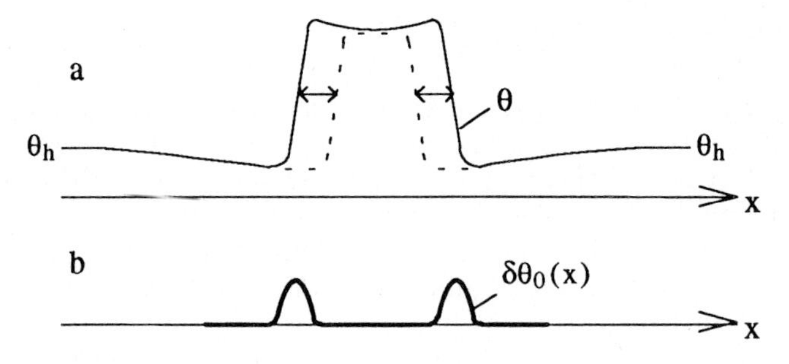

Fig. 4.9. (a) Schematic representation of the activator distribution (current density or electric field) in a pulsating wide autosoliton; (b) critical fluctuations causing the transition of a static into a pulsating wide autosoliton

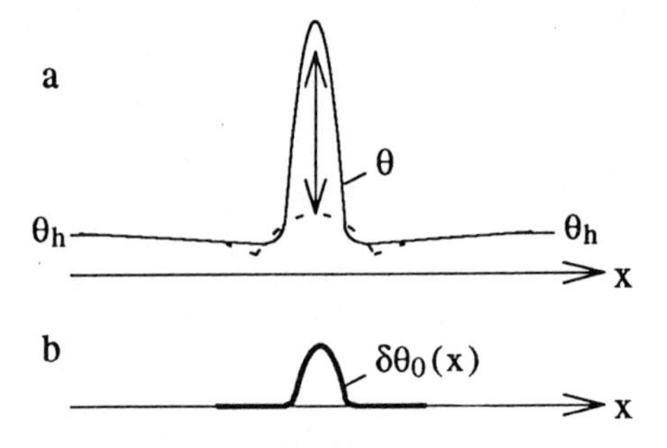

Fig. 4.10. (a) Schematic representation of the activator distribution (current density or electric field) in a pulsating spike autosoliton; (b) critical fluctuations causing the transition of a static into a pulsating spike autosoliton

(Fig. 4.10) autosolitons can be quite different [4.84]. This is due to the fact that the critical fluctuation of the activator $\delta\theta_0(x)cos(\omega_c t)$, the growth of which leads to the transition of a static autosoliton to a pulsating autosoliton, is practically localized only in the walls of the wide autosoliton (Fig. 4.9). A contrary situation is realized for spike autosolitons: the critical fluctuation of the activator $\delta\theta_0(x)cos(\omega_c t)$ is in this case localized practically in the whole spike of the autosoliton (Fig. 4.10) [4.84]. For these reasons, we may expect that pulsations of a wide autosoliton are oscillations mainly in the width of the autosoliton (Fig. 4.9). In contrary, the pulsating spike autosoliton can display oscillations in the amplitude or both in the amplitude and in the width of the spike (Fig. 4.10).

Depending on the parameters of a medium the onset of pulsation can be both soft (supercritical Hopf bifurcation) and hard (subcritical Hopf bifurcation). In the latter case the pulsation can be a relaxation oscillation with a period of the order of the characteristic relaxation time of the inhibitor τ_η and with an autosoliton-amplitude growth time of the order of the characteristic relaxation time of the activator τ_θ. This kind of pulsations with a quick appearance of the autosoliton and a subsequent slower relaxation process may occur due to the condition (4.39).

These predications concerning the nonlinear properties of pulsating autosolitons [4.84] were confirmed by subsequent theoretical, experimental and numerical investigations of various semiconductors [4.41, 54, 57, 70, 71, 84, 105–107]. Thus, the form of periodically pulsating autosolitons has been investigated in heated electron-hole plasmas [4.54]. Pulsating spike autosolitons in the form of pulsating field domains have been found in numerical investigations of electron-hole plasmas [4.57]. Possibly, the same kinds of pulsating autosolitons have been observed in experimental investigations of electron-hole plasmas in Ge [4.105]. A pulsating wide or a pulsating spike current filament can also spontaneously appear in semiconductors with an S-shaped characteristic [4.70, 84], when there is a large enough capacity in the external circuit of a sample. A detailed numerical nonlinear analysis of the latter case [4.70] has shown both the soft and the hard onset of the pulsations. In the case of the hard onset the pulsations are relaxation oscillations: during one cycle of the pulsation the spike current filament quickly appears in the medium and then during the slower relaxation process the filament almost disappears [4.70]. Also for an activator-inhibitor model qualitatively similar pulsating spike filaments have been found in numerical investigations [4.70] and thus confirmed the earlier predications of the theory of autosolitons concerning the possible behaviour of pulsating spike autosolitons [4.47, 84]. Such kind of pulsating spike current filaments have been experimentally found in *p-i-n* stuctures [4.41]. Pulsating current filaments of the discussed types [4.84] are possible also in *n*-GaAs and *p*-Ge at low temperatures [4.106, 107] and in *p-n-p-n* structures [4.71]. In the latter case numerical investigations have shown the appearance of current filaments pulsating both in the amplitude and in the width [4.71].

Rocking autosolitons. From the stability analysis of static autosolitons it is also known that besides symmetrical critical fluctuations of the activator, $\delta\theta_0(x)$ (Figs. 4.9 and 4.10), nonsymmetrical dangerous fluctuations $\delta\theta_1(x)$ (Figs. 4.11 and 4.12) are localized in the walls of a wide autosoliton or in the spike of a spike autosoliton [4.47, 85, 86]. The growth of such nonsymmetrical fluctuations which vary at some frequency ω_{c1} implies feasibility of a wide (Fig. 4.11) or a spike *rocking* autosoliton (Fig. 4.12) [4.47,86]. However, when a control parameter is varied the condition for a spontaneous growth of such a nonsymmetrical fluctuation is met later than for the symmetrical fluctuation $\delta\theta_0(x)$. In other words, a rocking autosoliton may be excited by way of a finite amplitude perturbation [4.47,86]. The hard excitation of a steadily rocking autosoliton by way of a finite deformation of a static autosoliton was found in numerical simulations of electron-hole plasmas [4.54].

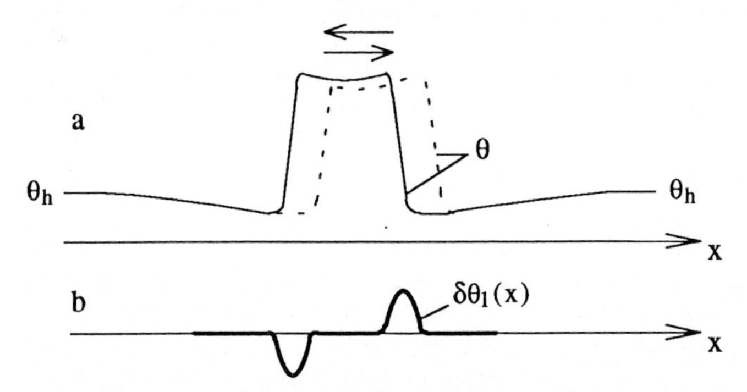

Fig. 4.11. (a) Schematic representation of the activator distribution (current density or electric field) in a rocking wide autosoliton; (b) critical fluctuation which may be responsible for the appearance of a rocking wide autosoliton

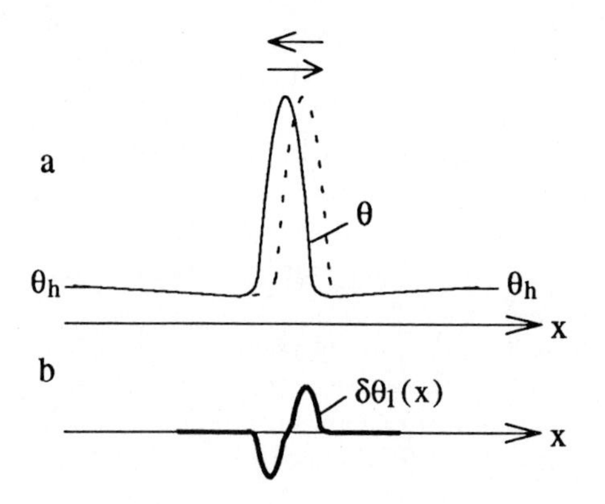

Fig. 4.12. (a) Schematic representation of the activator distribution (current density or electric field) in a rocking spike autosoliton; (b) critical fluctuation which may be responsible for the appearance of a rocking spike autosoliton

Spontaneous appearance of rocking autosolitons. Peculiarities of dynamics of autosolitons caused by parameters of the external circuit. The conclusion that a rocking autosoliton may be excited by way of a finite amplitude perturbation concerns the activator-inhibitor model described by a set of two equations for the activator and the inhibitor [4.47, 86]. Owing to the influence of the load resistor in the external circuit, a *spontaneous* appearance of rocking autosolitons becomes possible. This effect has been found out from experimental investigations of current filaments in *p-n-p-n* structures [4.42, 43] and it has been explained on the basis of qualitative analytical considerations of the dynamical behaviour of pulsating and rocking autosolitons [4.42].

The arguments for such a behaviour of autosolitons are considered here by means of the example of an autosoliton in the form of a current filament. A static current filament can transform into a pulsating current filament if the symmetrical perturbation of the current density $\delta j_0(x) \propto \delta\theta_0(x)$ (Figs. 4.8 and 4.9) grows. On the one hand, such a perturbation of the current density in a medium of a finite length L_x causes a change of the *total* current in the external circuit:

$$\delta I = L_z \int_0^{L_x} \delta j_0(x) dx. \tag{4.40}$$

On the other hand, in the limit case of a current source, i.e., if the load resistor is $R_L = \infty$, it is evident that the total current should be constant. For this reason, the growth of the symmetric fluctuation $\delta\theta_0$ is suppressed. In contrast to this, the perturbation of the current density $\delta j_1 \propto \delta\theta_1(x) cos(\omega_{c1} t)$, which describes the rocking motion, can appear. This is due to the fact that this perturbation is nonsymmetrical and therefore may be realized without a change in the total current, i.e., the nonsymmetrical perturbation can appear even in the case $R_L = \infty$, when the total current is constant. Thus a rocking current filament can occur instead of a pulsating one if the load resistor as well as the total current exceed a certain critical value [4.42]. These qualitative physical considerations have been confirmed by numerical simulations concerning the spontaneous appearance of rocking current filaments made in [4.108]. The analogous conclusions can be made for the spontaneous appearance of a rocking electric field domain in the case when the load resistor is smaller than a certain critical value [4.42].

In the frame of the concept of activator and inhibitor these effects can be explained by the existence of two types of inhibitors: the local inhibitor and the global inhibitor. The latter corresponds to the total voltage drop across the sample in the case of an autosoliton in the form of a current filament or the total current in the case of an autosoliton in the form of an electric field domain.

From these examples one can see the important influence of the parameters of the external circuit (load resistor, external capacity or inductance, or character of a dynamical resistivity of the source of energy) on the dynamics of autosolitons in semiconductors and devices. In other words, changes in the total current (or in the total voltage drop across the sample) which are caused by a time-dependent character of the distributions of current density (or electric field) can lead to new diverse phenomena of self-organization and turbulence in semiconductors.

Travelling autosolitons. Travelling autosolitons, i.e., autosolitons which move at a certain finite velocity without attenuation can appear in semiconductors under the condition (4.39). Such travelling autosolitons can occur even in the absence of "drift forces", which could also be responsible for a motion (see below). If the inhibitor is both slow (4.39) and short-range in comparison with the activator, then the travelling autosolitons of the FitzHugh - Nagumo (FHN) type can be realized (Fig. 4.13) [4.109, 110]. The characteristic property of travelling autosolitons of FHN type is that the collision of two such autosolitons leads, as a rule, to an annihilation of these autosolitons (for a review see [4.111]). Besides, in such activator-inhibitor systems, where the inhibitor is both slow and short-range in comparison with the activator (these systems are called Ω-systems), the other types of autosolitons, i. e., static, pulsating or rocking autosolitons, cannot appear as a rule (e. g., [4.50]).

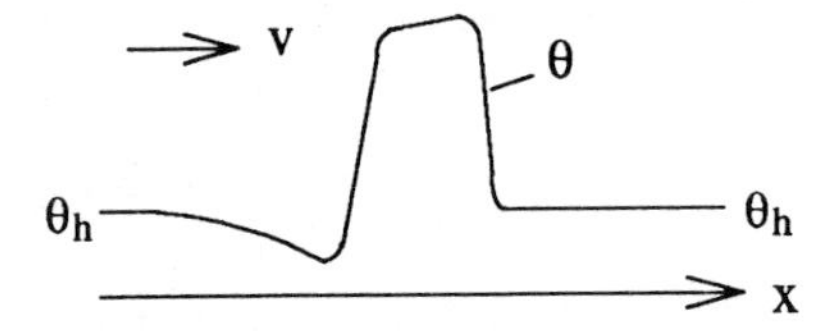

Fig. 4.13. Schematic representation of the activator distribution (current density or electric field) in a travelling autosoliton of the FitzHugh - Nagumo type [4.109, 110]

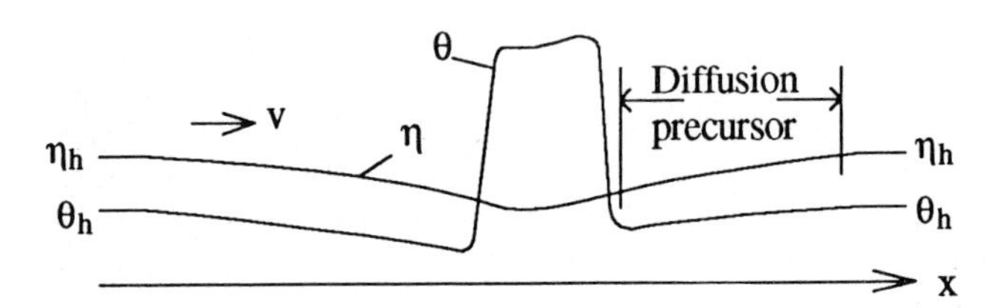

Fig. 4.14. Schematic representation of the activator distribution (current density or electric field) in a travelling autosoliton when a diffusion precursor in front of the first wall of the autosoliton is realized [4.47, 85, 86]

All possible types of autosolitons can occur in such a medium, where the inhibitor is both slow and long-range in comparison with the activator ($K\Omega$-systems) [4.47, 85]. In this case travelling autosolitons have an important peculiarity: in front of the first wall of the autosoliton there exists a diffusion precursor of the inhibiting component (Fig. 4.14) [4.47, 85]. Owing to this diffusion precursor two travelling autosolitons may repel each other or may transform to another type of autosoliton (see below) [4.47, 85, 86]. When the travelling autosoliton reaches the boundary of the medium it can repel from the boundary due to the diffusion precursor which acts as a feeler of the autosoliton [4.42].

The spontaneous appearance of travelling autosolitons in semiconductors, which occurs due to a transition of a rocking autosoliton into a travelling one, has been found out in experimental investigations of current filamentation in *p-n-p-n* structures [4.42]. The observed travelling current filaments show the properties of travelling autosolitons with diffusion precursor. In particular, such a travelling current filament does not reach the surface of the sample, but repels in a certain distance of it [4.42]. The mathematical model of the *p-n-p-n* structure proposed can be written as activator-inhibitor model along with the equation (Kirchhoff's rule) for the global inhibition [4.42]. Detailed numerical investigations of static, pulsating, rocking and travelling autosolitons in this model made in [4.71] permitted to investigate peculiarities of different types of autosolitons in *p-n-p-n* structures and also to confirm the general conclusions and predictions of the theory of autosolitons in $K\Omega$-systems, where the inhibitor is both slow and long-range in comparison with the activator [4.47, 85]. In particular, numerical investigations [4.71] show that travelling autosolitons with diffusion precursor really repel from the boundary of the sample, as it had been proposed [4.42] based on the results of the theory of autosolitons [4.47]. Travelling autosolitons have also been found in numerical investigations of a model for semiconductors in the regime of impurity impact ionization [4.107] and in both analytical and numerical investigations of a model for semiconductor heterostructures [4.67–69].

Autosolitons moving due to "drift forces". Electric field domains and current filaments, moving due to "drift forces" are well-known in semiconductors with N-shaped or S-shaped current voltage characteristics (e.g., [4.1, 2, 6, 8, 20–22]). This effect can also be realized in semiconductors with a single-valued characteristic. For example, autosolitons in the form of electric field domains have been discussed in Subsect. 4.2.4 with reference to electron-hole plasmas in which electrons and holes had the same parameters. In such "symmetric" electron-hole plasmas the autosolitons are static. In a real electron-hole plasma heated by a permanent electric field, i.e., in a plasma in which the parameters of electrons and holes are not the same, autosolitons, naturally, begin to move [4.60, 74, 112].

We emphasize that in these cases the reason for the motion of the autosolitons is absolutely different compared to the travelling of autosolitons discussed above. Travelling autosolitons (Figs. 4.13 and 4.14) move with a certain finite velocity without attenuation due to the fact that the inhibition process is slower than the process of activation (see (4.39)). Such a motion occurs also if no "drift forces" in the direction of the motion of the autosoliton exists. In contrary, the discussed motion of electric field domains in electron-hole plasmas is linked with the presence of "drift forces" in the direction of motion. These forces appear owing to the additional "drift" terms proportional to $j\partial\theta/\partial x$ and $j\partial\eta/\partial x$ in the model of the plasma, which accounts for the drift of autosolitons in electric fields [4.60, 74]. Depending on the relation between the parameters of electrons and holes, autosolitons may drift along or counter to the direction of current lines [4.60, 112]. When the concentrations of electrons and holes are equal, autosolitons move in the direction of the drift of those carriers whose mobility is higher (electrons, as a rule). If the concentrations differ considerably, autosolitons move with those carriers whose concentration is smaller [4.60, 112]. Note that the drift terms, which may have an influence on the autosoliton's motion, also appear in an other model for autosolitons in semiconductor heterostructures [4.67–69].

Likewise, autosolitons in the form of current filaments can begin to move due to the "drift forces". Such a situation, for example, may occur due to inhomogeneous heating of the sample in the direction transverse to the current lines, due to other inhomogeneities in this direction, or in crossed electric and magnetic fields due to the appearance of the Lorentz force. Current filaments moving in crossed electric and magnetic fields have analytically been investigated in a model of semiconductors with S-shaped current-voltage characteristic in [4.9]. Recently current filaments moving in crossed electric and magnetic fields have numerically been investigated in a model of a semiconductor with S-shaped current-voltage characteristic, where both stationary moving filaments and a complex dynamical behaviour of current filaments have been observed [4.113].

4.5 Some Properties of Autosolitons

Up to now we have discussed the forms and types of autosolitons in semiconductors. In this section we try to answer questions about mechanisms of the spontaneous appearance of autosolitons in semiconductors, about possible scenarios of the evolution of autosolitons, about transitions of an autosoliton of one type into an autosoliton of another type and about the occurrence of spatiotemporal chaos connected with the existence of autosolitons in semiconductors.

4.5.1 Main Types of Self-Organization Phenomena

We can distinguish the following types of self-organization phenomena connected with the spontaneous appearance and further evolution of autosolitons in one-dimensional systems [4.47–50, 94]:

1. Spontaneous appearance of a localized critical fluctuation acting as "nucleation centre" for the formation of an autosoliton.
2. Local breakdown effect: A *local* avalanchelike increase (or decrease) of the activator value in a medium. The local breakdown effect can cause a spontaneous appearance of autosolitons in the medium. The local breakdown effect is a *dynamical* effect, i.e., it may also be realized in the absence of fluctuations. The presence of fluctuations may give rise to the local breakdown before the critical point for the realization of this effect in the absence of fluctuations is reached. The local breakdown effect in media can be realized
   - near a small local inhomogeneity in a semiconductor sample,
   - in the centre of an autosoliton or between autosolitons,
   - in the oscillatory tails of an autosoliton,
   - in the case of a slightly inhomogeneous parameter distribution.
3. Repelling of autosolitons which appear owing to the splitting of an autosoliton.
4. Activator repumping between autosolitons. This effect results in a spontaneous destruction of one autosoliton or several autosolitons.
5. Annihilation of colliding travelling autosolitons of FHN type. This effect has already been discussed in Subsect. 4.4.2.

6. Spontaneous transition of an autosoliton into an autosoliton of another type when a control parameter is changed.
7. Spontaneous transitions of an autosoliton into an autosoliton of another type due to collisions between autosolitons.
8. Spontaneous appearance of a leading centre, i. e., the local source of travelling autosolitons.
9. Self-destruction of autosolitons.
10. Random recurrent processes of appearance and disappearance of autosolitons in a medium.

Some of these effects will briefly be considered in the next sections. Note that a competition between these effects may give rise to turbulence in semiconductors.

4.5.2 "Nucleation Centre" for the Spontaneous Formation of Autosolitons in Semiconductors

An autosoliton can spontaneously appear due to the occurrence of a *finite-amplitude localized critical fluctuation*, i. e., a localized fluctuation which causes the appearance of an autosoliton with the *maximal* probability. As a rule, autosolitons in experiments are more likely to arise near small local inhomogeneities which are always present in real media. Both localized critical fluctuations and local inhomogeneities play the role of "nucleation centres" for the spontaneous formation of autosolitons (for reviews, see [4.50, 94]).

Critical localized fluctuations in ideally homogeneous media. In an ideally homogeneous medium there is a finite probability that the activator distribution will *locally* reach a level sufficient for the spontaneous formation of an autosoliton even when the control parameter A is smaller than the value A_c which is critical for a *global instability* of the whole medium corresponding to the growth of nonhomogeneous fluctuations. As such localized critical fluctuations appear randomly, autosolitons may appear randomly both in time and in space.

Let us consider the probability of occurrence of a growing finite-amplitude localized fluctuation, P, in an ideally homogeneous medium at $A < A_c$ for the case of such media, where at $A = A_c$ the Turing instability is realized [4.114,115]:

$$P = K \frac{L_x}{l_c} \frac{\gamma^3}{G\tau} exp\left(-\frac{16}{3}\frac{\gamma^{3/2}}{G}\right). \tag{4.41}$$

In (4.41) L_x is the length of the medium; $K = const, \tau = const$; $l_c \sim k_c^{-1}$, k_c is the wave number of the critical global fluctuation at the threshold of the Turing instability; G is the dispersion of fluctuations which have been proposed to be delta-correlated;

$$\gamma = \frac{A_c - A}{A_c}. \tag{4.42}$$

Equation (4.41) is valid for $A < A_c$, precisely when

$$exp\left(-\frac{16}{3}\frac{\gamma^{3/2}}{G}\right) \ll 1. \tag{4.43}$$

The probability of the spontaneous formation of an autosoliton increases according to (4.41) steadily with the size of the medium L_x. This implies that in an infinite medium an autosoliton will spontaneously arise somewhere for a certainty even before the value of the control parameter A reaches the critical point A_c [4.114].

The form of the localized critical fluctuation of the activator $\Delta\theta(x)$ of a finite amplitude, the growth of which causes a spontaneous formation of an autosoliton in the ideally homogeneous media, is given by [4.115]

$$\Delta\theta(x) = K_1\gamma^{1/2}u_c(x)exp(ik_cx), \tag{4.44}$$

where $K_1 = const$ and the function $u_c(x)$ is:

$$u_c(x) = \frac{2^{1/2}exp(i\phi)}{cosh[\gamma^{1/2}l_c^{-1}(x-x_0)]}. \tag{4.45}$$

In (4.45) x_0 and ϕ are constants.

Equations (4.41) and (4.45) [4.115] are similar to the results of the theory of critical fluctuations in *equilibrium* media with first-order phase transitions developed in [4.116,117]. However, in *nonequilibrium* media under consideration the order parameter u_c (4.45) is complex rather than real. Besides, the localized critical fluctuation (4.44) oscillates in space in nonequilibrium media rather than fading in space monotonously. The oscillating character of the critical fluctuation (4.44) may lead to the nontrivial consequences in the evolution of this fluctuation. In particular a complex localized structure of many interacting autosolitons can spontaneously appear in the medium [4.47–50].

Note that the probability of an occurrence of a localized critical fluctuation of large amplitude in a model of homogeneous bistable nonequilibrium media has earlier been found in [4.118].

The influence of small local inhomogeneities. Real media always contain small inhomogeneities which play the role of a "seed" for the spontaneous appearance of autosolitons [4.47,104,119]. The latter is linked with the fact that at a certain value of the control parameter

$$A = A_c^- < A_c \tag{4.46}$$

in the neighbourhood of the inhomogeneity a *local breakdown* occurs, i.e., a local increase (or decrease) in the activator in an avalanche fashion. The influence of an inhomogeneity is comparable to the effect of the boundary of the semiconductor, if the distributions of the activator and/or the inhibitor near the boundary are inhomogeneous.

As a result of the local breakdown, static, pulsating, rocking, travelling or more complex autosolitons can spontaneously appear in the medium. We emphasize that the local breakdown is a dynamical effect which is not related to the presence of fluctuations in the medium [4.47,119]. The critical value $A = A_c^-$ defines the threshold of a dynamical rearrangement of the medium. The critical value A_c^- has been calculated in [4.120]. Note that in the presence of fluctuations there is a finite probability for a finite-amplitude fluctuation localized near the inhomogeneity to grow before the critical point A_c^- is reached [4.121]. Therefore, the fluctuations can lead to a random appearance of autosolitons near the inhomogeneity.

The influence of slightly inhomogeneous distributions of semiconductor parameters. Some parameters of a semiconductor as the lattice temperature, the effective carrier temperature, the carrier concentration, the current density, etc., can be slightly inhomogeneous even in semiconductors without local inhomogeneities, for example, due to external influences or due to some transition processes in the semiconductor. Note that a local breakdown effect causing a spontaneous appearance of autosolitons may also occur in this case. This is linked with the possibility of self-formation of localized critical perturbations in some parts of the slightly inhomogeneous distributions in the semiconductor. The occurrence of self-formed localized critical perturbations in slightly inhomogeneous media [4.122] may be a dynamical effect which is not linked with the presence of fluctuations. Apparently, this effect may explain the random appearance of spike autosolitons found in slightly inhomogeneous electron-hole plasmas in numerical investigations [4.55].

4.5.3 Effects Determining the Evolution of Autosolitons

Splitting of autosolitons. Splitting of autosolitons, i. e., the division of one autosoliton into two autosolitons, is a deterministic effect which is not associated with the presence of fluctuations in the medium [4.47–50, 119]. Splitting of autosolitons occurs at a certain value of the bifurcation parameter $A = A_d$ and is linked with the local breakdown in the centre of the autosoliton. The features of the splitting process depend on the form of the autosoliton, i. e., whether the autosoliton is a wide or a spike autosoliton. The theory of this effect and also the results of experimental and numerical investigations can be found in [4.47–50].

Repelling of autosolitons. Two autosolitons which appear owing to the splitting of an autosoliton (Fig. 4.15) move in opposite directions (for a review, see [4.50]). The mechanism of such a repelling of two autosolitons has been considered in [4.94]. The repelling of autosolitons is linked with a considerable difference between the inhibitor value in the walls (or spikes) of the autosolitons appearing at the first moment after the splitting and the characteristic inhibitor value for static autosolitons. It causes the motion of autosolitons in opposite directions even when the condition (4.39) is not satisfied. In the latter case the velocity of the repelling autosolitons decreases as the distance between them increases.

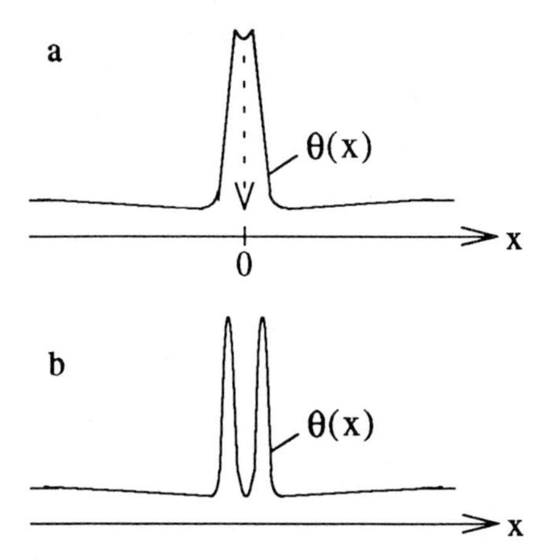

Fig. 4.15. (a) Schematic representation of the splitting of a static autosoliton: The activator distribution at the initial moment after reaching the critical point $A = A_d$ (a) and directly after the local breakdown in the centre of the autosoliton (b); the arrow in (a) schematically shows the local breakdown, i. e., a local decrease of the activator in the centre of the autosoliton in an avalanche fashion

Local breakdown between autosolitons and in the tails of autosoliton. Maximal possible distance between autosolitons. The spontaneous appearance of new autosolitons can also be caused by local breakdown effects between autosolitons and in the tail of an autosoliton (for a review, see [4.50]). These effects together with the splitting of autosolitons determine a maximal distance between autosolitons L_{max}: if the distance L_p between autosolitons is larger than L_{max}, $L_p > L_{max}$, the local breakdown effect starts and new autosolitons appear in the medium. Therefore, the given number of autosolitons in the medium can exist when the distance between them is

$$L_p < L_{max}. \tag{4.47}$$

The value L_{max} depends on the control parameter. There is some range of the control parameter, where $L_{max} = \infty$, i. e., a single autosoliton or a few autosolitons located arbitrarily far from each other can exist.

"Pumping" of the activator between autosolitons. Minimal possible distance between autosolitons. Let us consider two autosolitons which are located at some distance L_p. In this case one of the dangerous fluctuations of the activator can be approximately represented as the antibinding combination of the functions $\delta\theta_0$ (Fig. 4.10) of each of the "isolated" autosolitons:

$$\delta\theta_{1,0}(x) \simeq \delta\theta_0\left(x + \frac{L_p}{2}\right) - \delta\theta_0\left(x - \frac{L_p}{2}\right). \tag{4.48}$$

It follows from the form of this critical fluctuation $\delta\theta_{1,0}(x)$ (Fig. 4.16) that the growth of this fluctuation leads to an increase of the amplitude (or the width) of one autosoliton (left in Fig. 4.16) and to a decrease of the amplitude (or the width) of a neighbouring autosoliton. In other words, $\delta\theta_{1,0}(x)$ describes a pumping of the activator between autosolitons. As a result of the growing of this fluctuation one of the autosolitons (the right in the Fig. 4.16) gradually disappears.

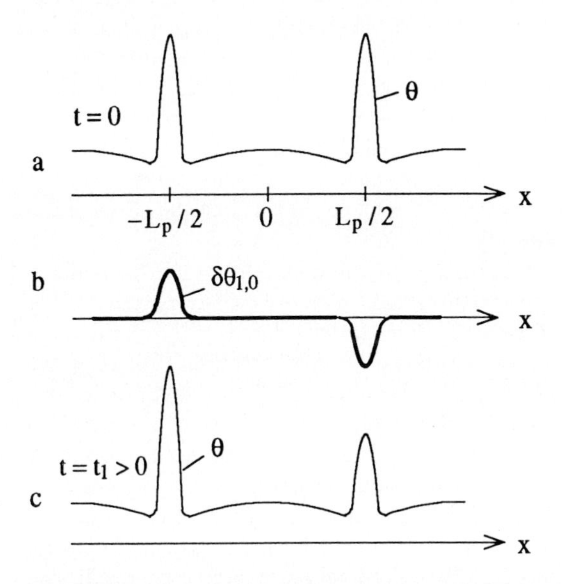

Fig. 4.16. Illustration of the activator "pumping" between two autosolitons: (a) schematic form of the autosolitons located at some distance L_p; (b) form of the critical activator fluctuation $\delta\theta_{1,0}(x)$; (c) intermediate activator distribution realized during the growth of the fluctuation $\delta\theta_{1,0}(x)$

The growth of the fluctuation $\delta\theta_{1,0}(x)$ is damped by a change in the inhibitor. The damping effect of this change in the inhibitor is reduced by its diffusive spreading, which is more pronounced the smaller the distance L_p between autosolitons is. Therefore, there is a critical distance between autosolitons $L_p = L_{min}$: Autosolitons which are located at a distance $L_p < L_{min}$ are unstable with respect to the growth of the fluctuation $\delta\theta_{1,0}(x)$ of the activator (4.48). Therefore, two or more autosolitons can only be stable if the distance between them is

$$L_p > L_{min}. \tag{4.49}$$

The theory of the pumping activator effect can be found in [4.48,50].

Self-destruction of autosolitons. On the one hand, for a spontaneous appearance of an autosoliton the growth of a local perturbation of the activator should be realized. The growth of the local perturbation of the activator occurs due to a competition between active and damping processes which exists in a medium in the *absence* of an autosoliton. On the other hand, after the formation of an autosoliton of large amplitude the competition between active and damping processes in the vicinity of the autosoliton's centre may change qualitatively. This can occur because in the vicinity of the centre of an autosoliton of large amplitude some new nonlinear processes, which appear in a medium in the *presence* of the autosoliton, are possible. Therefore, the appearance of an autosoliton of large amplitude in the medium can lead to considerable changes in some parameters of the medium in the region of the autosoliton's location. Effects of a self-destruction of the autosoliton can be realized if the autosoliton changes the parameters of the medium so much that the conditions of the existence of the autosoliton cannot be fulfilled any more. As a result, the autosoliton disappears. Let us consider two possible different mechanisms of the effect of the self-destruction of the autosoliton.

The first mechanism of self-destruction of the autosoliton can be realized due to a possible dependence of the threshold value A_b on the parameters of the medium. Indeed, we have already mentioned, that there is a threshold value A_b of the control parameter for the existence of an autosoliton: For $A < A_b$ the autosoliton cannot exist [4.47,50]. The threshold value A_b is a characteristic parameter of a medium. It can depend on other parameters of the medium such as, for example, the temperature of a sample. In media, where A_b is an increasing function of a parameter of the medium, which grows in the presence of an autosoliton, a self-destruction of the autosoliton can occur. In such media, after the autosoliton has spontaneously appeared, A_b begins to grow. After some delay time it can reach the value of the given control parameter. As a result, the autosoliton disappears.

This mechanism of self-destruction of autosolitons has been found in numerical investigations of the formation of microplasmas (spike autosolitons of huge amplitude) in Si *p-n* structures [4.123]. In this case the control parameter A corresponds to the voltage drop V across the structure, i.e., $A = V$. The threshold value V_b of the voltage drop across the structure for the existence of the autosoliton is an increasing function of the temperature of the Si *p-n* structure. After the microplasma has appeared, the temperature of the structure in a vicinity of the microplasma begins to grow. Therefore, the threhold value V_b begins to rise, too. As a result, the autosoliton disappears when V_b reaches the given value of the control parameter. In the case under consideration local heating of a sample, which appears only in the *presence* of the autosoliton in the semiconductor, is the new nonlinear process determining the realization of the effect of the self-destruction of the autosoliton.

The second mechanism of self-destruction of autosolitons can be realized in semiconductors, where spike autosolitons appear [4.54, 55]. When considering spike thermodiffusion autosolitons in electron-hole plasmas (Subsects. 4.2.3 and 4.2.4), we have emphasized that the effective carrier temperature in the centre of autosolitons increases when the control parameter is increased. In some semiconductors this may bring about a situation where in a small neighbourhood of the centre of the autosoliton intensive interband impact ionization of carriers occurs. This new process, which appears only in the *presence* of autosolitons in the semiconductor, can lead to a value of the generation rate of carriers which is so large that the thermodiffusion ejection mechanism will no longer cope with carriers resulting from impact ionization in the centre of the autosoliton. As a result, the autosoliton disappears [4.54,55].

4.5.4 Processes of Random Appearance and Disappearance of Autosolitons

The nonlinear processes discussed above can lead to the occurrence of different recurrent random processes of appearance and disappearance of autosolitons [4.47,50]. This is linked with the fact that some of the effects of appearance and of disappearance of autosolitons can be realized at the *same* value of the control parameter. Recurrent processes of the appearance and disappearance of autosolitons in semiconductors may occur, for example, in the following cases:

- An autosoliton appears near an inhomogeneity of a sample at $A > A_c^-$ (4.46) and after the autosoliton has been formed it disappears due to one of the mechanisms of self-destruction of autosolitons. An example of such a kind of recurrent processes is the appearance and disappearance of microplasmas in Si *p-n* structures which has been observed in numerical simulations [4.123].
- An autosoliton appears due to a slightly inhomogeneous parameter distribution of a medium and after the autosoliton has been formed it disappears due to one of the mechanisms of self-destruction of autosolitons. Apparently, such a kind of recurrent processes can explain the appearance and disappearance of thermodiffusion autosolitons in electron-hole plasmas found in numerical investigations [4.55].

We emphasize that both the appearance of an autosoliton near an inhomogeneity and the self-destruction of autosolitons are dynamical processes, i.e., they might also be realized in the absence of fluctuations. Obviously, the fluctuations which cause a random appearance or a random disappearance of autosolitons before the critical values are reached, can be responsible for the occurrence of recurrent random appearance and disappearance of autosolitons in semiconductors. Autosolitons randomly appearing and disappearing can also arise without fluctuations. This may occur, for example, if the appearance of autosolitons is caused by a slightly inhomogeneous parameter distribution of a medium and, on the other hand, this distribution is sensible to the existence of autosolitons in the medium.

4.5.5 Transitions between Different Types of Autosolitons

The effects of local breakdown, repelling of autosolitons, activator pumping and self-destruction of autosolitons determine scenarios of self-organization and turbulence in many cases. If the condition (4.39) is fulfilled, then, besides of these effects, many additional diverse scenarios of self-organization, processes of random appearance and disappearance of autosolitons and turbulence connected with the processes of appearance and disappearance of pulsating, rocking, or travelling autosolitons and with transitions of autosolitons into autosolitons of another type can occur [4.47,50].

Coexistence of different types of autosolitons. In media, where the inhibitor is slow and long-range compared with the activator, static, pulsating, travelling, rocking and other types of autosolitons may simultaneously coexist (see the condition of such coexistence in [4.47,50,85]). In particular, this implies that in different areas of the same sample it may be possible to excite autosolitons of different types, by adjusting the parameters of the exciting pulse.

Nonlinear interaction between autosolitons. In media, where various types of autosolitons can coexist, collisions between autosolitons may lead to the transition of autosolitons into autosolitons of another type [4.47,50,85]. For example, the collision between two travelling autosolitons due to a "diffusion precursor" running ahead of their front walls (Fig. 4.14) may lead to a repelling of these autosolitons, or they may transform into a static or a pulsating autosoliton.

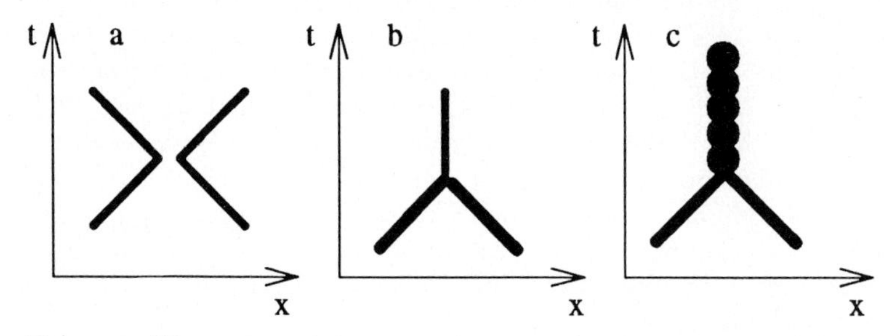

Fig. 4.17. Illustration of the transitions of autosolitons into autosolitons of another type upon collision of two travelling autosolitons: repelling of autosolitons (a), transition of two travelling autosolitons into a static (b), or a pulsating autosoliton (c)

Evolution of autosolitons. A transition of an autosoliton into an autosoliton of another type can also be realized owing to a loss of the stability of an autosoliton, or due to a dynamic rearrangement of the initial state. These effects can occur when the parameters of an autosoliton in the process of its evolution, which is realized by varying the control parameter, reach some critical values (for a review, see [4.47,50]). We have already mentioned the possible transitions of static autosolitons into pulsating or rocking autosolitons in Subsect. 4.4.2. Also the reverse transitions may be possible. Furthermore, the transitions travelling ⇔ pulsating autosoliton, rocking ⇔ pulsating autosoliton or rocking ⇔ travelling autosoliton may supposed to appear when the initial parameters of an autosoliton change in the process of its evolution. Some of such transitions, static ⇔ rocking autosoliton and rocking ⇔ travelling autosoliton, have already been found in experimental investigations of current filaments in p-n-p-n structures [4.42,43,45,46].

4.5.6 On Mechanisms of Spatiotemporal Chaos (Turbulence) in Semiconductors

Temporal chaos in the external circuit of a semiconductor sample, i. e., irregular oscillations of the total current through the sample or of the voltage drop across the sample, accompanied by an existence of autosolitons in the form of current filaments have experimentally been discovered in a number of semiconductors: in n-GaAs, in p-Ge [4.32, 35–40, 124] and in p-n-p-n Si structures [4.44]. In the latter case the correlation of irregular spatial motions of the current filament and chaotic oscillations of the device voltage has been demonstrated for the case of a rocking filament by using a streak camera. However, in many other cases the understanding of the correlations of the experimentally observed *temporal chaos* (irregular oscillations of the voltage or total current) to the spatial behaviour of the current density distribution is far from being complete.

In semiconductors *spatiotemporal chaos*, i. e., turbulence, has been discovered in numerical investigations of spike autosolitons in electron-hole plasmas heated by an electric field [4.55–57]. In this case, in particular, the process of random appearance and disappearance of spike autosolitons in different parts of the medium has been observed. The conclusion about the occurrence of turbulence in an electron-hole plasma followed from the behaviour of both the time and space correlation functions of patterns in the electron-hole plasma: The observed oscillations are not correlated either in time or in space [4.57].

Recently period-doubling cascades leading to chaos have been found in numerical investigations of pulsating current filaments in the model of p-n-p-n Si structures [4.71] and also in the activator-inhibitor model for elements with S-shaped current-voltage characteristic [4.70].

Mechanisms of spontaneous appearance of temporal or spatiotemporal chaos (turbulence) in systems of autosolitons have been proposed based on the results of the theory of autosolitons [4.47–50, 84, 99, 104]. Turbulence is associated with a spontaneous random formation and subsequent evolution of one or several autosolitons. Scenarios of the development of turbulence are thus determined by nontrivial properties of autosolitons (random appearance of static, pulsating, rocking or travelling autosolitons), by the complex dynamic interaction between autosolitons, by transitions of autosolitons into autosolitons of another type and by the random appearance and disappearance of autosolitons in different parts of a medium. It has also been emphasized that the spontaneous appearance of temporal spatiotemporal chaos (turbulence) can most probably be expected in systems of *spike* autosolitons due to their nontrivial properties [4.47, 84, 99, 104]. This supposition, following from the theory of spike autosolitons (for reviews see [4.47–50]), has first been confirmed by numerical investigations of electron-hole plasmas [4.55–57] and later by numerical investigations of p-n-p-n Si structures [4.71] and of an activator-inhibitor model [4.70].

The mechanisms of chaos in a system of autosolitons can be divided into two groups [4.47–50, 84, 94]:

1. Mechanisms of chaos, which are linked with the existence of a *time-oscillatory* behaviour of some nonhomogeneous stationary states of a medium. This situation can be realized in media, for example, under the condition (4.39),

when time-oscillatory nonhomogeneous stationary states, such as pulsating or rocking autosolitons, can appear. In other words, *local* self-oscillators in active media may serve as sources of the onset of irregular inhomogeneous oscillations in a system [4.84].

2. Mechanisms of spatiotemporal chaos, where the *local breakdown effect* plays a decisive role [4.48–50, 94]. In these cases the spatiotemporal chaos may also be realized in such media, where the time-oscillatory behaviour of both nonhomogeneous and homogeneous stationary states of a medium does not exist.

The first group of mechanisms of spatiotemporal chaos in semiconductors may mathematically be described by the methods developed for dynamical systems. In particular, one can expect that the well-known period-doubling scenario leading to chaos (e. g., [4.125]) can be realized in this case. Indeed, under the conditions (4.39) the period-doubling scenario leading to chaos has been found both in experimental (e. g., [4.44]) and numerical investigations of autosolitons in semiconductors [4.70, 71]. In this case it may also be expected for some models of semiconductors by an analogy with hydrodymamical systems [4.126] that it may be possible instead of partial differential equation, describing spatial distributions of parameters of a medium, to write a set of ordinary differential equations which can describe temporal changes in the parameters of autosolitons, such as maximal and minimal values of activator or inhibitor in autosolitons, the width of autosolitons, etc. However, it is necessary to note that in many cases the processes of spontaneous appearance and disappearance of autosolitons, splitting of autosolitons and a lot of other processes, which can be important for the development of chaos in a system of autosolitons, may be very sensitive even to very small variations of the spatial parameter distributions of the medium.

The second group of mechanisms of spatiotemporal chaos includes:

- Mechanism of spatiotemporal chaos in the whole medium connected with the competition of the local breakdown and the activator repumping effects [4.50, 84, 99]. This mechanism can be possible in media, where spike autosolitons appear, when in some range of the control parameter the maximal possible distance between autosolitons is less than the minimal one [4.84]:

$$L_{max} < L_{min}. \tag{4.50}$$

- Mechanism of local spatiotemporal chaos connected with the competition of three processes: splitting of autosolitons, repelling of autosolitons and activator repumping between autosolitons [4.94].
- Mechanism of spatiotemporal chaos connected with the spontaneous appearance and self-destruction of spike autosolitons [4.55, 56].

4.6 Conclusions

In this review we have briefly considered some results concerning the investigations of autosolitons in the form of current filaments and electric field domains in semiconductors and semiconductor devices. Already in the simplest one-dimensional case autosolitons in semiconductors may show intricate properties. We observe spontaneously appearing wide or spike static, pulsating, rocking and travelling autosolitons, complex processes of evolution of autosolitons, collisions between autosolitons and transitions of autosolitons of one type into autosolitons of another type, random processes of spontaneous appearance and disappearance of autosolitons, temporal or spatiotemporal chaos in systems of autosolitons.

In experiments autosolitons usually appear near small inhomogeneities of a sample. After an autosoliton has been formed, its properties do not, as a rule, depend on the parameters of the inhomogeneity, which acts as nucleus for the formation of autosolitons of large amplitude. In some cases autosolitons of gigantic amplitude may develop, where the current density or the electric field in the centre of the autosoliton exceeds the values on the periphery of the autosoliton by many orders of magnitude.

On the one hand, the investigations of autosolitons in semiconductors may help to understand important physical properties of semiconductors. On the other hand, they may give ideas for the development of new semiconductor devices with better characteristics. We believe that the discussion of properties of autosolitons in semiconductors made in this review will be useful for further experimental, theoretical and numerical investigations of autosolitons in semiconductors and semiconductor devices.

References

[4.1] J.B. Gunn: Solid State Commun. **1**, 88 (1963)
[4.2] B.K. Ridley: Proc. Phys. Soc. **82**, 954 (1963)
[4.3] T. Fukami, J. Homma: Jap. J. Appl. Phys. **2**, 535 (1963)
[4.4] I. Melngailis, A.G. Milnes: J. Appl. Phys. **33**, 995 (1962)
[4.5] A. M. Barnett, A.G. Milnes: J. Appl. Phys. **37**, 4215 (1966)
[4.6] B.W. Knight, G.A. Peterson: Phys. Rev. **155**, 393 (1967)
[4.7] A.V. Volkov, S.M. Kogan: Sov. Phys.-JETP **25**, 1095 (1967)
[4.8] E. Pytte, H. Thomas: Phys. Rev. **179**, 431 (1969)
[4.9] A.K. Zvezdin, V.V. Osipov: Sov. Phys.-JETP **31**, 90 (1970)
[4.10] V.V. Osipov, V.A. Kholodnov: Mikroelektronika **2**, 529 (1973) (in Russian)
[4.11] Z.S. Gribnikov: Sov. Phys. Semicond. **6**, 1204 (1973)
[4.12] B.K. Ridley: Rep. Prog. Phys. **54**, 169 (1991)
[4.13] F. Pacha, F. Paschke: Electron. Commun. **32**, 235 (1978)
[4.14] K. Hess, H. Morkoc, H. Shichijo, B.G. Streetman: Appl. Phys. Lett. **35**, 469 (1979)
[4.15] K. Aoki, K. Yamamoto, N. Mugibayashi, E. Schöll: Solid-St. Electron. **32**, 1149 (1989)
[4.16] A. Kastalsky, M. Milshtein, L.G. Shantharama, J. Harbison, L. Florez: Solid-St. Electron. **32**, 1841 (1989)
[4.17] L.L. Bonilla, S.W. Teitsworth: Physica D **50**, 545 (1991)
[4.18] N. Balkan, B.K. Ridley, A. Vickers (eds.): *Negative Differential Resistance and Instabilities in 2-D Semiconductors* (Plenum Press, New York 1993)
[4.19] M.P. Shaw, V.V Mitin, E. Schöll, H.L. Grubin: *The Physics of Instabilities in Solid State Electron Devices* (Plenum Publishing Corp., New York 1992)

[4.20] A.V. Volkov, S.M. Kogan: Sov. Phys.-Uspekhi **11**, 881 (1969)
[4.21] V.L. Bonch-Bruevich, I. Zvyagin, A.G. Mironov: *Domain Electrical Instabilities in Semiconductors* (Consultants Bureau, New York 1975)
[4.22] E. Schöll: *Nonequilibrium Phase Transitions in Semiconductors* (Springer, Berlin, Heidelberg, New York 1987)
[4.23] B.S. Kerner, V.V. Osipov: Mikroelektronika **6**, 337 (1977) [Engl. transl.: Sov. Microelectronics **6**, 256 (1977)]
[4.24] B.S. Kerner, V.F. Sinkevich: JETP Lett. **36**, 436 (1982)
[4.25] B.S. Kerner, N.A. Kozlov, A.M. Nechaev, V.F. Sinkevich: Sov. Phys. Semicond. **17**, 1235 (1983)
[4.26] B.S. Kerner, N.A. Kozlov, A.M. Nechaev, V.F. Sinkevich: Mikroelektronika **12**, 217 (1983) (in Russian)
[4.27] B.S. Kerner, D.P. Litvin, V.I. Sankin: Sov. Tech. Phys. Lett. **13**, 342 (1987)
[4.28] K.M. Mayer, J. Parisi, R.P. Huebener: Z. Phys. B **71**, 171 (1988)
[4.29] V.A. Vashchenko, B.S. Kerner, V.V. Osipov, V.F. Sinkevich: Sov. Phys. Semicond. **24**, 1065 (1990)
[4.30] K. Aoki, T. Kobayashi, K. Yamamoto: J. Phys. Soc. Japan **51**, 2373 (1982)
[4.31] K. Aoki, K. Miyamas, T. Kobayashi, K. Yamamoto: Physica B+C **117/118**, 570 (1983)
[4.32] K. Aoki, K. Yamamoto: Phys. Lett. A **98**, 72 (1983)
[4.33] J. Peinke, A. Mühlbach, R.P. Huebener, J. Parisi: Phys. Lett. A **108**, 407 (1985)
[4.34] K.M. Mayer, J. Peinke, B. Röhricht, J. Parisi, R.P. Huebener: Physica Scripta T **19**, 505 (1987)
[4.35] U. Rau, K.M. Mayer, J. Parisi, J. Peinke, W. Clauss, R.P. Huebener: Solid State Electron. **32**, 1365 (1989)
[4.36] J. Spangler, A. Brandl, W. Prettl: Appl. Phys. A **48**, 143 (1989)
[4.37] A. Brandl, M. Völcker, W. Prettl: Solid State Commun. **72**, 847 (1989)
[4.38] A. Brandl, W. Prettl: In *Festkörperprobleme (Advances in Solid State Physics)* Vol. 30, ed. by U. Rössler (Vieweg, Braunschweig 1990) pp. 371 - 385
[4.39] U. Rau, W. Clauss, A. Kittel, M. Lehr, M. Bayerbach, J. Parisi, J. Peinke, R.P. Huebener: Phys. Rev. B **43**, 2255 (1991)
[4.40] J. Peinke, J. Parisi, O.E. Rössler, R. Stoop: *Encounter with Chaos* (Springer, Berlin, Heidelberg 1992)
[4.41] R. Symanczyk, S. Gaelings, D. Jäger: Phys. Lett. A **160**, 397 (1991)
[4.42] F.-J. Niedernostheide, B.S. Kerner, H.-G. Purwins: Phys. Rev. B **46**, 7559 (1992)
[4.43] F.-J. Niedernostheide, M. Arps, R. Dohmen, H. Willebrand, H.-G. Purwins: Phys. Status Solidi (b) **172**, 249 (1992)
[4.44] F.-J. Niedernostheide, M. Kreimer, H.-J. Schulze, H.-G. Purwins: Phys. Lett. A **180**, 113 (1993)
[4.45] F.-J. Niedernostheide, M. Kreimer, B. Kukuk, H.-J. Schulze, H.-G. Purwins: Phys. Lett. A **191**, 285 (1994)
[4.46] F.-J. Niedernostheide: this volume, chapter 8
[4.47] B.S. Kerner, V.V. Osipov: Sov. Phys.-Uspekhi **32**, 101 (1989)
[4.48] B.S. Kerner, V.V. Osipov: Sov. Phys.-Uspekhi **33**, 679 (1990)
[4.49] B.S. Kerner, V.V. Osipov: *Autosolitons* (Nauka, Moscow 1991) (in Russian)
[4.50] B.S. Kerner, V.V. Osipov: *Autosolitons: A New Approach to Problems of Self-Organization and Turbulence* (Kluwer Academic Publishers, Dordrecht, Boston, London 1994)
[4.51] A.L. Dubitsky, B.S. Kerner, V.V. Osipov: Sov. Phys. Semicond. **20**, 755 (1986)

[4.52] V.V. Gafiichuk, B.S. Kerner, V.V. Osipov, A.G. Yuzhanin: Sov. Tech. Phys. Lett. **13**, 543 (1987)
[4.53] V.V. Gafiichuk, B.S. Kerner, V.V. Osipov, A.G. Yuzhanin: Sov. Phys. Semicond. **22**, 1298 (1988)
[4.54] V.V. Gafiichuk, V.E. Gashpar, B.S. Kerner, V.V. Osipov: Sov. Phys. Semicond. **22**, 1161 (1988)
[4.55] Z.I. Vasyunik, V.V. Gafiichuk, B.S. Kerner, V.V. Osipov: Sov. Phys.-Solid State **31**, 1875 (1989)
[4.56] V.V. Gafiichuk, B.S. Kerner, V.V. Osipov, Z.I. Vasyunik: In *Noise in Physical Systems. 10th International Conference*, ed. by A. Ambrozy (Budapest 1989) p. 609
[4.57] V.V. Gafiichuk, B.S. Kerner, V.V. Osipov, T.N. Shcherbachenko: Sov. Phys. Semicond. **25**, 1020 (1991)
[4.58] B.S. Kerner, V.V. Osipov, M.T. Romanko, V.F. Sinkevich: JETP Lett. **44**, 77 (1986)
[4.59] A.L. Dubitsky, B.S. Kerner, V.V. Osipov: Sov. Phys.-Solid State **28**, 725 (1986)
[4.60] V.V. Gafiichuk, B.S. Kerner, V.V. Osipov, I.V. Tysluk: Sov. Phys.-Solid State **31**, 46 (1989)
[4.61] V.A. Vashchenko, B.S. Kerner, V.V. Osipov, V.F. Sinkevich: Sov. Phys. Semicond. **23**, 857 (1989)
[4.62] V.V. Gafiichuk, B.I. Datsko, B.S. Kerner, V.V. Osipov: Sov. Phys. Semicond. **24**, 455 (1990)
[4.63] V.V. Gafiichuk, B.I. Datsko, B.S. Kerner, V.V. Osipov: Sov. Phys. Semicond. **24**, 806 (1990)
[4.64] M.N. Vinoslavskii: Sov. Phys.-Solid State **31**, 1461 (1989)
[4.65] M.N. Vinoslavskii, B.S. Kerner, V.V. Osipov, O.G. Sarbey: J. Phys. - Condens. Matter **2**, 2863 (1990)
[4.66] V.A. Vashchenko, Yu.A. Vodakov, V.V. Gafiichuk, B.I. Datsko, B.S. Kerner, D.P. Litvin, V.V. Osipov, A.D. Roenkov, V.I. Sankin: Sov. Phys. Semicond. **25**, 730 (1991)
[4.67] D. Döttling, E. Schöll: Physica D **67**, 418 (1993)
[4.68] D. Döttling, E. Schöll: In *Negative Differential Resistance and Instabilities in 2-D Semiconductors*, ed. by N. Balkan, B.K. Ridley, A. Vickers (Plenum Press, New York 1993) p. 179
[4.69] D. Döttling, E. Schöll: Solid-State Electron. **37**, 685 (1994)
[4.70] A. Wacker, E. Schöll: Z. Phys. B **93**, 431 (1994)
[4.71] F.-J. Niedernostheide, M. Ardes, M. Or-Guil, H.-G. Purwins: Phys. Rev. B **49**, 7370 (1994)
[4.72] B.S. Kerner, V.V. Osipov: JETP Lett. **18**, 70 (1973)
[4.73] B.S. Kerner, V.V. Osipov: Microelectronics **1**, 5 (1974)
[4.74] B.S. Kerner, V.V. Osipov: Sov. Phys.-Solid State **21**, 1348 (1979)
[4.75] B.S. Kerner, V.V. Osipov: Sov. Phys. Semicond. **13**, 523 (1979)
[4.76] C.G. Thornton, C.D. Simmons: IEEE Trans. on Elect. Dev. **5**, 6 (1958)
[4.77] R.M. Scarlett, W. Shockley: IEEE Intern. Conv. Rec. **11**, 3 (1963)
[4.78] F. Bergmann, D. Gerstner: Arch. Electr. Übertr. **17**, 467 (1963)
[4.79] S.M. Sze: *Physics of Semiconductor Devices*, 2nd ed. (Wiley, New York 1981)
[4.80] A.R. Verma, P. Krishna: *Polymorphism and Polytypism in Crystals* (Wiley, New York 1966)
[4.81] A.P. Dmitriev, A.O. Konstatinov, D.P. Litvin, V.I. Sankin: Sov. Phys. Semicond. **17**, 686 (1983)
[4.82] R.A. Suris, B.S. Shchamkhalova: Sov. Phys. Semicond. **18**, 738 (1984)

[4.83] K. Seeger: *Semiconductor Physics* (Springer, Berlin, Heidelberg, New York 1973)
[4.84] B.S. Kerner, V.V. Osipov: Sov. Phys.-JETP **56**, 1275 (1982)
[4.85] B.S. Kerner, V.V. Osipov: Mikroelektronika **12**, 512 (1983) (in Russian)
[4.86] B.S. Kerner, V.V. Osipov: Sov. Phys.-JETP **62**, 337 (1985)
[4.87] E. Meron: Phys. Rep. **218**, 1 (1992)
[4.88] M. Cross, P.C. Hohenberg: Rev. Mod. Phys. **65**, 3 (1993)
[4.89] H. Haken: *Introduction to Synergetics* (Springer, Berlin, Heidelberg, N.Y. 1977)
[4.90] H. Haken: *Advanced Synergetics* (Springer, Berlin, Heidelberg, N.Y. 1983)
[4.91] G. Nicolis, I. Prigogine: *Self-Organization in Nonequilibrium Systems* (Wiley, New York, London, Sydney, Toronto 1977)
[4.92] J.D. Murray: *Mathematical Biology* (Springer, Berlin 1989)
[4.93] B.S. Kerner, V.V. Osipov: Sov. Phys.-JETP **52**, 1122 (1980)
[4.94] B.S. Kerner: In *Experimental and Theoretical Advances in Biological Pattern Formation*, ed. by H.G. Othmer, P.K. Maini, J.D. Murray, NATO ASI Series A (Life Sciences) Vol. 259 (Plenum Press, New York 1993) pp. 191-210
[4.95] B.S. Kerner, V.V. Osipov: Sov. Phys.-JETP **47**, 874 (1978)
[4.96] B.S. Kerner, V.V. Osipov: Sov. Phys. Semicond. **13**, 424 (1979)
[4.97] H.-G. Purwins, Ch. Radehaus, Th. Dirksmeyer, R. Dohmen, R. Schmeling, H. Willebrand: Phys. Lett. A **136**, 480 (1989)
[4.98] Th. Dirksmeyer, R. Schmeling, J. Berkemeier, H.-G. Purwins: In *Patterns, Defects and Materials Instabilities*, ed. by D. Waldgräf, N.M. Ghoniem, NATO ASI Series E (Applied Sciences) Vol. 183 (Kluwer Academic Publishers, Dordrecht 1990) pp. 91-107
[4.99] B.S. Kerner, V.V. Osipov: Sov. Phys.-Doklady **28**, 485 (1983)
[4.100] A.L. Dubitsky, B.S. Kerner, V.V. Osipov: Sov.Phys.-Doklady **34**, 906 (1989)
[4.101] B.S. Kerner, V.V. Osipov: JETP Lett. **41**, 473 (1985)
[4.102] B.S. Kerner, V.V. Osipov: Sov. Phys.-Doklady **32**, 43 (1987)
[4.103] B.S. Kerner, V.V. Osipov: In *Nonlinear Waves 1: Dynamics and Evolution*, ed. by A.V. Gaponov-Grekhov, M.I. Rabinovich, J. Engelbrecht (Springer, Berlin, Heidelberg, New York 1989) pp. 126-150
[4.104] B.S. Kerner, V.V. Osipov: Mikroelektronika **14**, 389 (1985) (in Russian)
[4.105] K.M. Aliev, R.I. Bashirov, M.M. Gadzhialiev: Semiconductors **28**, 524 (1994)
[4.106] E. Schöll, D. Drasdo: Z. Phys. B **81**, 183 (1990)
[4.107] G. Hüpper, K. Pyragas, E. Schöll: Phys. Rev. B **47**, 15515 (1993)
[4.108] F.-J. Niedernostheide, R. Dohmen, H. Willebrand, B.S. Kerner, H.-G. Purwins: Physica D **69**, 425 (1993)
[4.109] R. FitzHugh: Biophys. J. **1**, 445 (1961)
[4.110] I. Nagumo, S. Arimoto, S. Yoshizawa: Proc. IRE **50**, 2061 (1962)
[4.111] A.S. Mikhailov: *Foundations of Synergetics I*, Springer Series of Synergetics Vol. 51 (Springer, Berlin 1990)
[4.112] A.L. Dubitsky, B.S. Kerner, V.V. Osipov: In *Abstracts of Reports of 1st Soviet Conference on Physics and Physicochemical Principles of Microelectronics* (Vilnius 1987) (in Russian)
[4.113] G. Hüpper, K. Pyragas, E. Schöll: Phys. Rev. B **48**, 17633 (1993)
[4.114] B.S. Kerner, S.L. Klenov: Verhandl. DPG (VI) **27**, 463 (1992)
[4.115] B.S. Kerner, S.L. Klenov: unpublished (1992)
[4.116] J.S. Langer: Annals of Physics **41**, 108 (1967)
[4.117] J.S. Langer: Annals of Physics **54**, 258 (1969)
[4.118] T. Ohta: Prog. Theor. Phys. Suppl. **99**, 425 (1989)
[4.119] B.S. Kerner, V.V. Osipov: Sov. Phys.-Doklady **27**, 484 (1982)

[4.120] B.S. Kerner, S.L. Klenov: Sov. Tech. Phys. Lett. **16**, 731 (1990)
[4.121] B.S. Kerner, S.L. Klenov: JETP Lett. **52**, 326 (1990)
[4.122] The appearance of the localized critical perturbation, which is formed by itself and causes the formation of an autosoliton, has been found out in numerical investigations of a traffic jam formation in a slightly inhomogeneous traffic flow; see Figs. 2 - 4 in B.S. Kerner, P. Konhäuser: Phys. Rev. E **50**, 54 (1994)
[4.123] B.I. Datsko, V.V. Gafiichuk, B.S. Kerner: unpublished (1991)
[4.124] R. Huebener: In *Festkörperprobleme (Advances in Solid State Physics)* Vol. 30, ed. by U. Rössler (Vieweg, Braunschweig 1990) pp. 387 - 401
[4.125] H.G. Schuster: *Deterministic Chaos*, 2nd ed. (VCH Verlag, Weinheim 1988)
[4.126] E.N. Lorenz: J. Atmos. Sci. **20**, 130 (1963)

5 Pattern Formation of the Electroluminescence in AC ZnS:Mn Devices

Ch. Goßen, F.-J. Niedernostheide, H.-G. Purwins

Institut für Angewandte Physik, Universität Münster, Correnstr. 2/4
D-48149 Münster, Germany

Abstract. Spatial inhomogeneous distributions of the electroluminescence radiation density are studied in ac driven thin film ZnS:Mn electroluminescent (ACTFEL) devices. Apart from stationary microfilaments, dynamical patterns occur on a timescale ranging from tenths of a second to several seconds. In particular, moving filaments and strings, fluctuating networks of strings, autowaves, or global oscillations may arise. The spatiotemporal structures are extremely sensitive to the conditions of preparation and the driving parameters of the samples.

5.1 Introduction

The occurence of 'structures without a constructor', thus pattern formation in distributed media, is of great interest, on the one hand for basic research of self-organization in nonlinear dynamical dissipative systems, on the other hand with regard to applications in medicine or technology.

Besides microfilaments, the ZnS:Mn system preferably shows dynamical, wave-like structures, both in ac driven [5.1,2] and in dc driven systems [5.3]. This wave patterns, e. g., rotating spiral waves or expanding concentric ringwaves, are very similar to those found in Belousov-Zhabotinskii systems, as already pointed out by Beale et al. [5.3]. Reports on pattern formation in ACTFEL (alternating current thin film electroluminescence) devices are quite rare in view of the numerous investigations of thin film luminescent devices with inherent memory in the seventies and eighties, especially the ZnS:Mn devices. While in former times the main attention has been focused on the application of memory devices with multifilamentary states for flat display panels with inherent grayscale addressing and the formation of wave patterns in ZnS:Mn electroluminescent devices has been considered as a curiosity not being suitable for applications, recently, the pattern forming ZnS:Mn ACTFEL device has been rediscovered as an examplary system of the synergetics of pattern formation in nonlinear dissipative systems [5.4,5].

As pattern formation in ZnS:Mn devices is not well-understood until now, the aim of this article is to get a better understanding particularly with respect to synergetic processes. Starting with a description of the experimental set-up (Sect. 5.2) and a short elucidation of the basics of the device physics (Sect. 5.3),

Springer Proceedings in Physics, Vol. 79 **Nonlinear Dynamics and Pattern Formation in Semiconductors and Devices**
Editor: F.-J. Niedernostheide

we give a survey of typical patterns arising in ZnS:Mn ACTFEL devices with Y_2O_3 as dielectric material (Sect. 5.4). Most important control parameters are the amplitude and the frequency of the driving voltage and the device temperature. By varying these parameters, stationary or mobile filaments, mobile luminescent strings and string networks, autowaves, and global spatiotemporal oscillations can be observed. Finally, in (Sect. 5.5) the experimental results are discussed using concepts of synergetics. It turns out that a description of pattern formation observed in such devices in terms of reaction-diffusion systems is very promising.

5.2 Experimental Set-Up

The ZnS:Mn ACTFEL device consists of Mn-doped ZnS as semiconducting active layer (S), symmetrically sandwiched between two insulating layers (I), with Y_2O_3 as suitable insulating material. This stack of thin films is vacuum-deposited sequentially on a ITO-precoated glass substrate by means of thermal coevaporation of ZnS-granules from a Mo-boat and Mn from a Al_2O_3-crucible for the active layer, and electron gun evaporation techniques for the I-layers, respectively. The ITO-precoating acts as transparent metallic electrode (M) on the glass-side, and the device is completed by depositing Al-dots as metallic counterelectrodes (M) on the opposite side. The resulting MISIM-device is shown schematically on the right hand side of Fig. 5.1. Optional postdeposition annealing is done, e. g., a 700 K heat treatment in ambient air was applied for 1 h to the Z10-sample to improve its electroluminescent characteristics. The parameters of samples refered to in this article are listed in Table 5.1.

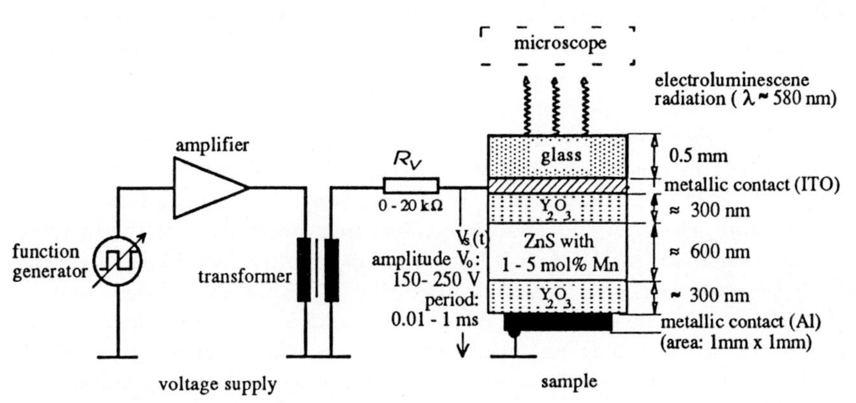

Fig. 5.1. Experimental set-up. The electroluminescence radiation is observed through the transparent ITO electrode by using a microscope. The transformer is required only for thicker film stacks (schematic, not to scale)

In Fig. 5.1 the complete experimental set-up is shown schematically. The power supply consists of a function generator, an amplifier and a transformer, the latter is used for thick film stacks to achieve sufficiently high voltage amplitudes, and provides the device with bipolar pulse trains or sinusoidal driving voltages with amplitudes V_0 in the range of 150 - 200 V, depending on the thickness of the device. Usually the devices are operated at room temperature.

The electroluminescence radiation is observed through the ITO-coated glass-substrate from the top (Fig. 5.1) by using a microscope and either a simple CCD-camera for a rough view or an image-intensifier camera for a detailed view.

Table 5.1. Data of samples used in the experiments

Sample	Z01	Z08	Z10
width of upper Y_2O_3 layer in Fig. 5.1 in nm	580	300	not measured
width of ZnS layer in nm	313	650	400
width of lower Y_2O_3 layer in Fig. 5.1 in nm	0	300	not measured
Mn concentration of the ZnS layer determined after [5.6] in mol%	not measured	1	2
annealing process after evaporation	none	none	1 h, 700 K

5.3 Electroluminescence Basics

With a suitable concentration of the Mn-dopant in the range of 1 - 5 mol%, the ZnS:Mn ACTFEL device shows an intense yellow phosphorescence glow if the electrical field in the active layer exceeds a threshold of about 2 MV/cm. In this high-field mode of operation, the response of the device is no longer purely capacitive, i.e., in addition to the displacement current a dissipative, ohmic current does exist. In region a) of Fig. 5.2 the origin of this current is illustrated for the case that the right hand side of the MISIM-device is the momentary cathode. For sufficiently high amplitudes of the driving voltage electrons from localized states of the SI-interface are injected into the S-region by tunneling through a barrier of approximately triangular shape having a field dependent width and a hight of about 1 eV. Phonon-loss-free acceleration of the injected electrons takes place in the S-region, with the possibility to impact-excite the Mn^{2+} luminescence centres on Zn^{2+} substitutional sites from the groundstate to the first excited state after having gained more than 2.2 eV from the field (region b) in Fig. 5.2). Since the de-excitation of the excited Mn^{2+}-centre by an optical transition to the groundstate is highly forbidden, the de-excitation luminescence, peaking at about 580 nm, is characterized by a typical phosphorescence time constant on the order of 0.1 ms to approximately 1 ms, depending on the Mn-concentration. Once calibrated, this dependence can be used for the estimation of the Mn-concentration, since the relation between phosphorescence time constant and the Mn-concentration is definite [5.6, 8, 9].

Having gained more than 3.6 eV - the gap-energy of ZnS - the hot electrons are able to impact-ionize the lattice, resulting in carrier multiplication by the generation of electron-hole pairs (region c) in Fig. 5.2). The generated free

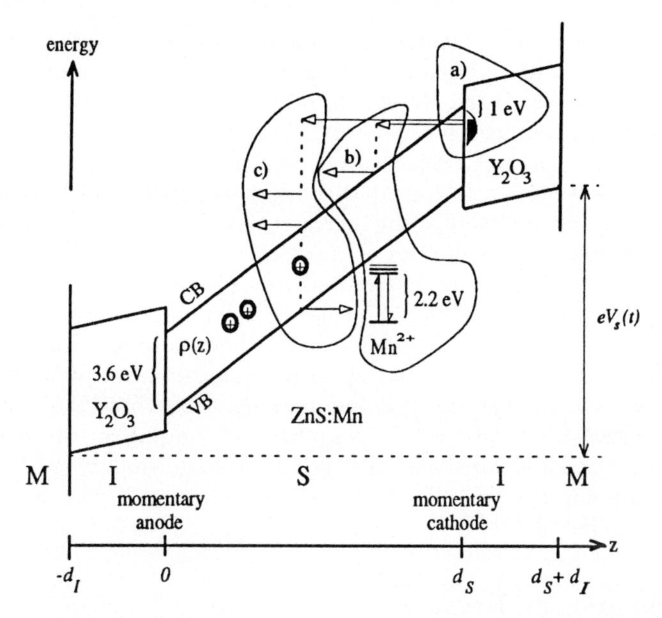

Fig. 5.2. Schematic visualization of hot electron processes within the S-layer of the film structure under high-field excitation $\geq$2 MV/cm: Tunnel emission of electrons from interface states of the IS-interface at the cathode (a), followed by the phonon-loss-free acceleration up to energies sufficient for impact excitation of the Mn^{2+} luminescent centres (b) or for impact ionization of the host lattice, eventually in combination with the build-up of space charge due to hole-trapping (c) (energy-band diagram drawn from [5.7])

holes either drift in the field towards the momentary cathode, or, according to an extended model introduced by Howard et al. [5.10], they are captured instantaneously by hole-traps. This takes place at sites within the S-layer where unoccupied hole-traps are available, leading to the build-up of positive space charge. Electrons that finally reach the momentary anode are trapped again in IS-interface states. With progressing charge transfer within one half-cycle, the accumulating negative interface charge at the anode decreases the field at the cathode so that (within a time of about 10 μs) further carrier-injection rapidly is decreased, too, resulting in a gradually disappearing dissipative current and in a limitation of the total transferred charge. Now, when changing the polarity of the applied voltage at the beginning of the new half-cycle, the former field-lowering interface charge becomes field-strengthening, until the amount of transferred charge again inhibits further charge transfer within this half-cycle, so that in steady state operation the device runs in a charge-controlled breakdown mode. The light output or brightness of the device is in good approximation proportional to the total transferred charge [5.11, 12], so that the luminescence radiation density can be looked upon as a measure for the dissipative current density. Ignoring any influence of the build-up of space charge for the present, the total transferred charge increases nearly linearly with rising voltage amplitude above the threshold voltage for tunnel emission, in the experimental case [5.13] as well as for numerical considerations [5.14]. This linear characteristic is clearly disturbed if the Mn-concentration is about 2 mol% [5.9]. Then a feedback mechanism becomes possible, with the space charge acting as mediator: Any positive space charge strengthens the field at the cathode and lowers the width of the tunnel barrier. In consequence, the strongly field-dependent tunnel emission and the dissipative current increases. At sufficiently high dissipative currents the averaged generation of space charge becomes more effective than its recombination with free-flowing electrons from the conduction band, and the higher dissipative current leads to a further build-up of space charge. This autocatalytic process continues until all available hole-traps are occupied and a new equilibrium steady state is reached. In this 'high'-state of total transferred charge the feedback mechanism stops and the 'high'-state branch of the charge-transfer voltage characteristic can be considered as the shifted extrapolation of the 'low'-state branch towards a smaller driving voltage amplitude.

More details of the physics of ZnS:Mn ACTFEL devices and their technology can be found in [5.7,9,13].

5.4 Experimental Results

5.4.1 Stationary Microfilaments and Hysteresis

The relationship between integrated radiation density and driving voltage, firstly discovered by Marello et al. [5.15], is seen in Fig. 5.3 for the sample Z01 and parameters as given in the figure caption. The hysteresis is quite evident: Figure 5.4

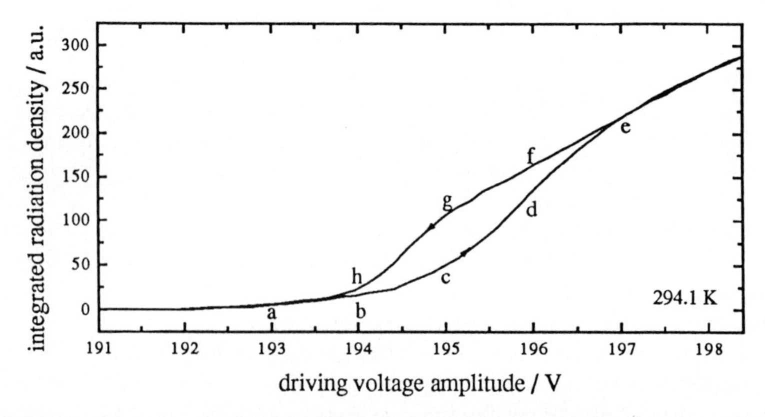

Fig. 5.3. Integrated radiation density as a function of the driving voltage of a ZnS:Mn electroluminescence device. The positions on the curve marked with letters b-c-d and h-g-f refer to the corresponding pictures in Fig. 5.4. Parameters: Sample Z01, pulsed excitation, pulse width 13.5 μs, f=5.3 kHz, R_V=5 kΩ, Θ_0=294.1 K

shows the photographs of the radiation distribution. We observe well-defined localized regions of high radiation density with a diameter of $\leq 1\,\mu$m. The number

Fig. 5.4. Photographs of the multifilamentary 'low' states (upper row) and 'high' states (lower row) refering to the hysteresis loop in Fig. 5.3. The letters in the photographs correspond to those of Fig. 5.3. The parameters are as in Fig. 5.3

of the luminous localized regions increases with increasing applied voltage and vice versa. The letters b, c, d and h, g, f indicate radiation density distributions which correspond to points b, c, d and h, g, f in Fig. 5.3 for increasing and decreasing applied voltage amplitude, respectively. Figure 5.3 results from radiation density patterns like those of Fig. 5.4 by integrating the emitted radiation of individual filaments being switched on and off while increasing and decreasing the applied voltage, respectively. The hysteresis of the brightness-voltage characteristic (Fig. 5.3) is a consequence of the hysteresis of the number of filaments as is clearly indicated in Fig. 5.4. The number is smaller when approaching a certain driving voltage amplitude from below than from above.

With increasing frequency, the once stationary filaments are getting more and more fidget, until they begin to move around the plane in a stepwise way at frequencies on the order of some 10 kHz. At the same time the walls of filaments become increasingly sharp at higher excitation frequencies.

5.4.2 Frontpropagation

At sufficiently low excitation frequencies when bistability is still present, the 'on'-region appears with a dense, almost homogeneous luminescence. Starting from the 'off'-state and increasing the voltage amplitude, primary activity arises at material inhomogeneities within the film plane where the threshold for switching to the 'on'-state is a little bit lower than in the surroundings. From this preferential sites the growth of circular expanding 'high'-state domains starts, see Fig. 5.5. The 'on'-region is characterized by a visible granular luminescence distribution, indicating the presence of closely condensed microfilaments. When the voltage amplitude is decreased, single diffusely bordered microfilaments can clearly be resolved.

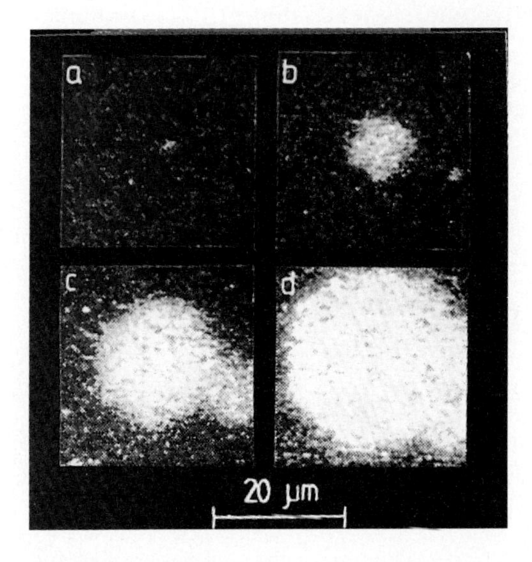

Fig. 5.5. Growth of two circular luminescence domains. Their overlapping area shows the same luminescence density as a single isolated domain. Parameters: Sample Z10, sine-wave excitation, V_0=117.6 V, f=1 kHz, R_V=5 kΩ, Θ_0=293.4 K, time between consecutive pictures 0.4 s; the order of consecutive pictures is indicated by the letters a - d

The velocity of the propagating front marking the borderline between the 'on' and the 'off'-region becomes nearly independent of the domain radius when the domain size is sufficiently high. As can be taken from Fig. 5.6a, the growth of the domain radius with increasing time is the higher the greater the radius is. This

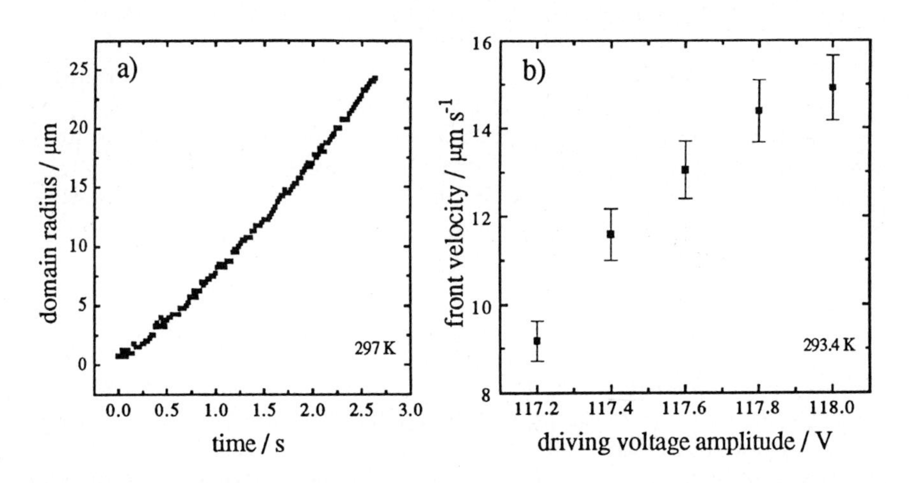

Fig. 5.6. Radius vs. time for an expanding circular domain (a) and mean front velocity vs. voltage amplitude (b). Sample and parameters as in Fig. 5.5, except V_0=116.6 V and Θ_0=297 K for (a)

can be understood if one considers a point in the frontline of the propagating front: Assuming that the activation spreads out by diffusion, the switching of such a point from the 'off'-state to the 'on'-state is supported by an increasing portion of 'on'-state sites in the surroundings of this point with increasing domain radius; this results in a faster frontpropagation. The mechanism of activation is discussed in Sect. 5.5. The domain expansion is strongly voltage dependent, as can be seen in Fig. 5.6b: A small variation of the voltage amplitude on the order of tenths of a volt changes the mean propagation velocity drastically. A similar strong dependence of the front velocity on the driving voltage amplitude is valid in the case of autowaves (Sect. 5.4.3).

5.4.3 Autowaves

Travelling waves with a self-sustaining profile arise in certain samples within the first 20 - 30 minutes of operation. The aging (under electrical excition) is characterized by an increase of the threshold voltage for the outset of luminescence, connected with a decrease of the hysteresis width. The width of the waves is typically 5 μm, the wave velocity is on the order of several ten μm/s. With decreasing voltage amplitude, the velocity and, above all, the width of a wave decreases strongly. The formation of autowaves takes place at the same sites where circular expanding domains can arise from. Starting with the formation of a circular domain, the medium is reset to the 'off'-state at its centre regeneratively when the domain diameter has reached a critical value (Fig. 5.7). From the formation

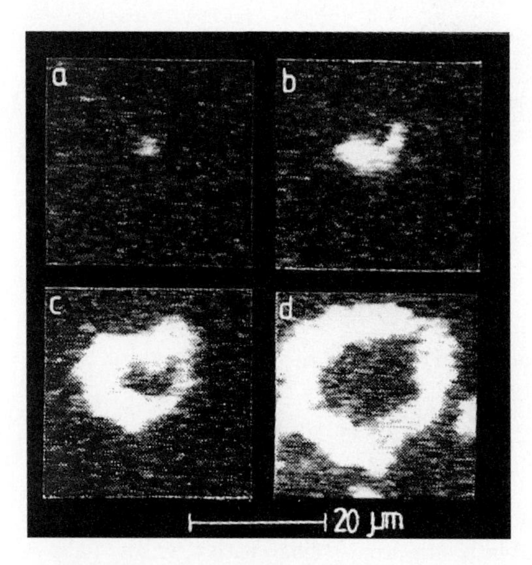

Fig. 5.7. The formation of an autowave from a growing domain. When the domain reaches a certain size, the luminescence extinguishes at the domain centre so that an expanding ring with fixed width is left behind. Parameters: Sample Z08, pulsed excitation, pulse width 13.5 μs, V_0=262 V, f=1 kHz, R_V=5 kΩ, Θ_0=293.4 K, time between consecutive pictures: 4.5 s; the order of consecutive pictures is indicated by the letters a - d

of successively generated autowaves at the same organizing centre a pattern of expanding concentric autowaves ensues. No substructure of the bright central region of the waves is discernable, except some microfilaments in the wake of the waves. These separated filaments exist up to several tenths of a second until they extinguish. In contrast to this, Rüfer et al. [5.1] observed travelling walls that obviously consist of numerous single filaments. Onton and Marello [5.2] observed that the motion of the filament wall appears to be accomplished primarily by the ignition of new filaments along the leading edge and filament decay at the trailing edge of the wall, i.e., the wall motion is supposed to be associated with the propagation of a filament-switching thermal wave rather than with the coherent, cooperative movement of combined filaments. The waves exhibit a characteristic behaviour in the interaction with boundaries or with themselves, respectively: When two waves collide, their touching parts annihilate. The remaining parts form a cusp which smoothes down with further propagation, like the two colliding ringwaves in Fig. 5.8. Those parts of a wave that reach a boundary are not

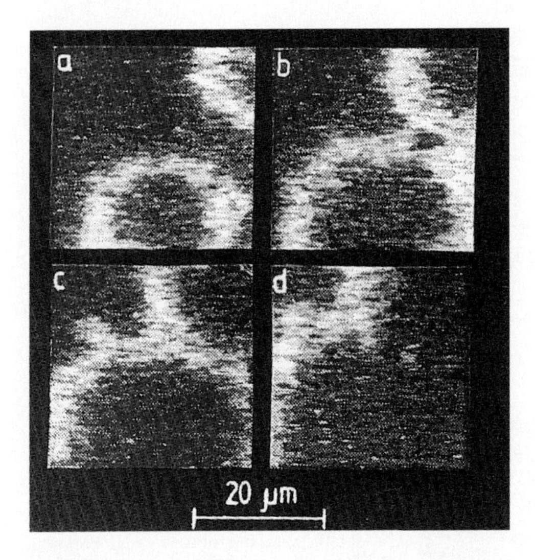

Fig. 5.8. Annihilation of two colliding autowaves. The parts left behind form a new waveline that smoothes down with further propagation. Parameters: Sample Z08, pulsed excitation, pulse width 13.5 μs, V_0=262 V, f=1 kHz, R_V=5 kΩ, Θ_0=293.4 K, time between consecutive pictures: 3.46 s; the order of consecutive pictures is indicated by the letters a-d

reflected but disappear. In a sample with a circular inhomogeneity, a wave that reaches the boundary of this inhomogeneity, splits up into two parts at the point of first contact with the boundary, and the two parts glide along the borderline until they reunite at the opposite point. When a wave reaches such a circular boundary just with its single free end, this end glides around the boundary continuously, forming a rotating spiral arm. Free ends not being influenced by any boundary tend to roll up.

5.4.4 Mobile Filaments and Strings

With increasing operating time the maximum wave width decreases. After an operating time of about 1/2 hour thin luminescent strings with a width of about 1 μm have been developed. Note that the width remains constant with further device operation. Very often we observed that the luminescent strings coexist with microfilaments. The diameter of the filaments is approximately equal to the string width and appears to be independent of the driving voltage as does the string width. One could suppose that a luminescent string consists of one-dimensionally coupled filaments, but this substructure was not resolvable yet. Both strings and filaments are not stationary but quite mobile. At lower frequencies, the strings have a higher tendency to meander or to wobble, whereas at higher frequencies they prefer to bend perpendicular to their axis (Fig. 5.9). The mobility of the strings and the filaments increases with increasing frequency. Higher mobility also means a higher mean velocity, both of strings and filaments, whereby the motion of the filaments is characterized by an abrupt random hopping, eventually superposed by a drift if a significant gradient of the film thickness is present.

The luminescent strings are most likely comparable with the 'moving threads' that Rüfer et al. [5.1] reports on.

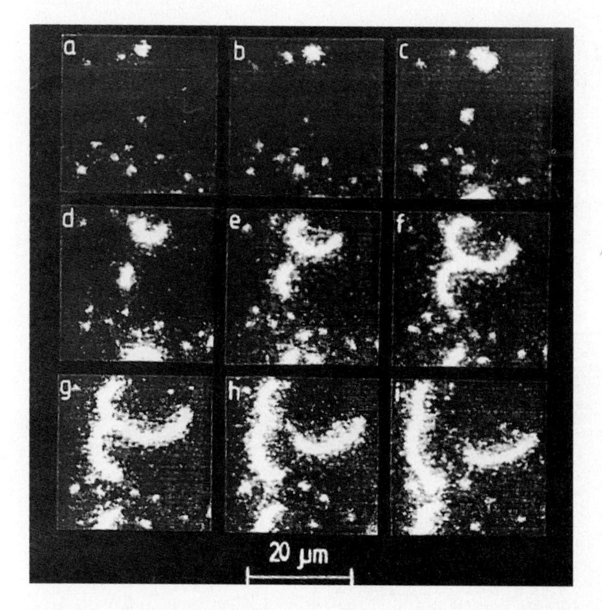

Fig. 5.9. The burst of two mobile filaments into two luminescence strings. Strings can merge together (f) or break up into parts (h). Parameters: Sample Z10, sine-wave excitation, V_0=157.5 V, f=54.5 kHz, R_V=0, Θ_0=313.6 K, time between consecutive pictures: 100 ms; the order of consecutive pictures is indicated by the letters a - i

Evolution of filaments and strings. The experiment described above and further investigations reveal that the generation and disappearance of moving filaments and strings can take place in different ways:

- Filaments are generated and extinguished spontaneously.
- Filaments are spawned at a generation centre, inhomogeneities act as filament sources.

- A hopping filament can split spontaneously into two different filaments, or two filaments can transform into a single one when they encounter accidentally. The character of randomness of the filament hopping seems to be independent of the distance of filaments from each other. Contrary to this, Rüfer et al. [5.1] speaks about a close-range repulsive and long-range attractive interaction between filaments.
- Filaments separate from a free end of a luminescent string, or a string entirely decays into single filaments.
- A moving filament bursts into a highly mobile string. This can be seen in Fig. 5.9: Two mobile filaments form two strings at different sites.
- Strings separate from other ones, or touching strings merge together, as can be seen in Fig. 5.9: The two growing strings melt together, separating a part at the same time.

Influence of the driving voltage. When moving strings and filaments exist (Fig. 5.10b), a lowering of the voltage amplitude causes the disappearance of the strings while single mobile filaments are still present (Fig. 5.10a). On the other hand, a variation of the voltage amplitude towards higher values enlarges the portion of strings compared to the portion of filaments. At even higher values of the voltage, a compact filament-free network of fluctuating strings is formed (Fig. 5.10c). The mean width of the meshes decreases with further voltage increase.

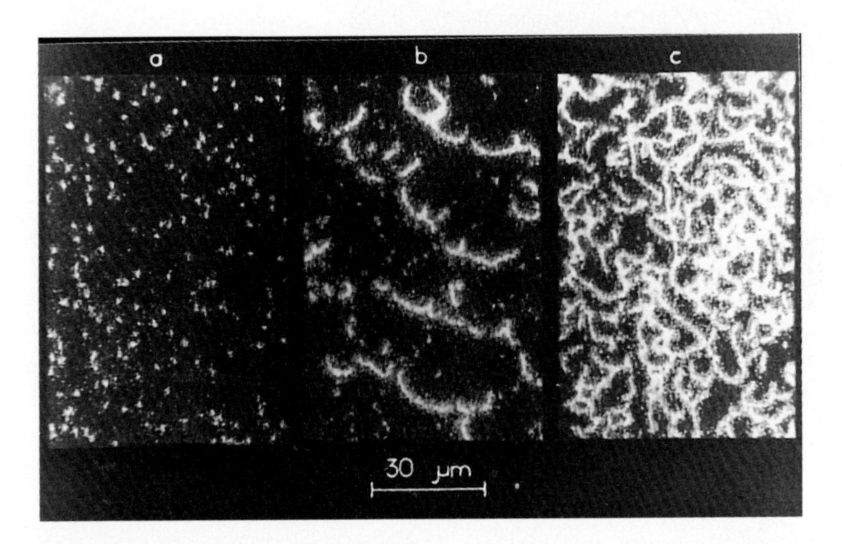

Fig. 5.10. Electroluminescence patterns for different values of the voltage amplitude (a) single mobile filaments, V_0=169.7 V, (b) coexisting mobile filaments and strings, V_0=170 V, (c) string network, V_0=170.3 V. Other parameters: Sample Z10, pulsed excitation, pulse width 13.5 μs, f=27.8 kHz, R_V=5 kΩ

5.4.5 Global Spatiotemporal Oscillations

When driven at higher frequencies (preferable above 50 kHz), the occuring luminescence is restricted to a spotlike region with a diameter of several hundred μm. The spot diameter becomes greater or smaller when the driving voltage amplitude is increased or decreased, respectively. The spot undergoes an oscillating movement if the ratio of the spot area to the total contacted area is smaller than a critical value. In the simplest case, the spot just moves back and forth on the contacted area. The more the voltage amplitude is decreased, the more complicated the spot motion can be.

Macroscopic view. Figure 5.11 shows an example of a plain rotating motion of a luminescence spot along the boundary of the basic region. The periodic motion is counterclockwise, as indicated by the four phases in Fig. 5.11. The time series in Fig. 5.12 was obtained by detecting the luminescence radiation emitted from an area of $156 \times 52\,\mu\mathrm{m}^2$ located in the upper left quadrant of the contact area. The time trace reveals the periodicity as well as the luminescence degration due to aging. The signal is correlated with the motion of the rotating luminescent spot: When the luminescent spot passes the upper left quadrant of the contact area, the maximum integrated luminescence density is measured, whereas the minimum is reached when the spot passes the lower right quadrant. The aging

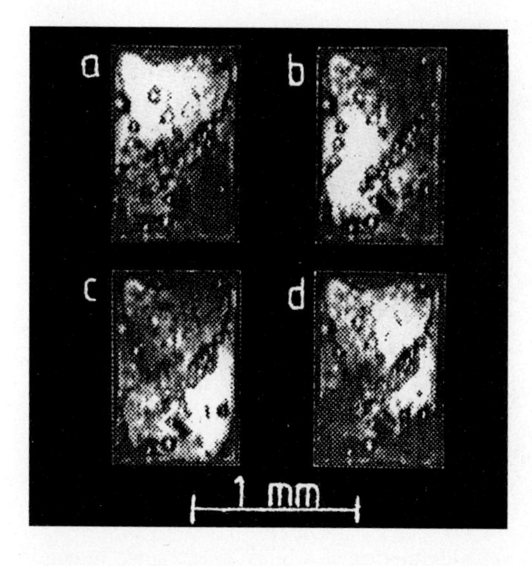

Fig. 5.11. Different stages of a simple rotation of a luminescence spot. Parameters: Sample Z10, sine-wave excitation, V_0=140 V, f=71 kHz, R_V=20 kΩ, Θ_0=294.6 K, time between consecutive pictures: 1.04 s; the order of consecutive pictures is indicated by the letters a - d

behaviour, well-known for Y_2O_3 as dielectric material and caused by changes in the IS-interfaces [5.16, 17], is related to an increase of the threshold voltage for the luminescence activity, thus the driving voltage amplitude has to be increased to keep the brightness at a constant value. In this way, the aging behaviour can be interpreted as a continuous decrease of the driving voltage amplitude of a non-aging device. Consequently, the brightness or, more precisely, the spot

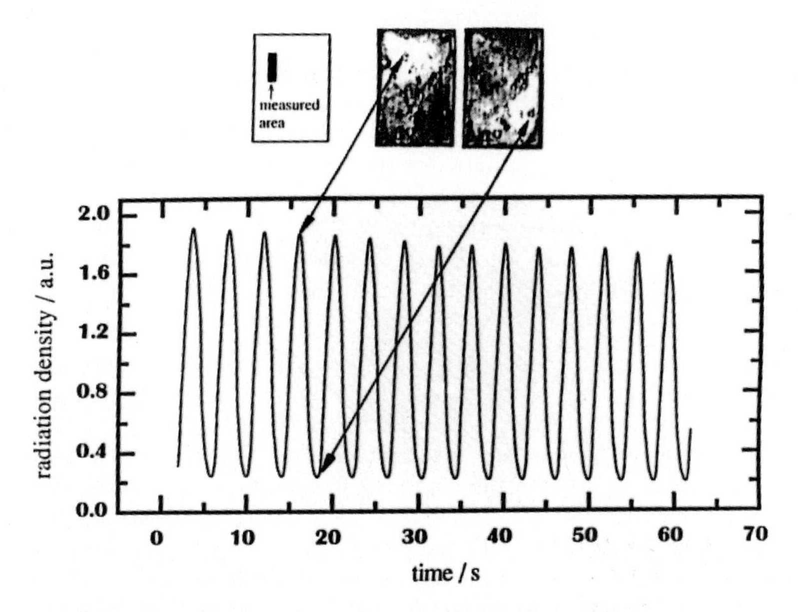

Fig. 5.12. Time series of the luminescence radiation emitted from an area of 156 × 52 μm^2 located in the upper left quadrant of the contact area. Parameters as in Fig. 5.11

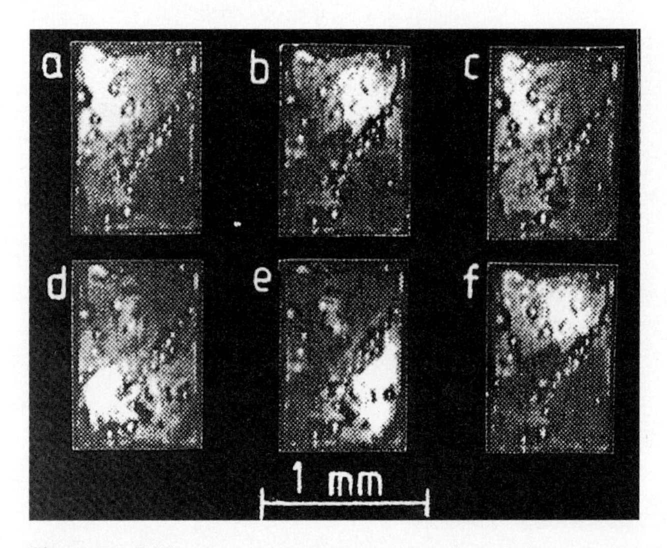

Fig. 5.13. Different stages of a more complex motion of a luminescent spot. In comparison with the plain rotation of Fig. 5.11, the spot motion has become more complex: A short clockwise movement at the upper border of the contacted area is embedded in the original counterclockwise rotation; parameters as in Fig. 5.11, but the sample is aged by one minute; time between consecutive pictures is 1.12 s, 0.64 s, 0.8 s, 0.64 s, and 1.28 s, respectively; the order of consecutive pictures is indicated by the letters a-f

diameter decreases with proceeding time, quite as the period increases. At a critical diameter, the regularity is disturbed spontaneously by a random motion, until the spot movement locks in a new periodic motion, as indicated in Fig. 5.13: The primary plain rotation period is increased by an additional modulation, i. e., the luminescent spot undergoes a reverse movement from the upper left to the upper right quadrant, before the original counterclockwise motion is continued. The more complex motion is reflected also in the time series of the integrated luminescence radiation density as illustrated in Fig. 5.14. In comparison with Fig. 5.12 one makes out an additional local minimum caused by the modulation.

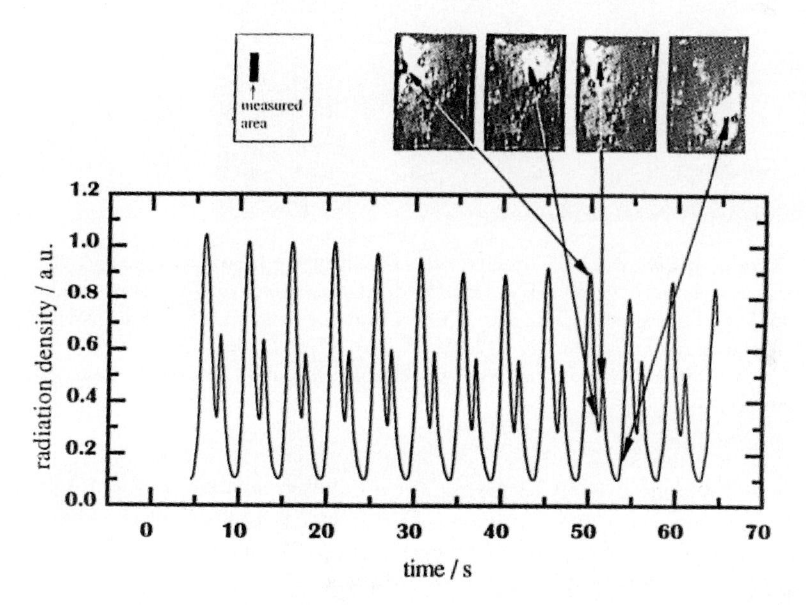

Fig. 5.14. Time series of the luminescence radiation emitted from an area of 156 × 52 μm^2 located in the upper left quadrant of the contact area; parameters as in Fig. 5.13

Microscopic view. If viewed on a microscopic scale, some substructures of the moving luminescent spot are disclosed. Figure 5.15 represents the range of substructures, obtained from a certain part of the total contact area where the outer parts of the luminescent spot pass by: If the spot centre is far away, the basic region is nearly dark with only some single hopping filaments. The main luminescent activity is signalized by the appearance of some additional randomly moving strings and filaments (Fig. 5.15a-f). The maximum of the light density is connected with the formation of an extreme bright, homogeneously lightening area that reveals no substructure (Fig. 5.15h). The great brightness then dissolves to a fluctuating network of strings for a short time (Fig. 5.15i, j), until the

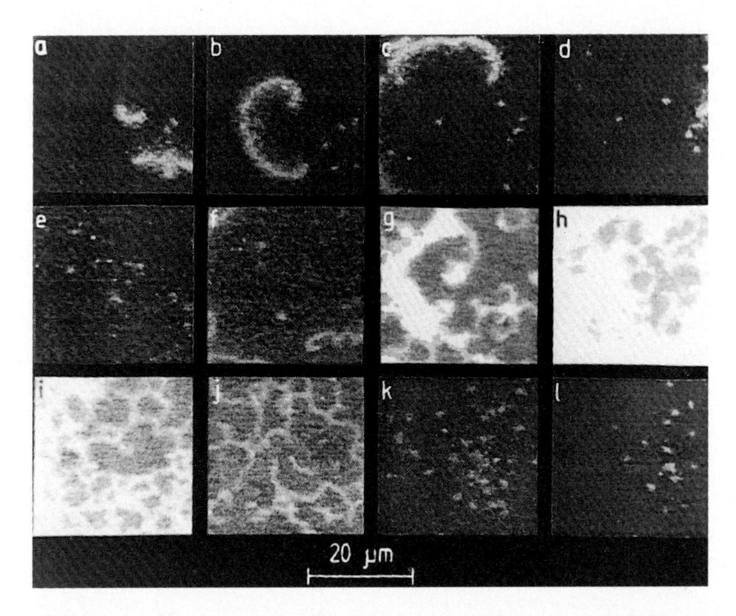

Fig. 5.15. Microscopic view on the outer regions of a rotating luminescence spot. The passing luminescence activity reveals a variety of substructures, ranging from mobile filaments (k, *l*) and luminescent strings (b, c) to dense string networks (i, j). Parameters: Sample Z10, sine-wave excitation, V_0=170 V, f=52.6 kHz, R_V=5 kΩ, Θ_0=294.6 K, time between consecutive pictures: 180 ms; the order of consecutive pictures is indicated by the letters a - *l*

luminescence diminishes to some remaining moving filaments anew (Fig. 5.15k, *l*), so the whole process restarts. In this way, a rich variety of filament and string behaviour can be observed in a single scenario. We point out that the luminescence motion of a moving spot does not appear to be a formation of collectively moving substructures, but instead, the substructures seem to be created by the high activity of the passing (unstructured) luminescence spot centre, i. e., the substructures do not move coherently with the spot centre. That means, that, e. g., the fluctuating network stays at the place of its formation during its lifetime.

5.4.6 Summary of Experimental Results

In Fig. 5.16 a survey of the experimentally observed patterns, arranged with respect to the amplitude and the period of the driving voltage, is given. Generally, the patterns become more complex and mobile with increasing frequency: At low frequencies just frontpropagation is observable, whereas at higher frequencies autowaves, mobile luminescent strings and filaments, and global spatiotemporal oscillations are present. An increase of the driving voltage amplitude means an increase of the front velocity, both for the simple frontpropagation and the autowaves; in the latter case additionally the width of the waves increases. The width of microfilaments or luminescent strings remains approximately constant when varying the applied voltage, whereas the number of structure elements per unit area changes in this case. At highest frequencies, in the mode of global oscillations, the diameter of the macroscopic luminescent spot decreases with decreasing voltage amplitude, and a more complex oscillating behavior becomes possible.

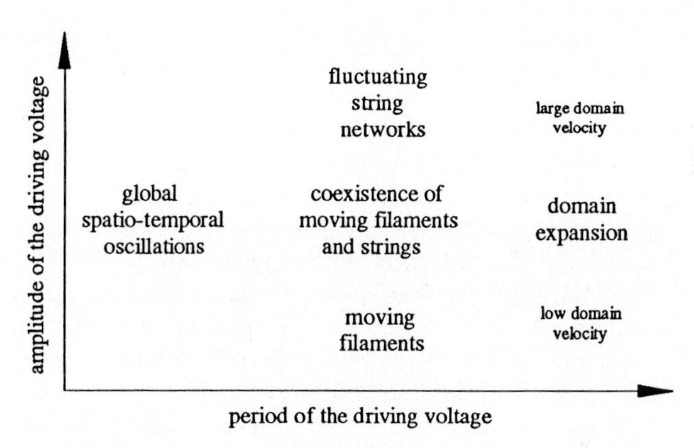

Fig. 5.16. Classification of the electroluminescence patterns with respect to the amplitude and the period of the driving voltage

5.5 Discussion

5.5.1 Formation of Microfilaments

As the active ZnS layer is polycristalline with grain sizes up to 100 nm, the question arises whether microfilaments are induced or stabilized by geometric effects due to field peaks at the grain boundaries, or whether they can be understood as dissipative structures self-organized in the sence of synergetics. Onton et al. [5.2] sees an indication of the influence of grain boundaries in the hopping nature of mobile filaments. However, a very similar discontinuous filament motion can be observed in ac driven gas discharge panels, with obviously no influence of inhomogeneities [5.18], because the filament diameter there is on the order of millimeters, whereby the dielectric layer is smooth if viewed on this space scale. The common opinion in the literature is that the influence of grain boundaries can not be ruled out, however the predominant effect for filamentary conduction is caused by an inherent bistable behaviour of the ZnS:Mn bulk material (e. g., [5.2, 15]).

An explanation for the stabilization of filaments could be the following: Consider a filament at sine-wave excitation, so the outset of dissipative current takes place at first at the filament centre where the trapped interface charge has its highest value. Due to the field enhancement of this interface charge at the beginning of any charge transfer, the required field for tunnel injection is reached there at first when the field increases at the momentary cathode. With proceeding charge transfer, the enhancement effect decreases and tilts over to a self-limiting, weakening effect when half the total transferred charge has been flown. This field weakening is not active in the central current channel only, but also hinders charge transfer in the filament environment due to the field influence of the central charge separation on the surrounding field distribution. Thus, the mechanism of stabilization of the filament's lateral extension occurs as an interplay between the prefered ignition at the current channel centre due to field enhancement of the trapped interface charge and the subsequent lateral field weakening effect of the transferred charge in the channel centre on its surroundings, so that, e. g., a neighbouring current channel is kept at a certain distance. Basically, this ideas could be proved by including one lateral dimension in a numerical model based on the lateral zero-dimensional model of Howard et al. [5.10]; however, the vectorial character of currents and the electrical field makes the numerical implementation much more complicated and time-consuming.

Filamentary conduction is also known for dc driven devices with a *monocristalline*, epitaxial ZnS:Mn layer and a ZnSe control layer [5.19], confirming the interpretation of microfilaments as self-organized structures generated by synergetic processes. It is worth to note that in systems constructed in a similar way and containing an active layer, in which an autocatalytic process occurs, and a passive layer that counteracts the autocatalytic process, filamentary conduction can be understood within the framework of a two-component reaction-diffusion system (e. g., [5.20,21]). The basis for this kind of modelling is that the reaction-diffusion components evolve time-continuously, so that their evolution can be described by differential equations. Although a time-continuous state evolution within a half-cycle of the ac excitation can reflect the bistable behaviour of ac driven ZnS:Mn devices properly [5.10,14,22], they clearly differ in details from dc driven systems since the momentary roles of the interface electrodes swap each half-cycle.

5.5.2 Comparison with Reaction-Diffusion Systems

The strong similarity of the phenomena observed in ac ZnS:Mn with those in chemical systems suggest to treat ac ZnS:Mn systems in terms of reaction-diffusion systems; for this description the use of amplitudes or averaged quantities is necessary. For a further discussion we explain the reaction-diffusion concept in more detail. Considering a two-component system of activator-inhibitor type, the evolution of the system shall be determined by

$$\tau_v \frac{\partial v}{\partial t} = L_v^2 \Delta v + f(v, w), \tag{5.1}$$

$$\tau_w \frac{\partial w}{\partial t} = L_w^2 \Delta w + g(v, w), \tag{5.2}$$

where L_v (L_w) means the diffusion length of the activator v (inhibitor w), τ_v (τ_w) the corresponding time constant, Δ the Laplacian, and $f(v,w)$ ($g(v,w)$) describe the local kinetics. Here we consider the case where the activator has (in a certain range) the property to enhance itself as well as the inhibitor, whereas the inhibitor damps both activator and inhibitor. For $\tau_w > \tau_v$, dynamical structures are the prefered ones, e. g., for thyristor-like semiconductor devices [5.23] or gas-discharge systems [5.24] repulsively interacting structure elements are known for $L_v < L_w$, whereas the well-known annihilating reaction-diffusion waves in Belousov-Zhabotinskii systems belong to the contrary category with $L_v > L_w$ (see, e. g., [5.25]). Since charge transfer in ZnS:Mn devices is autocatalytic in the region of bistability of the total transferred charge, a charge-related variable, e. g., the interface charge or the space charge, is a promising aspirant to the activating component. Presupposing the presence of a second, inhibiting component for the ZnS:Mn system, the properties of both components appear to be dependent on the driving conditions. At low excitation frequencies when front propagation is predominant, it is expected that $L_v > L_w$ and $\tau_w < \tau_v$. Increasing the excitation frequency, the appearance of annihilating autowaves indicates that τ_w becomes greater than τ_v at higher frequencies, whereby L_v is expected still to be greater than L_w, thus the inhibitor has to react slowly and is to be short-ranging by nature. On the other hand, in the case of mobile strings and fluctuating string networks a situation with $L_v < L_w$ appears to be favoured, because the strings reveal some behaviour that is rather different from the properties of annihilating waves: The strings can wobble back and forth or meander, respectively, whereas the autowaves always keep to a once prefered direction of propagation. Above this, strings can form crossings, like they do in the mode of fluctuating networks, and their width is nearly independent of the driving voltage amplitude, whereas the width of the autowaves is strongly voltage-dependent. So, the mechanisms of the string dynamics appear to be related rather to that of mobile filaments with $L_v < L_w$ than to those of autowaves, i. e., the inhibiting mechanism for strings and filaments on the one hand and annihilating waves on the other hand is assumed to be of different quality.

The fact that the mobility of strings and filaments increases with increasing excitation frequency indicates that the ratio τ_w / τ_v increases, too. If τ_w depends only weakly on the excitation frequency, an increase of this ratio can be expected because an increase of the excitation frequency leads to an increase of the total transferred charge per unit time and, consequently, to a decrease of the relaxation time constant τ_v of the activator.

Up to now, the inhibiting mechanism is not fully clear. However, there are some indications that temperature effects play an important role. Subsequently, we sum up some essential results concerning the influence of temperature and discuss a possible mechanism for mobile structure elements.

As far as autowaves are concerned, their appearance is discussed in the context of thermal effects in the literature. Besides a thermal model for filament motion that is based on the presence of a critical temperature where filaments extinguish, Rüfer et al. [5.1] explains the sequential formation of concentric ring waves at a certain organizing centre as follows: The organizing centre initiates a growing luminescent domain that extinguishes at its centre when the temperature has reached a critical value there, so that the expanding autowave cannot be followed up by another autowave formation until the centre has cooled down sufficiently. The nature of the presence of the critical temperature is not specified further.

Onton et al. [5.2] proposes that moving waves are related with accompanying thermal waves where filaments switch on along the leading edge and off along the trailing edge. Likewise, Beale et al. [5.3] supposes that local heating at sites of high luminescence, thus high power dissipation, is the reason for the mobile luminescence.

In a recent report on pattern formation in ZnS:Mn devices Beletskii et al. [5.4] points out a connection between the presence of travelling waves and the temperature dependence of the brightness-voltage hysteresis. More specifically, the decrease of total transferred charge with increasing temperature is seen as inhibiting process. The interface charge in the IS-interfaces is addressed as activating component in the sense that the spread of interface charge out of 'on'-sites into 'off'-sites causes them switching to the 'on'-state, too, if the device is driven in a bistable mode. A possible mechanism leading to activity waves with a self-sustaining profile (autowaves) is based on the temperature-dependence of the hysteresis loops of single filaments. As can be seen in Fig. 9 of [5.11], the bistability vanishes as temperature increases. Equilibrium states of the total transferred

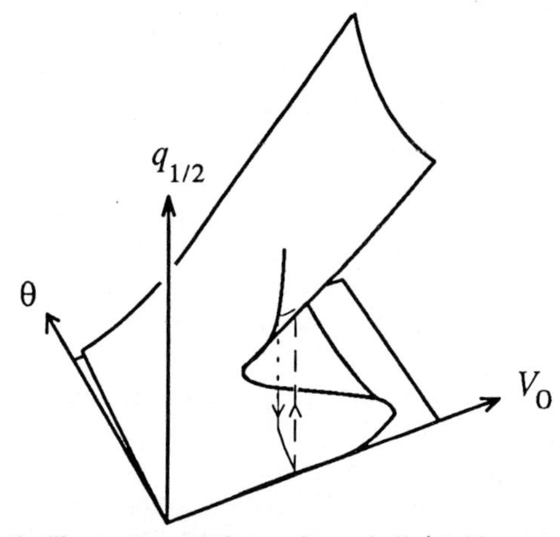

Fig. 5.17. Schematic illustration of the surface of all (stable or unstable) equilibrium states of the total transferred charge $q_{1/2}$, inspired by the single filament characteristics of [5.11]. The marked route at constant driving voltage amplitude, but varying temperature Θ is explained in the text

charge as a function of Θ and V_0 might look like the fold sketched in Fig. 5.17. With a driving voltage amplitude to the left of the cusp (indicated in Fig. 5.17), the following scenario is conceivable:

- The 'off'-state at a certain position is excited by neighbouring 'on'-sites (dashed line in Fig. 5.17).
- This excited state of high luminescence activity is supposed to be metastable, because the power consumption at 'high'-state sites heats up the material, so that the 'high'-state gradually shifts to higher temperatures (upper solid line in Fig. 5.17).

- When a critical temperature is reached, the damping effect of the temperature on the autocatalytic process prevails, resulting in a decrease of the charge transfer and, consequently, to a rapid decrease of the luminescence activity to a 'low'-state (dotted line in Fig. 5.17).
- In view of the little power consumption of this 'low'-state, the system regeneratively settles down to the only equilibrium rest state (lower solid line in Fig. 5.17).

To obtain the above-mentioned metastable niveau, the ratio of heat production within the phosphor layer to heat transfer to the environmental heat reservoir has to be sufficiently high; in this case, the film stack may heat up due to power consumption at luminescent sites so that no second common equilibrium state of temperature Θ and transferred charge is possible. Furthermore, the time constant for the autocatalytic processes involved in the excitation should be small with respect to the time constant of thermal relaxation, and, according to the above-introduced diffusion lengths, the diffusion length of the inhibiting temperature should be small with respect to that of the charge-related activating component. By that, it is achieved that the accompanying inhibiting thermal wave is mainly located in the wake of the excitation wave, so that, when such autowaves collide, they do not 'perceive' each other, until the delayed inhibiting thermal waves meet the oncoming autowave.

5.6 Conclusions

The ZnS:Mn system reveals pattern formation on different space and time scales, with microfilaments and mobile luminescent strings as the smallest structures on a space scale on the order of 1 μm and a time scale down to tenths of a second, up to moving luminescent spots with a diameter of several hundred μm and a time scale on the order of several seconds. In general, the structure elements are preferable dynamical ones, whereby the annihilating waves are of particular interest because of their exciting similarity to reaction-diffusion waves, e. g., autowaves in Belousov-Zhabotinskii systems. This great similarity between the wave patterns in ZnS:Mn devices and reaction-diffusion waves in two-component reaction-diffusion systems confirms the supposition that the main aspects of those patterns can be modelled in terms of reaction-diffusion equations, however, details about the dynamics of the presupposed components are not finally clear. So, the ZnS:Mn device offers a promising system for the study of autowave phenomena, with the advantage that wave patterns in a non-aging device can be kept at a constant distance from thermodynamic equilibrium without any extra effort, in contrast to chemical systems where products and educts have to be removed or added.

References

[5.1] H. Rüfer, V. Marello, A. Onton: J. Appl. Phys. **51**, 1163 (1980)
[5.2] A. Onton, V. Marello: In *Advances in Image Pickup and Display*, Vol. **5** (Academic Press, New York 1982) p. 177
[5.3] M.I.J. Beale, J. Kirton, M. Slater: In *Electroluminescence*, ed. by S. Shionoya, H. Kobayashi, Springer Proceedings in Physics Vol. 38 (Springer, Berlin, Heidelberg 1989) pp. 296 - 300
[5.4] A.I. Beletskii, N.A. Vlasenko: Pis'ma Zh. Tekh. **19**, 33 (1993) [Engl. transl.: Tech. Phys. Lett. **19**, 13 (1993)]
[5.5] F.-J. Niedernostheide, Ch. Goßen, H.-G. Purwins: "Dynamics of current filaments studied in two examplary systems: dc Si p-n-p-n devices and ac ZnS:Mn films", in: *Advances in Synergetics*, Vol. 2, ed. by V. Kuvshinov, C. Krylov (Belarusian State University Press, Minsk 1994) in print
[5.6] J. Benoit, P. Benalloul, A. Geoffroy: In *Electroluminescence*, ed. by S. Shionoya, H. Kobayashi, Springer Proceedings in Physics Vol. 38 (Springer, Berlin, Heidelberg 1989) pp. 85 - 88
[5.7] G.O. Müller: Phys. Stat. Sol. (A) **81**, 597 (1984)
[5.8] J.M. Hurd, C.N. King: J. Elec. Mat. **8**, 879 (1979)
[5.9] R. Mach, G.O. Müller: Phys. Stat. Sol. (A) **69**, 11 (1982)
[5.10] W.E. Howard, O. Sahni, P.M. Alt: J. Appl. Phys. **53**, 639 (1982)
[5.11] W. Rühle, V. Marello, A. Onton: J. Elec. Mat. **8**, 839 (1979)
[5.12] Y.A. Ono, H. Kawakami, M. Fuyama, K. Onisawa: Japan. J. Appl. Phys. **26**, 1482 (1987)
[5.13] G.O. Müller, R. Mach, B. Selle, G. Schulz: Phys. Stat. Sol. (A) **110**, 657 (1988)
[5.14] K.A. Neyts: IEEE Trans. ED-**38**, 2604 (1991)
[5.15] V. Marello, W. Rühle, A. Onton: Appl. Phys. Lett. **31**, 452 (1977)
[5.16] R. Mach, G.O. Müller: In *Electroluminescence*, ed. by S. Shionoya, H. Kobayashi, Springer Proceedings in Physics Vol. 38 (Springer, Berlin, Heidelberg 1989) pp. 264 - 272
[5.17] R. Mach: In *Acta Polytechnica Scandinavica Appl. Phys. Ser.*, Vol. 170, ed. by M. Leskelä, E. Nykänen (1990) p. 85
[5.18] E. Ammelt: Private communication (1994)
[5.19] A.P.C. Jones, A.W. Brinkman, G.J. Russel, J. Woods, P.J. Wright, B. Cockayne: Semicond. Sci. Technol. **2**, 621 (1987)
[5.20] Ch. Radehaus, K. Kardell, H. Baumann, D. Jäger, H.-G. Purwins: Z. Phys. B - Condensed Matter **65**, 515 (1987)
[5.21] Ch. Radehaus, T. Dirksmeyer, H. Willebrand, H.-G. Purwins: Phys. Lett. A **125**, 92 (1987)
[5.22] J.M. Jarem, V.P. Singh: IEEE Trans. ED-**35**, 1834 (1988)
[5.23] F.-J. Niedernostheide, B.S. Kerner, H.-G. Purwins: Phys. Rev. B **46**, 7559 (1992)
[5.24] H. Willebrand, M. Or-Guil, M.Schilke, H.-G. Purwins, Yu.A. Astrov: Phys. Lett. A **177**, 220 (1993)
[5.25] J.J. Tyson, P.C. Fife: J. Chem. Phys. **73**, 2224 (1980)

6 Structure Formation in Charge Density Wave Systems

M.J. Bünner, G. Heinz, A. Kittel, J. Parisi

Physical Institute, University of Bayreuth
D-95440 Bayreuth, Germany

Abstract. We investigate organic and anorganic crystals with a quasi-one-dimensional electronic structure. It is well established that these systems are able to undergo a Peierls transition. Below the transition temperature the electrons increasingly condensate to a charge density wave. The charge density wave is able to contribute to the charge transport above a threshold electrical field, leading to nonlinear conductivity. The motion of the condensate is accompanied by narrow-band noise and broad-band noise. The metastability of the charge density wave defects leads to hysteresis and memory effects. Self-organized criticality and switching oscillations have been found near the threshold electrical field. The observed complex behavior in the time domain makes it plausible to expect charge density wave systems also to organize in space.

6.1 The Peierls Transition

As has been pointed out by Peierls [6.1], a one-dimensional metal is unstable against perturbations with the wave vector $2k_F$. This can be already understood using the simple model of non-interacting electrons in a periodic lattice potential with the periodicity a. The band structure is cosinus-like with an energy gap at the edge of the first Brillouin zone. The electrons fill the energy states up to the Fermi vector k_F. The electronic density $\rho(x)$ is spatially homogeneous (Fig. 6.1a). Introducing then a periodic lattice distortion with the wave vector $2k_F$, the electronic system lowers its total energy by opening an additional energy gap just at k_F. Assuming repulsive forces between the atoms of the lattice, the lattice distortion with $2k_F$ will have a finite amplitude balancing the energy costs for the lattice distortion with the energy gain of the electrons. In this state the electronic density has the additional periodicity π/k_F (Fig. 6.1b). In general, the lattice period a and the Fermi wave vector k_F are not related. We distinguish between a commensurate and an incommensurate Peierls distortion. In the commensurate case, the quotient of the lattice wave vector k_a and the Fermi wave vector k_F is rational. Otherwise, the distortion is called incommensurate. In the latter case, the one-dimensional metal loses its periodicity. The unit cell equals the system size.

Solving the quantum-mechanical eigenvalue equation for the coupled electron-phonon system in one dimension [6.2] using the mean-field approximation, it

Springer Proceedings in Physics, Vol. 79 **Nonlinear Dynamics and Pattern Formation in Semiconductors and Devices**
Editor: F.-J. Niedernostheide

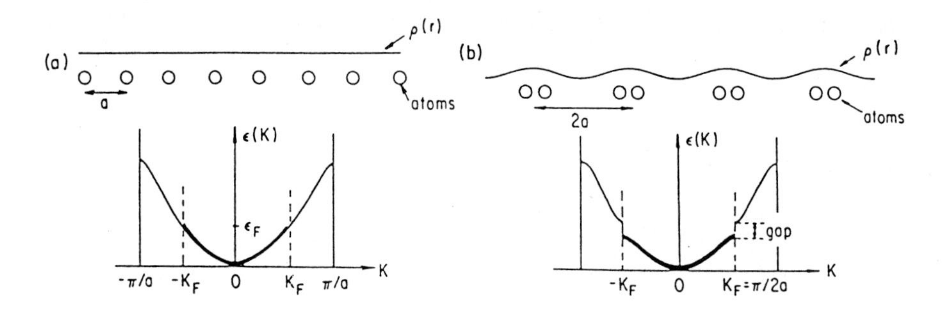

Fig. 6.1. a) Electronic density and band structure without Peierls distortion, b) with a commensurate Peierls distortion: $k_F = \pi/2a$

turns out that for the temperature $T = 0$ the ground state is not one-particle-like, but consists of electron-hole pairs, which are coupled by $2k_F$-phonons. The electron-hole pairs are bosons and condense into a coherent, macroscopic quantum state, the Charge Density Wave (CDW). The nature of the electronic system is radically different from either three-dimensional metals or semiconductors. We must realize at this point that the Born-Oppenheimer approximation, where the electron and the phonon system can be treated separately, breaks down for one-dimensional metals. Instead we have to deal with a strongly coupled electron-phonon system. It turns out that in materials with a CDW state the coherence of the CDW is not strong enough to form a coherent CDW in a single crystal. A given crystal consists of several CDW domains as has been observed in $NbSe_3$ [6.3,4] and TaS_3 [6.5].

To describe the temperature dependence of the properties of one-dimensional metals, it is essential to introduce a three-dimensional coupling [6.6]. We call a quasi-one-dimensional metal a three-dimensional lattice of slightly interacting parallel chains of metallic conductivity. It turns out that quasi-one-dimensional metals undergo a metal-semiconductor transition at the Peierls temperature. Above the Peierls temperature, the system is metallic. Below the Peierls temperature, the system is able to form a CDW. The phase of the CDW plays the role of an order parameter. Because not all the electrons condense into the CDW for non-vanishing temperature, it seems appropriate to describe the electronic properties with the help of a two-fluid model, distinguishing between electrons, which are in single-particle states, and electrons, which are condensed into the CDW. The physical properties connected with the CDW state have been reviewed several times [6.7–9].

6.2 Materials

Experimental work on CDW systems started as soon as crystals with quasi-one-dimensional electronic structure were available. The inorganic materials $NbSe_3$ and the blue bronze $K_{0.3}MoO_3$ are well-investigated [6.10,11]. The structure of the blue bronze $K_{0.3}MoO_3$ is to be seen in Fig. 6.2a. The Mo- and O-ions form octaeders, which line up in chains. The chains are separated by K-ions, which, because of their high electro-negativity, donate the MoO_3-chains. The axis with metallic conductivity is along the MoO_3-chains. CDW states have also been found in organic crystals, like the TTF-TCNQ [6.12] and recently the Fluoranthene radical cation salts (FARKS) [6.13, 14]. The structure of the FARKS is shown in Fig. 6.2b. The planar Fluoranthene molecules form staples with high electrical conductivity. These staples are separated by PF_6^--molecules.

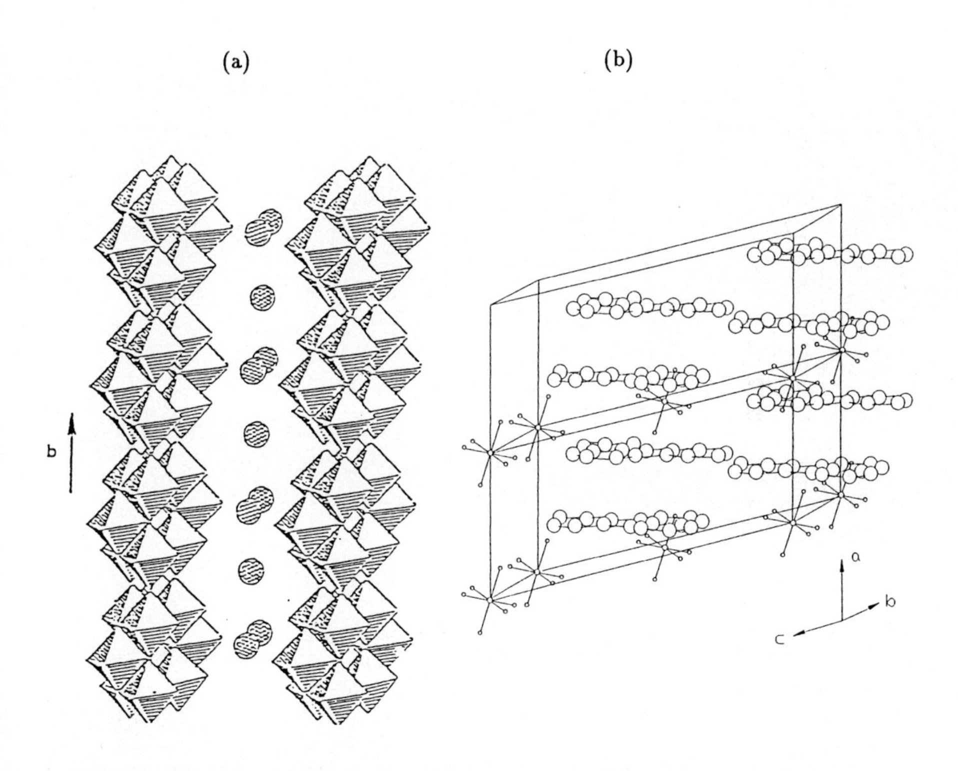

Fig. 6.2. a) Crystal structure of $K_{0.3}MoO_3$, b) crystal structure of $(Fluoranthene)_2PF_6$. Taken from [6.10, 14]

6.3 The Electronic Transport in CDW Systems

We discuss the electronic transport in CDW systems below the Peierls temperature with the help of the two-fluid model outlined above, where one clearly distinguishes between electrons, which are in single-particle states, and electrons, which are condensed in the CDW. The relative densities of the two phases are strongly temperature-dependent. More and more electrons condense into the CDW for decreasing temperature. The Peierls gap dominates the transport of the electrons in the single-particle states. The conductivity of these electrons is not field-dependent. We therefore call this phase "ohmic phase". The conductivity of the ohmic phase is thermally activated (Fig. 6.3).

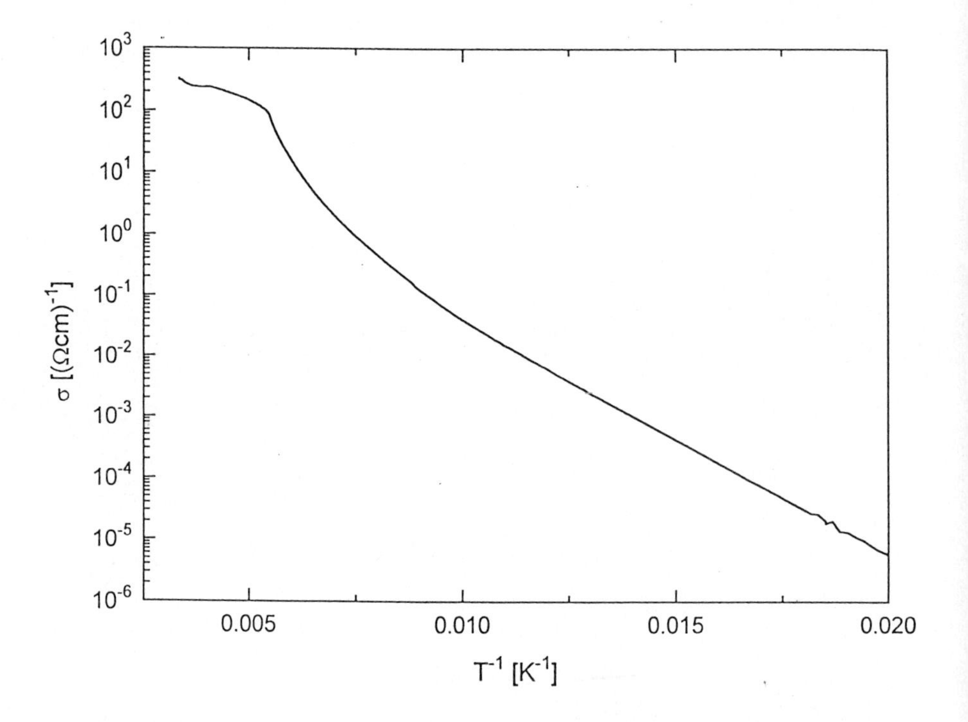

Fig. 6.3. Temperature dependence of the conductivity of the ohmic phase of $(Fluoranthene)_2PF_6$. Taken from [6.14]

In contrast a CDW domain is a coherent, macroscopic quantum state and can therefore only move as a whole. The movement of a CDW domain can be inhibited by pinning forces. Local pinning forces are provided by the following mechanisms:

- interaction of CDW defects, for instance domain walls, with the regular lattice.
- interaction of defects of the regular lattice with the CDW.
- interaction of CDW defects with defects of the regular lattice.

For a commensurate CDW the regular lattice exerts a nonlocal pinning force on the CDW (commensurate pinning). The CDW starts to contribute to the electrical transport as soon as the applied electrical field exceeds a threshold value E_{th} and is then strong enough to compensate the pinning forces. This leads to nonlinear conductivity. In Fig. 6.4 the contributions of the ohmic phase and the CDW phase to the electrical transport can be clearly distinguished. The specific shape of the current-voltage (I-V) characteristics depends on the sample and the electrical and thermal history as will be outlined below. The onset of nonlinear conduction is accompanied by $1/f$-noise and small voltage oscillations, whose frequency is proportional to the contribution of the CDW to the total current.

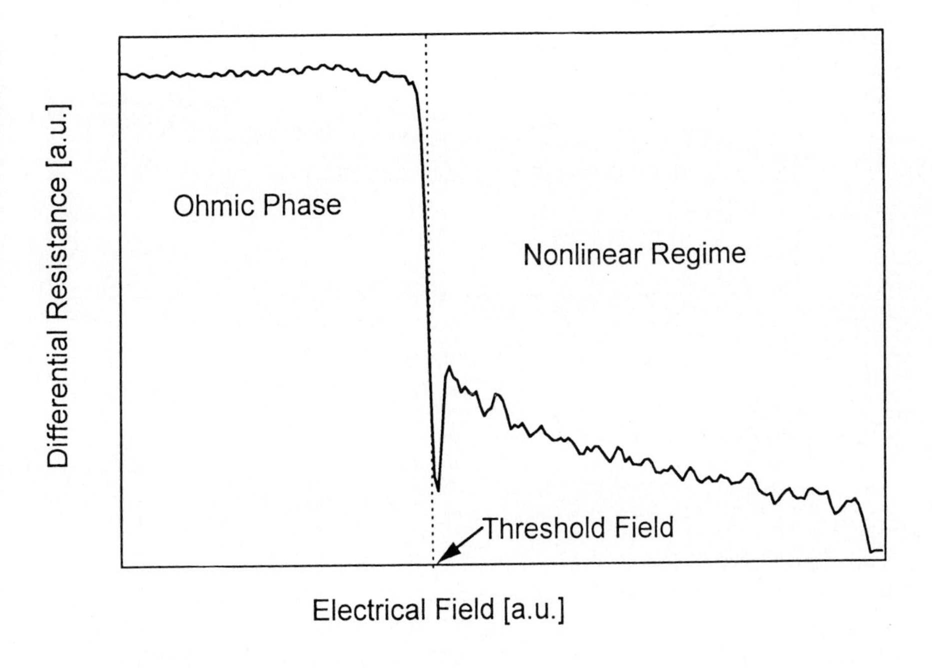

Fig. 6.4. Differential resistance vs. electrical field of a blue bronze sample at 77 K

6.4 Structure Formation of CDW Systems

In this section, we summarize the findings, which make it plausible to expect spatial degrees of freedom to be important for the electronic transport in CDW materials.

6.4.1 Metastability, Memory, and Hysteresis

As outlined above, the electrical transport of the CDW depends strongly on the pinning forces and therefore the defects of the CDW and the regular lattice. The defects act as local pinning centers and show extremely long relaxation times. CDW states with different spatial distribution of defects are metastable.

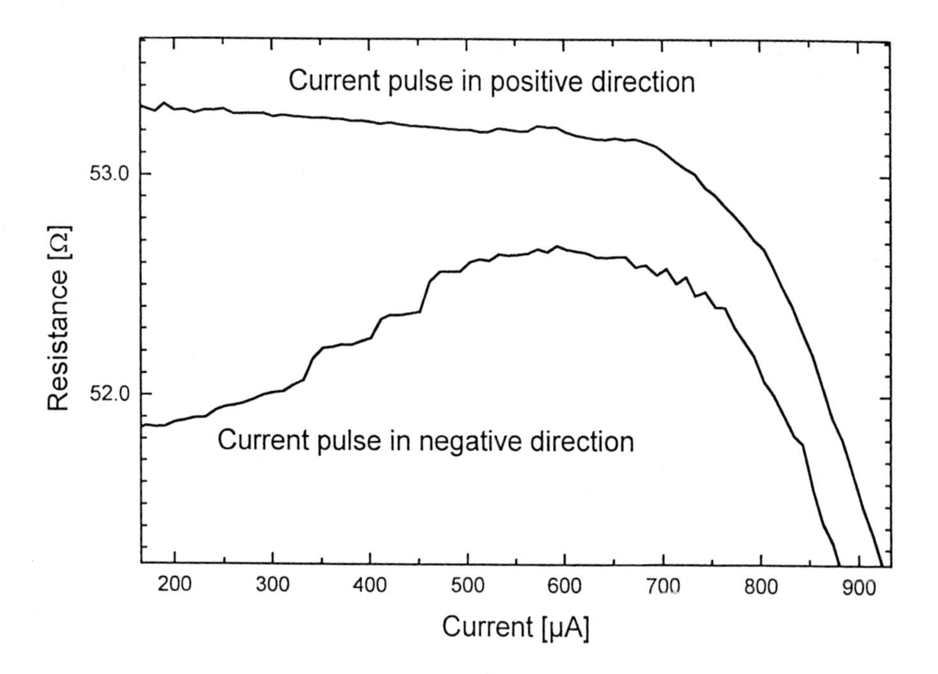

Fig. 6.5. Resistance vs. current of a blue bronze sample at 77 K after application of a current pulse in either positive or negative direction

Therefore, CDW systems must be viewed as disordered systems with random distribution of pinning centers and a large number of metastable states. This leads to a glassy behavior of the CDW. As a result of the existence of metastable states, several physical properties of CDW systems depend on the history of the sample leading to memory effects and hysteresis. Thermal memory effects have been explored elsewhere [6.10]. We concentrate on electrical memory effects. Measurements were performed on the blue bronze $K_{0.3}MoO_3$ at 77 K. The ohmic resistance of a sample measured below the threshold electrical field has been analyzed after application of a current pulse. The ohmic resistance depends on the duration, height, and sign of the current pulse. This is interpreted as a pulse-induced switching between different metastable states of the CDW. With the same technique we were able to obtain structures in the I-V characteristics, which appear as steps in the current-dependent resistance. Measuring the current-dependent resistance after having applied a current pulse with a height well above the threshold electrical current, we observe no peculiarities in the resistance-current characteristic, if the resistance is measured in the same direction as the pulse has been applied. If we measure the resistance in the direction opposite to the current pulse, we observe a few steps just below threshold (Fig. 6.5). We attribute these steps to single field-induced transitions between metastable states of the CDW, where the system loses its memory of the current pulse applied in opposite direction. Different pulse memory effects have been already reported [6.15]. We claim that the observation of spatial degrees of freedom is necessary to understand the nature of the metastable states and the hysteresis of the ohmic resistance.

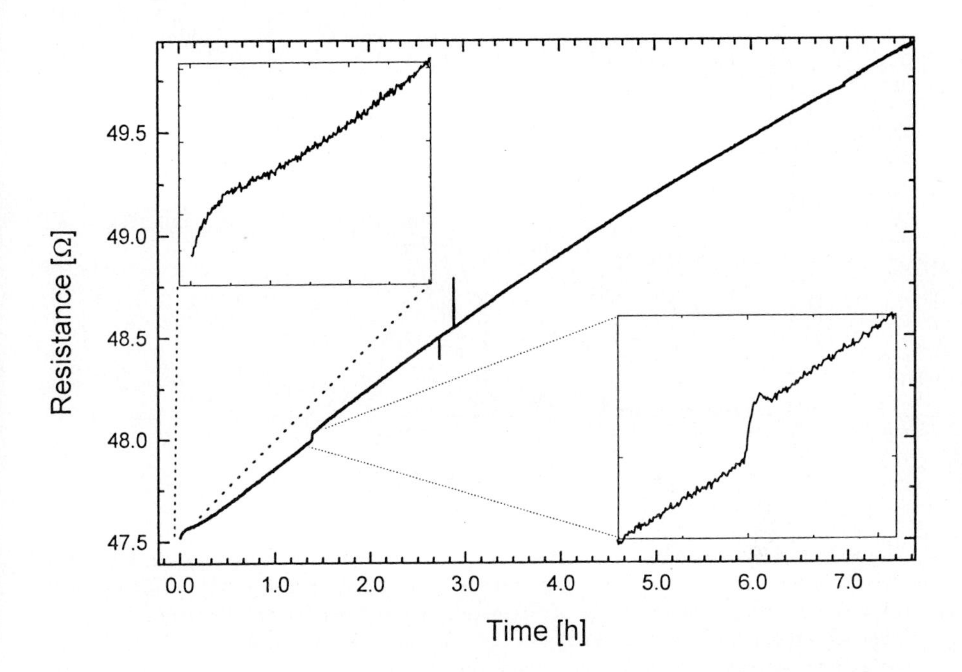

Fig. 6.6. Resistance of a blue bronze sample at 77 K measured as a function of time in the nonlinear regime. The upper inset magnifies the fast decay at the beginning of the measurement. The lower inset shows a blow-up of a jump of the resistance

The glassy behavior of the CDW can be observed by measuring the time-dependent resistance with an applied current well above the threshold current (Fig. 6.6). Surprisingly, the resistance does not tend to an equilibrium value in laboratory time scales, but increases steadily. The resistance does not change continuously, but single jumps are observed, which are attributed to transitions between metastable states.

6.4.2 Intermittent Oscillations and *S*-Shaped Negative Differential Resistance

The *I*-*V* characteristic of the blue bronze at 77 K has a negative differential resistance with an *S*-shape (Fig. 6.7). The low-conducting state below the threshold electrical field is interpreted as a pinned state of the CDW. The high-conducting state above the electrical threshold then corresponds to a depinned state of the CDW. Close to the threshold electrical field, the system shows intermittent oscillations by switching between a pinned and a depinned CDW state. Peinke et

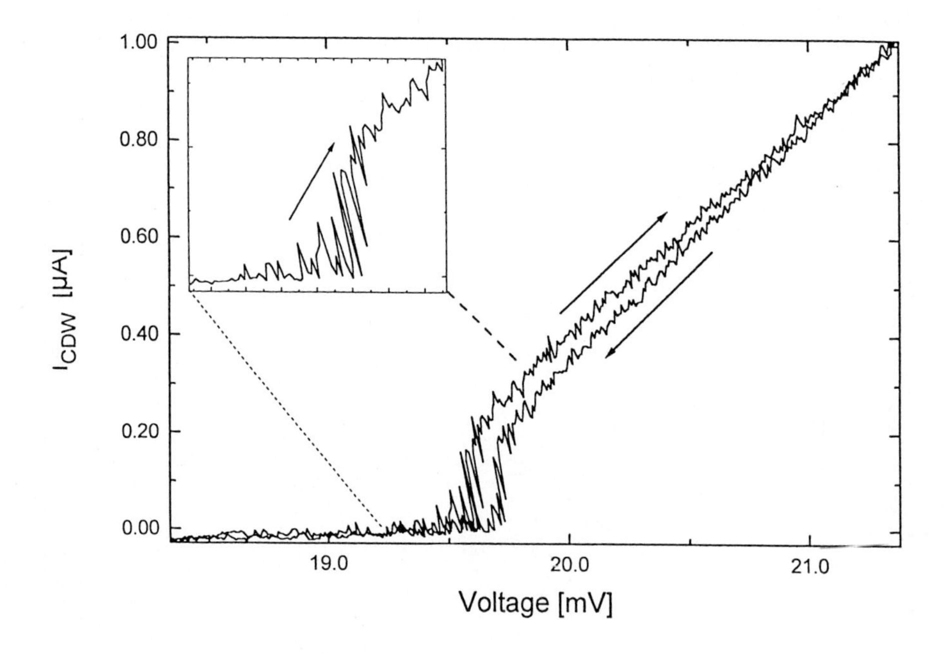

Fig. 6.7. *I-V* characteristic of a blue bronze sample at 77 K upon cycling (i. e., for increasing and afterwards decreasing current bias). I_{CDW} has been obtained by subtracting the current contribution of the ohmic phase from the total current. The different threshold fields for up- and down-cycling are clearly visible. The intermittent oscillations for electrical fields slightly below the threshold value can be seen in the inset

al. [6.16] analyzed the distribution of waiting times at 4.2 K and found a power law pointing towards intermittency [6.18] and self-organized criticality [6.17]. Whether the sample switches homogeneously to the depinned state or only certain domains switch to the depinned state, while the other ones remain in the pinned state, is still unclear. In semiconductors the *I-V* characteristics with *S*-shaped negative differential resistance could be clearly associated with spatial self-organization [6.19]. The question if one is able to connect the *S*-shaped *I-V* characteristics observed in CDW materials to spatial patterns is still open.

6.4.3 Narrow-Band Noise and Broad-Band Noise

The onset of nonlinear conductivity is accompanied by small voltage oscillations, the narrow-band noise (NBN) and broad-band noise (BBN). The frequency of the NBN is proportional to the contribution of the CDW to the total current. In some samples one observes several NBN peaks at different frequencies in the noise spectra (Fig. 6.8). The NBN peaks show up at different total currents. This is commonly attributed to several spatial domains contributing to the conductivity

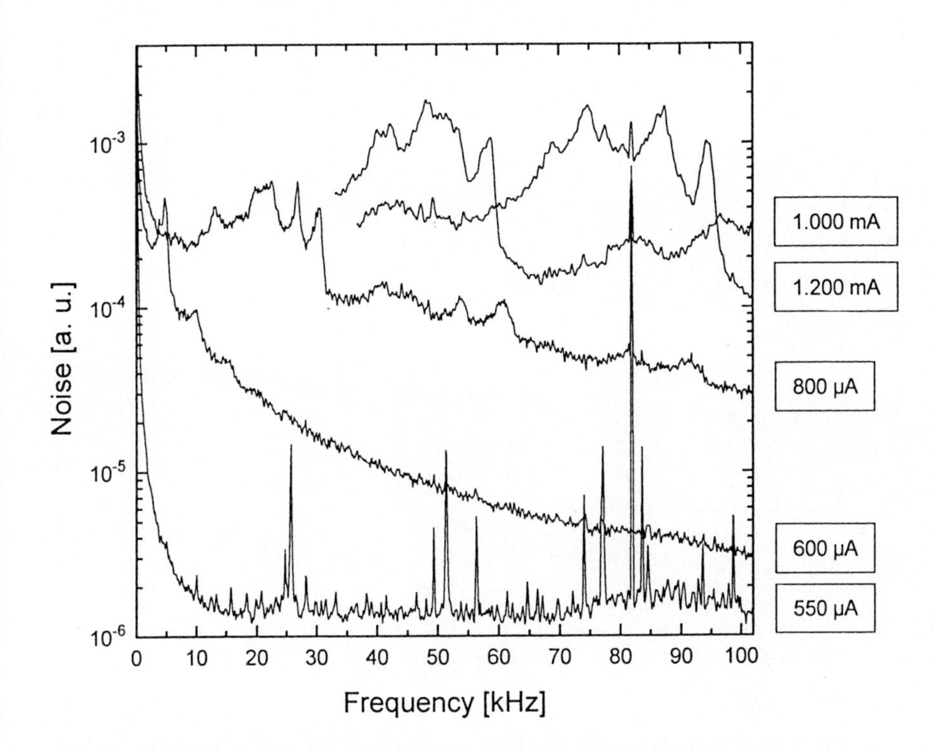

Fig. 6.8. Noise spectra of a blue bronze sample at 77 K for increasing total current (the current values are indicated on the r. h. s.). The threshold current obtained from the *I-V* characteristics is 700 μA. The onset of the BBN and the increasing frequencies of the NBN with increasing total current are clearly observed

at different threshold fields. The frequencies of the NBN peaks increase differently with increasing total current indicating different pinning forces. The BBN noise appears at the threshold field. The noise spectrum obeys approximately a power law with an exponent of -1.0 [6.16]. Maeda et al. [6.20, 21] attribute the noise to transitions among metastable states. At low temperature ($T < 20$ K), the noise observed before threshold is due to the intermittent oscillations.

6.4.4 Mode-Locking

Thorne et al. [6.22] observed plateau-like structures in the differential resistance of $NbSe_3$ in the presence of an a.c. electrical field (Fig. 6.9). This is interpreted as mode-locking [6.18] of the NBN with the external field. Generally, the synchronization of a nonlinear, low-dimensional oscillator with an external periodic signal leads to mode-locking. The findings of Thorne et al. imply that the depinned state of the CDW is governed by a few degrees of freedom. This is in disagreement, however, with the results outlined above, because of the metastability of the CDW, a huge number of degrees of freedom have to be taken into account.

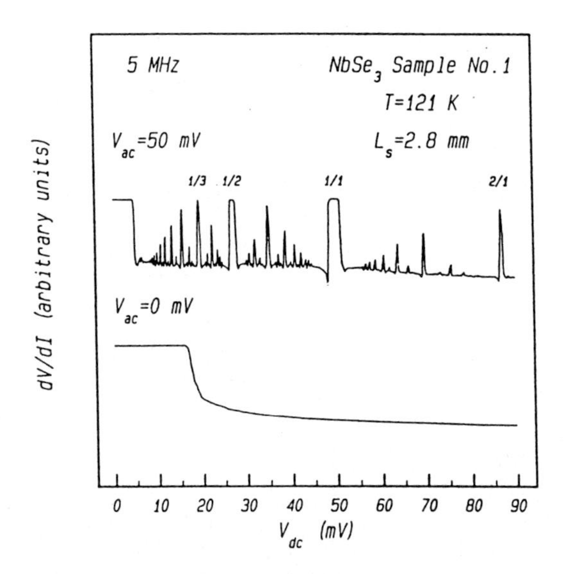

Fig. 6.9. Differential resistance vs. electrical field of a $NbSe_3$ sample without and with an a.c. electrical field (frequency 5 MHz, amplitude 50 mV). Taken from [6.22]

6.5 Conclusion and Outlook

We reviewed several findings which make it plausible to expect spatial degrees of freedom being of crucial importance to understand the unusual nonlinear transport in CDW materials. Emphasis was put on hysteresis and memory effects, intermittent oscillations, mode-locking and several noise phenomena. At the moment, we apply the EBIC (electron beam induced current) technique at liquid nitrogen temperature in order to investigate spatial degrees of freedom involved in the nonlinear electronic transport in CDW systems. Future developments will concentrate on different experimental techniques:

- low-temperature scanning-electron microscopy (at 4.2 K),
- low-temperature scanning-laser microscopy (at 4.2 K),
- low-temperature scanning-tunneling microscopy (at 4.2 K).

Acknowledgments. The work was supported financially by the *Deutsche Forschungsgemeinschaft*.

References

[6.1] R.F. Peierls: *Quantum Theory of Solids* (Clarendon, Oxford 1955) p. 108
[6.2] M.J. Rice, S. Strässler: Solid State Commun. **13**, 125 (1973)
[6.3] K.F. Fung, J.W. Steeds: Phys. Rev. Lett. **455**, 1696 (1980)
[6.4] C.H. Chen, R.M. Fleming, P.M. Petroff: Phys. Rev. B **27**, 4459 (1983)
[6.5] C.H. Chen, R.M. Fleming: Solid State Commun. **48**, 777 (1983)
[6.6] M.J. Rice, S. Strässler: Solid State Commun. **13**, 1389 (1973)
[6.7] G.A. Toombs: Phys. Rep. **40**, 181 (1978)
[6.8] P. Monceau: *Properties of Inorganic Quasi-One-Dimensional Metals*, Vol. 2 (Reidel, Holland 1986)
[6.9] G. Grüner: Rev. Mod. Phys. **60**, 1129 (1988)
[6.10] C. Schlenker, J. Dumas, C. Escribe-Filippini, H. Guyot: In *Low-Dimensional Electronic Properties of Molybdenum Bronzes and Oxides*, ed. by C. Schlenker (Kluwer, Dordrecht 1989) p. 159
[6.11] J. Dumas, C. Schlenker: Int. J. Phys. B, to be published
[6.12] A.J. Heeger: In *Highly Conducting One-Dimensional Solids*, ed. by J.T. Devreese (Plenum Press, New York 1979)
[6.13] W. Riess: Thermische Hysterese und Ladungsdichtewellen am quasi-eindimensionalen Leiter $(Fluoranthen)_2PF_2$. Ph. D. Thesis, University of Bayreuth (1992)
[6.14] W. Brütting: Peierlsinstabilität und Ladungsdichtewellentransport in Fluoranthen- und Perylenradikalkationensalzen. Diploma Thesis, University of Bayreuth (1992)
[6.15] R.M. Fleming, L.F. Schneemeyer: Phys. Rev. B **28**, 6996 (1983)
[6.16] J. Peinke, A. Kittel, J. Dumas: Europhys. Lett. **18**, 125 (1992)
[6.17] P. Bak, C. Tang, K. Wiesenfeld: Phys. Rev. A **38**, 364 (1988)
[6.18] H.G. Schuster: *Deterministic Chaos* (VCH, Weinheim 1988)
[6.19] J. Peinke, J. Parisi, O.E. Rössler, R. Stoop: *Encounter with Chaos* (Springer, Berlin 1992)
[6.20] A. Maeda, T. Furuyama, S. Tanaka: Solid State Commun. **55**, 951 (1985)
[6.21] A. Maeda, T. Furuyama, K. Uchinokura, S. Tanaka: Solid State Commun. **58**, 25 (1986)
[6.22] R.E. Thorne, W.G. Lyons, J.W. Lyding, J.R. Tucker, J. Bardeen: Phys. Rev. B **35**, 6360 (1987)

7 Current Filamentation in Dipolar Electric Fields

V. Novák[1] *and W. Prettl*[2]

[1] Institute of Electrical Engineering, Academy of Sciences, Dolejškova 5
182 02 Praha, Czech Republic
[2] Institut für Angewandte Physik, Universität Regensburg
93040 Regensburg, Germany

Abstract. In the first part of this contribution we report on filamentary current flow in epitaxial n-GaAs between two ohmic point contacts. A scanning laser microscope technique has revealed two types of filamentary structure: large-area filaments, typically appearing at higher sample currents, and bendable filaments, arising at low currents and becoming curved in perpendicular magnetic fields. In the case of large-area filaments self-organization is suppressed by the dipolar field distribution from the point contacts. Numerical simulations, based on the two-level model of Schöll, have been carried out for this case.
The second part discusses the response of the sample in the filamentary current regime to local interband illumination. Repetitive current pulses have been found to occur with a mean period strongly depending on the light intensity. Statistical properties of the pulse series on a large time scale reveal features, typical of a nonlinear dynamical system. A physical mechanism is proposed, which relates the pulse occurrence to a specific switching phenomenon in the region of the filament boundary.

7.1 Introduction

Current filamentation has been observed in a wide range of semiconductor materials and structures. Various methods have been used to visualize the filamentary structure of the current flow. Among them, non-invasive techniques seem to be preferable. However, technical difficulties (as in case of temperature or potential measurement [7.1]) or a low intensity of the corresponding physical process (as in case of recombination radiation measurement [7.2]) make them inapplicable in case of impurity breakdown induced current filamentation at low temperatures. Convincing results just under these conditions have been achieved by the following two methods: low temperature scanning electron microscopy [7.3–5] and its laser beam analogy [7.6, 7]. Both of them are based on some kind of interaction between the investigated structure and a testing device. The exact physical mechanism of this interaction is, however, not yet fully understood. It cannot be generally excluded that its influence is detrimental to the observed structure. Thus, the applicability of these strongly invasive imaging methods seems to be restricted to stationary and sufficiently stable filamentary patterns.

Such states have been found and investigated in epitaxial layers of n-GaAs with a fundamentally inhomogeneous electric field, imposed by some specific con-

Springer Proceedings in Physics, Vol. 79 **Nonlinear Dynamics and Pattern Formation in Semiconductors and Devices**
Editor: F.-J. Niedernostheide

tact geometry. In the first part of this contribution (Sect. 7.2) the investigation of a typical sample with a pair of point contacts is reported. At first, in Sect. 7.2.1, the principles of the laser beam imaging method are briefly explained. Then, in Sect. 7.2.2, the reconstructions of large-area filaments at higher sample currents are presented. In Sect. 7.2.3 the one-dimensional model of the cross-section of this filament is designed, based on the two-level model by Schöll [7.8, 10]. Numerical solutions of this model show satisfactory agreement with the experiment. Even some magnetic field related effects may be qualitatively reproduced by the model, as will be shown in Sect. 7.2.4. Some conclusions, analytically derived in Sect. 7.2.5, concern generally the interface region between the low and high conductive phases. It will be shown, however, that the described model does not yield an appropriate explanation of another type of filamentary structure, which is reported in Sect. 7.2.6.

The physical mechanism that underlies the imaging reaction induced by the focused illumination of the sample has not been analysed in detail up to now. The proposed interpretations of the light induced effect rely mostly on time-averaged or lock-in detected measurements [7.3, 7]. Unfortunately, the dynamic features of the phenomenon are thus concealed. Since the underlying physical mechanism seems to be of more general importance, we will analyse it separately in Sect. 7.3. At first, we will present the time resolved response to focused illumination of the sample in the filamentary regime, Sect. 7.3.1. We will then investigate some statistical properties of the response, which enables us in Sect. 7.3.2 to formulate a hypothesis concerning the actual physical mechanism of the phenomenon. In the last section the dynamic behaviour of the illuminated sample will be investigated on a large time scale. Some surprising features of the analysed signal will be pointed out that indicate a possible correspondence to previously reported experimental results of Aoki *et al.*, [7.11, 12].

7.2 Stationary Current Filaments in a Dipolar Electric Field

7.2.1 Experimental Setup

The experimental setup of the scanning laser microscope mentioned above is shown in Fig. 7.1. The spatially filtered beam of an interband light laser is scanned across the surface of the biased sample by a mechanical deflection unit. The beam is focused on the sample surface to a spot diameter of about 30 μm. Simultaneously with the scanning the sample current is recorded. Analogously to electron microscopy, the two-dimensional mapping of the light induced signal is reconstructed. It turns out that the interfaces between the regions of high and low conductivity are most sensitive to focused illumination. Accordingly, the two-dimensional image obtained reflects primarily the position of the boundaries of the filamentary structure. More detailed explanations concerning an underlying physical mechanism will be given in Sect. 7.3.

All experimental results reported here have been obtained with a sample of Si-doped n-conducting epitaxial layer of GaAs grown by LPE on semi-insulating

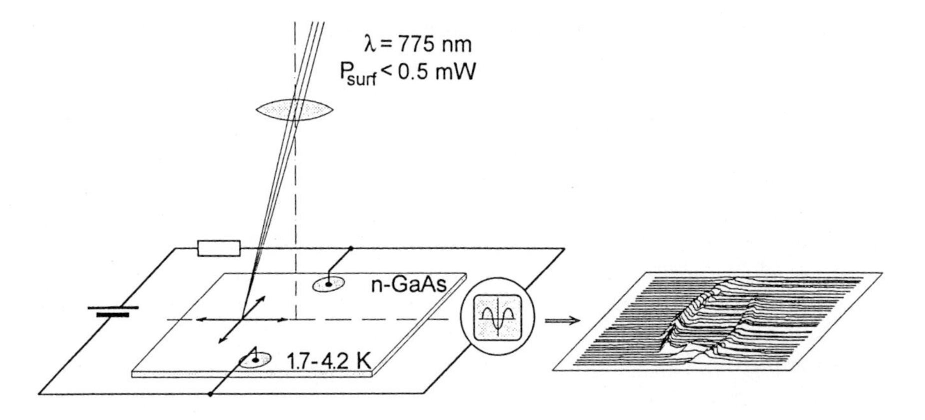

Fig. 7.1. Setup of scanning laser microscope

Cr-doped GaAs substrate. A pair of ohmic point contacts were alloyed into the epitaxial layer. The parameters of the sample are summarized in Table 7.1. It should be stressed, however, that they are in no way critical for the observed phenomena — similar results have been obtained with other n-GaAs samples, too.

The sample was mounted in an optical immersion cryostat and cooled to 1.8 K. Perpendicular to the epitaxial layer a magnetic field was applied by a superconducting magnet. The sample was supplied with a constant bias voltage in series with a load resistor.

Table 7.1. Geometry and material parameters of the sample

sample dimensions	-	4.1×3.7 mm
contact distance	c	3.5 mm
contact radius	R	< 0.5 mm
layer thickness	-	29μm
donor density	N_D	$11.9 \cdot 10^{14}$ cm^{-3}
compensation ratio	N_A/N_D	0.91
free electron density at 77 K	n	$9.6 \cdot 10^{13}$ cm^{-3}
electron mobility at 77 K	μ_n	$4.5 \cdot 10^4$ cm^2 V^{-1} s^{-1}

Figure 7.2 shows current–voltage characteristics of the sample. It can be seen that for low bias voltages the sample is practically nonconductive as all the impurity states are occupied. At a certain critical bias the avalanche breakdown of shallow impurities occurs, free electrons are multiplied and the sample becomes conductive. The back transition from conductive to nonconductive states is hysteretically shifted which gives rise to a broad region of bistability. The process of the transition between the two states is influenced by the value of the load resistor. For resistances greater than some critical value the system

becomes unstable and undamped relaxation oscillations arise in the breakdown regime [7.13].

A perpendicular magnetic field modifies the I–V curve in two ways: first, the conductivity of the sample is decreased at higher currents, which may be well attributed to the effect of magnetoresistivity. Secondly, the I–V curve is deformed in a complex way at lower currents and the boundaries of the bistability region are shifted. The explanation of this effect is not straightforward and will be discussed later.

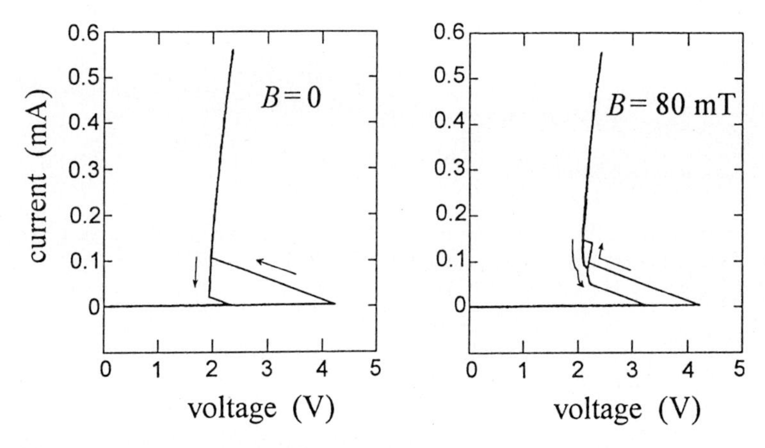

Fig. 7.2. Current–voltage characteristics of the sample without and with a magnetic field normal to the epitaxial layer

7.2.2 Large-Area Filament Reconstruction

The filament images shown in Fig. 7.3 have been recorded in a regime well above the threshold of breakdown. As has been mentioned above, the two ridge-like structures in each of the mappings correspond to the boundaries of a convex-shaped region with significantly higher conductivity. Not surprisingly, the width of this region rises with increasing current. This can be seen from Fig. 7.4, where only the central lines of filament reconstructions are plotted for a range of sample currents.

An important conclusion can be readily made: the shape of this current filament indicates the significant role of the electric field distribution. Unlike the effect of current contracting mechanisms, known e. g. from plasma physics, and unlike some previously observed filamentary structures [7.4,5], the width of the investigated filament reaches an apparent maximum halfway between the contacts, and the shape of the filament boundaries resembles the distribution of field lines in an undisturbed dipolar electric field from the sample contacts. The response of the filamentary structure to a perpendicular magnetic field yields additional support for the conclusion on the role of the electric field distribution, Fig. 7.3 bottom. While the symmetry in the *form* of the opposite filament

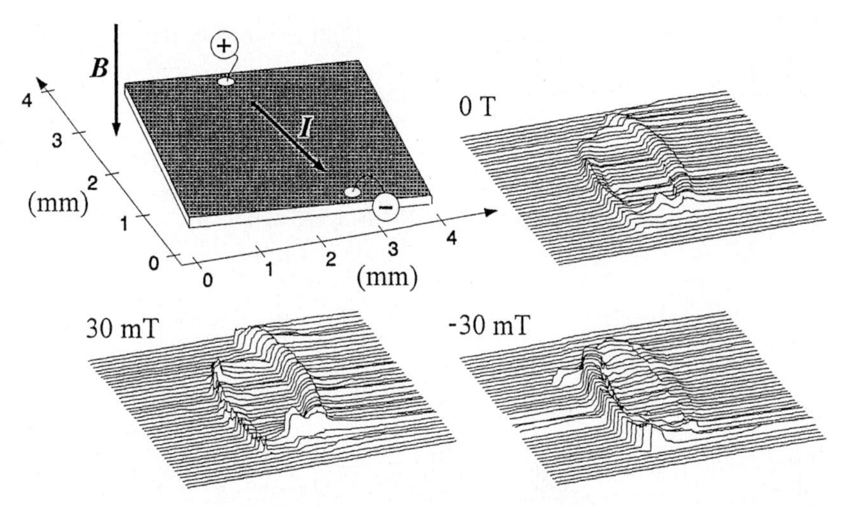

Fig. 7.3. Reconstructions of large-area filaments. The sample was biased via an 1 kΩ load resistor, the stationary current was 1 mA

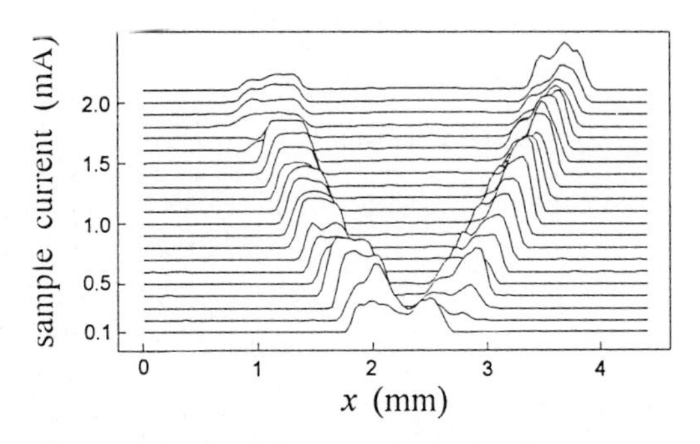

Fig. 7.4. Central lines of filament reconstructions at various sample currents. Each curve corresponds to the maximum filament width for the respective current

boundaries is lost, the filament itself exhibits no or negligible side shift, as could have been expected due to the Lorentz force.

Consequently, the dipolar electric field must be assumed to stabilize efficiently the macroscopic shape of the current filament in the investigated case. Moreover, the field distribution seems to be unsignificantly disturbed by the presence of a conductive area of the current filament. This fact establishes the possibility of reasonably simulating some of the filament related phenomena through a simple one-dimensional model.

7.2.3 One-Dimensional Model of a Large-Area Filament

Let us adopt the coordinate system according to Fig. 7.5, with the y-axis in the direction linking the two contacts and the x-axis lying in the middle of the filament. As discussed above, the following two assumptions are justified by the

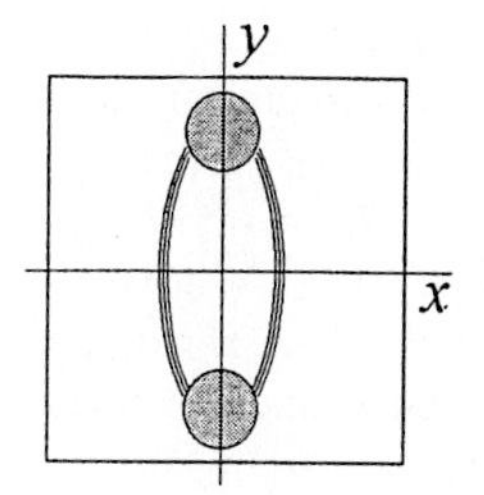

Fig. 7.5. Coordinate system of the model

experimental results: (i) the filament walls follow approximately lines of constant electric field, leaving thus essentially unperturbed the dipolar field distribution between the two point contacts; (ii) the filament is approximately symmetric with respect to reflection in the axis perpendicular to the contact dipole, i.e. with respect to the x-coordinate.

The symmetry about the x-coordinate implies that on this axis all derivatives (except the potential) in the y-direction tend to zero. Consequently, the y-components of the current density and the field strength vector vanish and an autonomous one-dimensional subsystem can be extracted from the physically two-dimensional problem: the cross-section along an axis transversal to the current flow. This fact enables us to adopt essentially the same one-dimensional mathematical apparatus, which was used by Schöll [7.8–10] to investigate the filamentary structures in a *homogeneous* E_y-field.

The governing equations for an n-type semiconductor in a steady state are Poisson's equation for the electric potential ψ,

$$\frac{\mathrm{d}^2\psi}{\mathrm{d}x^2} = -\frac{\rho}{\varepsilon_S} = -\frac{\mathrm{e}}{\varepsilon_S}\left(N_D^+ - N_A^- - n\right), \tag{7.1}$$

and the transport equation for the current density J_{nx} in the x-direction,

$$J_{nx} = -\mathrm{e}\,\mu_n n\frac{\mathrm{d}\psi}{\mathrm{d}x} + \mathrm{e}\,D_n\frac{\mathrm{d}n}{\mathrm{d}x}, \tag{7.2}$$

where ε_S is the permitivity of the semiconductor, N_D^+ and N_A^- are the densities of ionized donor and acceptor impurities, n is the free electron concentration, μ_n is the electron mobility and D_n the diffusion constant.

The continuity equation in a steady state degenerates into the requirement of spatially constant current density J_{nx}. Physical reasonableness leads then to the condition $J_{nx}(x) = 0$ along the x-axis. Upon integrating the transport equation the following simple relation between the free electron density n and the electric potential ψ can be derived,

$$n(x) = n_0 \exp\frac{\psi(x)}{u_T}, \tag{7.3}$$

provided that the Einstein relation $D_n = u_T\mu_n$ holds; u_T is the thermal potential, $u_T = k_B T/\mathrm{e}$, n_0 is the free electron density in a reference point x_0, $\psi(x_0) = 0$.

It has been shown that the presence of two levels of localized states in addition to the conduction band is necessary and sufficient to yield a system with up to three equilibrium states which differ in electron distribution [7.10]. Consequently, the material shows an S-shaped relation between current density and electric field strength (J–E characteristic), which is known to be typical for the occurrence of spatially inhomogeneous steady states in the form of current filaments (e.g. [7.14, 15]). In the case of n-type GaAs at liquid helium temperature the ground state and the first excited state of shallow donors are involved in the most probable transition processes, which are shown in Fig. 7.6. The transition rates

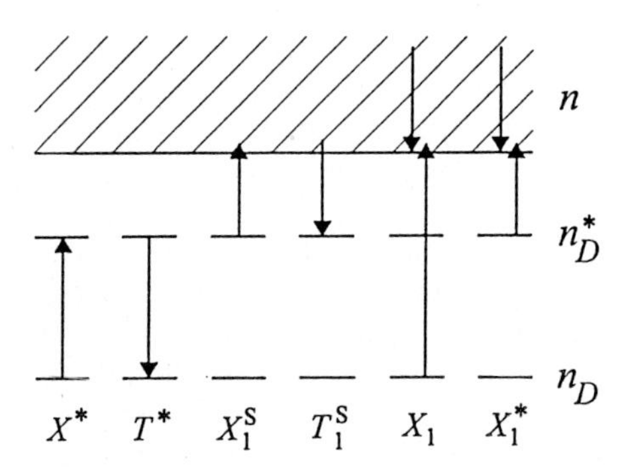

Fig. 7.6. Transition processes of two level model

can be described in analogy with chemical reaction kinetics in the following form [7.10]:

$$\begin{aligned} \dot{n}_D^* &= T_1^S N_D^+ n + X^* n_D - X_1^S n_D^* - X_1^* n_D^* n - T^* n_D^* \,, \\ \dot{n}_D &= T^* n_D^* - X^* n_D - X_1 n_D n \,, \end{aligned} \tag{7.4}$$

where n_D and n_D^* are the electron densities in the ground and excited states, respectively. The impact ionization rate coefficients X_1 and X_1^* depend strongly on the electric field. For the sake of simplicity these dependences are approximated by the semiempirical formula [7.16, 17]

$$X_1 = X_\infty \exp\left[-\left(\frac{E_{\text{av}}}{|\boldsymbol{E}|}\right)^\alpha\right] , \tag{7.5}$$

where X_∞, E_{av} and α are constants, $|\boldsymbol{E}|$ is the local electric field strength. An analogous relation holds for X_1^*.

In a steady state, the rate equations (7.4) become algebraic and both n_D and n_D^* can be expressed as rational functions of n. Since $N_D^+ = N_D - n_D - n_D^*$ and all the acceptor-like impurities are assumed to be occupied, (7.1) can be rewritten into the form

$$\frac{\mathrm{d}^2\psi}{\mathrm{d}x^2} = \frac{\mathrm{P}_3\left(\exp(\psi/u_T)\right)}{\mathrm{P}_2\left(\exp(\psi/u_T)\right)} , \tag{7.6}$$

where P_2 and P_3 for simplicity denote second and third order polynomials, respectively. This equation describes the distribution of an electric potential in the cross-section of the current filament.

It is noted that through the relation (7.5) some of the coefficients in the polynomials P_2 and P_3 are functions of the magnitude of the local electric field strength $|\boldsymbol{E}| = \left((\mathrm{d}\,\psi/\mathrm{d}\,x)^2 + E_y(x)^2\right)^{1/2}$. While the first derivative of the potential makes (7.6) nonlinear, the function $E_y(x)$ reflects the field distribution. With respect to assumption (i) from the beginning of this section, $E_y(x)$ in a steady state is presumed to be given only by the geometry of the contacts and by the applied voltage. Two basic configurations can thus be distinguished: parallel stripe-like contacts enforcing a spatially constant electric field, and the point or circular contacts with a dipolar field distribution. In the latter case the y-component of the electric field along the x-axis is described by the decreasing function [7.18] (as long as the finite boundary effect may be neglected)

$$E_y(x) = \frac{U/\delta}{1 + (x/\sigma)^2}, \tag{7.7}$$

where U is the applied voltage, δ and σ are constants reflecting the contact diameter R and distance c of the contact centers in a following way:

$$\delta = \sigma \cdot \ln \frac{\sigma + (c/2 - R)}{\sigma - (c/2 - R)},$$
$$\sigma = \sqrt{(c/2)^2 - R^2}\,.$$

It is physically reasonable to assume that sufficiently far from the filament the system tends to its homogeneous steady state. Such a homogeneous and thus charge neutral state corresponds to the trivial solution of (7.6):

$$P_3(n) = 0\,. \tag{7.8}$$

With the rate coefficients according to Table 7.2, the cubic equation (7.8) gives rise to the interval of bistability between E_h and E_{th} shown in Fig. 7.7.

Table 7.2. Numerical parameters of (7.6). The values of the rate coefficients are taken from [7.19] and correspond to a typical epitaxial layer of n-GaAs at liquid helium temperature

N_D	(m^{-3})	$5.7 \cdot 10^{20}$	N_A	(m^{-3})	$4.3 \cdot 10^{20}$
X^*	(s^{-1})	$2.34 \cdot 10^6$	T^*	(s^{-1})	$4.1 \cdot 10^7$
X_1^S	(s^{-1})	$1.16 \cdot 10^6$	T_1^S	$(\mathrm{m}^3\mathrm{s}^{-1})$	$5.0 \cdot 10^{-12}$
$X_{1\infty}$	$(\mathrm{m}^3\mathrm{s}^{-1})$	$4.4 \cdot 10^{-11}$	$X_{1\infty}^*$	$(\mathrm{m}^3\mathrm{s}^{-1})$	$2.2 \cdot 10^{-11}$
E_{av}	(Vm^{-1})	244	E_{av}^*	(Vm^{-1})	122
α	(-)	2	α^*	(-)	2
σ	(mm)	2.8	δ	(mm)	4.1

It can be seen that the values from the middle branch of the neutrality curve belong to the unstable homogeneous steady state, as its differential conductivity is negative. The remaining branches with $|\boldsymbol{E}| < E_{th}$ and $|\boldsymbol{E}| > E_h$ yield the values of electron density of a stable homogeneous state in a given electric

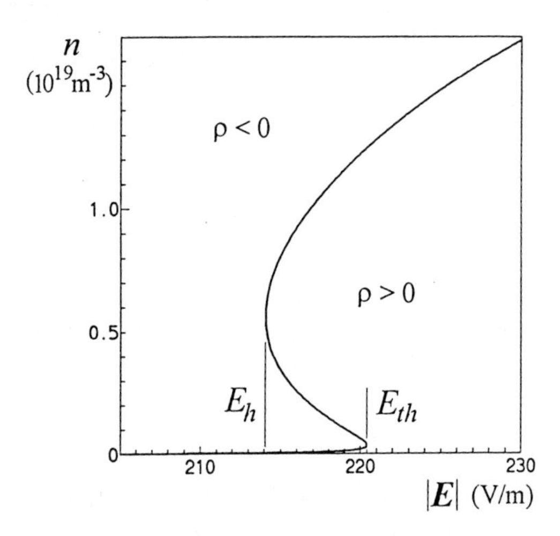

Fig. 7.7. Free electron density as a function of local electric field strength in a charge neutral state

field. Equations (7.7), (7.8) can be used also to define the Dirichlet boundary conditions for (7.6), assuming the boundary points lie sufficiently far from the filament walls.

Basic solutions of (7.6) for several values of the bias voltage are shown in Fig. 7.8. It can be seen that there exists a certain critical value of a *local* field strength: when the component E_y decreases below this critical value, the density of free electrons suddenly falls, forming thus an edge of the highly conductive region. This conclusion corresponds to the basic result of [7.10] about the coexistence of low and high electron density regions in a homogeneous field with a *globally* constant critical electric field — the coexistence field. Some general conclusions concerning the values of the coexistence field will be obtained in Sect. 7.2.5.

The position of the filament boundary in the modelled case coincides with the point (i. e. the line in the plane) of the coexistence field; it is shifted with changing bias. Assuming, in accordance with the presented model, some constant value E_{co} of this coexistence field, a simple square-root dependence of the filament width w on the applied voltage U can be derived through the inversion of (7.7):

$$w = 2\sigma\sqrt{\frac{U}{E_{co}\delta} - 1}\,. \tag{7.9}$$

Using σ and δ as free parameters, (7.9) has been fitted to the experimental data of Fig. 7.4. A good agreement with the real dimensions of the sample, Fig. 7.9, seems to justify indirectly both of the assumptions (i) and (ii) from the beginning of this section.

It may finally be concluded with reasonable justification that the shape of the filament as well as its position in a steady state are uniquely defined and stabilized by the field distribution. The role of self-organization is thus suppressed and confined only, on another spatial scale, to the formation of the phase interface — the filament boundary.

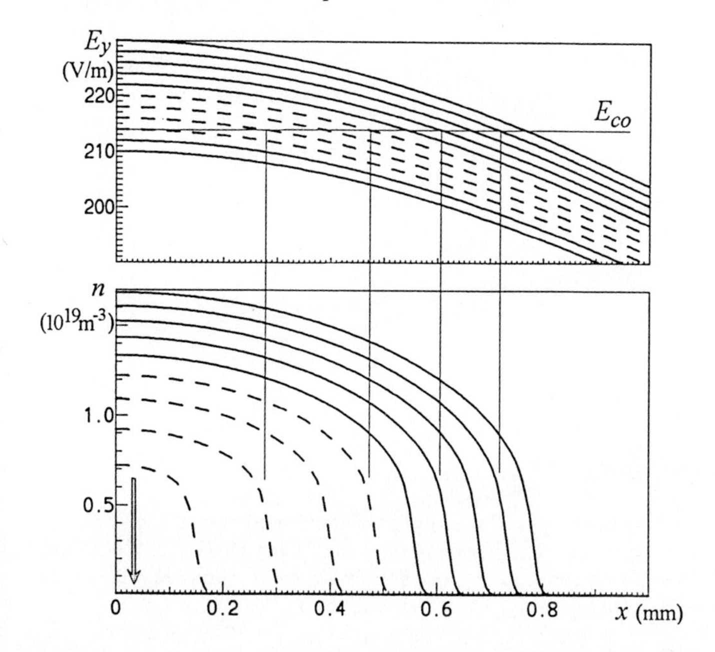

Fig. 7.8. The decay of the E_y component of the dipolar field in the transversal direction at several values of the bias voltage (top) and the respective free electron profiles given by the solution of equation (7.6) (bottom); $x = 0$ lies in the longitudinal symmetry axis of the filament. At this point the E_y-component of the electric field reaches its maximum, proportional to the applied voltage according to equation (7.7). The dashed lines depict the bistable states in which the maximum of E_y does not exceed the interval of bistability and two values of n are possible as boundary values. The corresponding second stationary state with a low electron density lies always on the x-axis within the resolution of the figure

7.2.4 Magnetic Field in the One-Dimensional Model

Introducing magnetic field effects into a *one-dimensional* model decreases in many aspects its reliability, especially in case of inhomogeneous solutions. The following results are therefore presented without any claim on quantitative agreement.

In a perpendicular magnetic field the Lorentz force leads to an additional transport of free electrons in the x-direction. This can be phenomenologically taken into account, within the above stipulation, by introducing a new term into the transport equation (7.2):

$$J_{nx} = \mathrm{e}\,\mu_n n E_x + \mathrm{e}\,D_n \frac{\mathrm{d}\,n}{\mathrm{d}\,x} + \mu_{\mathrm{H}} J_{ny}(x) B\,, \tag{7.10}$$

where J_{nx}, J_{ny} are components of the current density vector, μ_{H} is Hall mobility and B is the magnetic induction. In a stationary state, if it exists, the J_{nx}

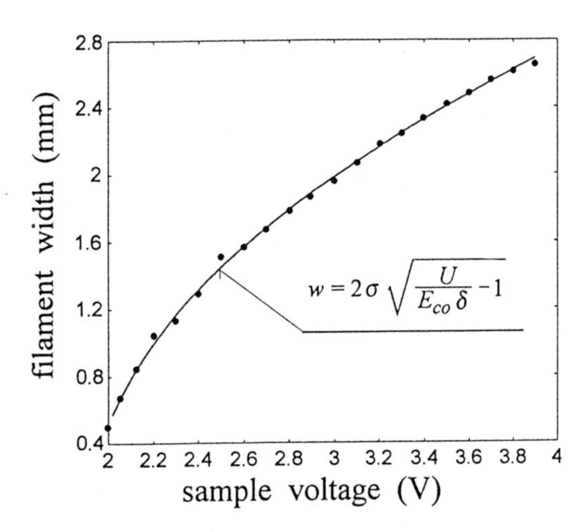

Fig. 7.9. Filament width as a function of the applied voltage. Full circles depict experimental data corresponding to Fig. 7.2. Parameters of the fitting curve (solid line) are $\sigma = 1.33$ mm and $E_{co}\delta = 1.93$ V, which correspond to the effective radius $R = 0.44$ mm and the distance between contacts $c = 2.8$ mm, *cf.* Table 7.1

component vanishes again, and integration of (7.10) yields an analogy to (7.3):

$$\psi(x) = u_T \ln\left(\frac{n(x)}{n_0}\right) - \int_{x_0}^{x} E_{\mathrm{H}}(x')\,\mathrm{d}x' \tag{7.11}$$

with the Hall field $E_{\mathrm{H}}(x) = -\mu_{\mathrm{H}} E_y(x) B$ and x_0 the reference point, $\psi(x_0) = 0$.

The interpretation of (7.11) is straightforward: through the modified charge distribution in the sample under a magnetic field the Hall field arises, compensating the Lorentz force, and a new stationary state is achieved. Under the simplifying assumptions of spatially constant E_{H} and electron mobility μ_n, the charge layers of a deficit or surplus of free electrons, establishing the Hall field, are formed along the sample edges. In such an idealized case it can be seen from (7.6) that the solution in the magnetic field free case remains valid even under a magnetic field, but is unsymmetrically modified in the regions of filament boundaries. The formation of the charge layers *might* be accompanied by the side shift of the whole of the filamentary structure over the distance proportional to the Hall field and to the reversal of electron density near the sample edges. It must be noted, however, that some previous calculations [7.17] indicate the instability of such a solution.

The above considerations can be adopted also in the case of a dipolar electric field with Hall field varying slowly according to (7.7). The solutions of (7.6) modified through the relation (7.10) are shown in Fig. 7.10. The loss of symmetry of the filament is apparent, as well as an otherwise small change in its position. A possible side shift during the Hall charge layers formation lies in the range of micrometers, but it is likely suppressed due to the dipolar contact geometry.

The assumption of constant electron mobility is probably unrealistic. Experiments suggest that inside the current filament the mobility achieves significantly

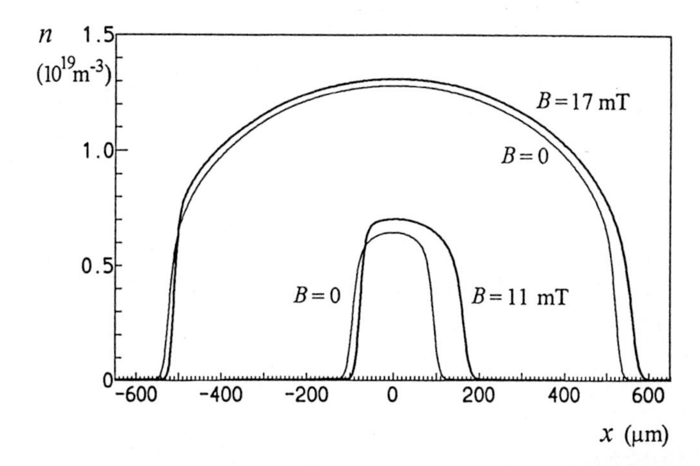

Fig. 7.10. Solutions of equation (7.6) in the presence of a magnetic field normal to the epitaxial layer. The thin lines show the solutions at the same bias but without magnetic field

higher values. The same result has been recently confirmed by a one-dimensional Monte Carlo simulation [7.20]. The Hall field building charge may then be distributed near the filament boundaries rather than along the sample edges. More complex changes in the character of the boundaries must be supposed in such a case.

7.2.5 Equal-Areas Rule for Coexistence Field

It has been shown in [7.10] that the phases of low and high free electron density can coexist, supposing the (overall constant) electric field reaches some specific value in the interval of bistability. A condition has been found, analogous to Maxwell's equal-areas rule in equilibrium thermodynamics, which specifies the value of this coexistence field. This integral condition may be directly used to compute the coexistence field under the assumption that the transversal component of the local electric field vector (E_x in our notation) does not contribute to impact ionization. In this section we will derive a specifying condition for the value of the coexistence field in the more general case, taking into account the full influence of the local electric field strength.

As has been discussed in Sect. 7.2.3, sufficiently far from the filament boundary the density of free electrons achieves its stable value from the upper or lower stable branch of the neutrality curve, (Fig. 7.7). In accordance with [7.10] we denote these two stable points by n_1 and n_3 in the schematic plot in Fig. 7.11. In the region of the filament boundary the electron density rises monotonically from n_1 to n_3 in a way, which can be described by the means of a trajectory in the phase plane of the variables $|E|$ and n. Some conclusions can easily be made about the course of this trajectory:

(i) $|\boldsymbol{E}| \geq E_{co} > E_h$, since $|\boldsymbol{E}|^2 = E_x^2 + E_{co}^2$;
(ii) $\rho(n, |\boldsymbol{E}|) < 0$ for $n_1 < n < n_2$, since E_x is rising;
(iii) in its rightmost point the trajectory crosses the neutrality curve since $\rho(n, |\boldsymbol{E}|) = 0$ if E_x reaches its maximum E_{xmax}. Obviously, $n = n_2$ in that point.

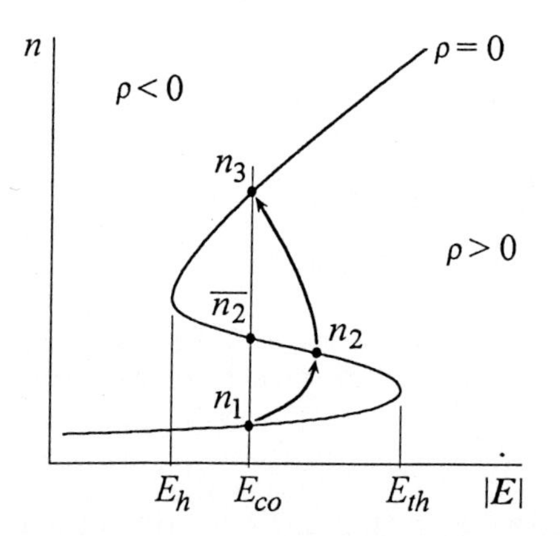

Fig. 7.11. Schematic plot of neutrality curve as phase plane

The conclusion (iii) implies the following relation:

$$E_{xmax}^2 + E_{co}^2 \leq E_{th}^2 \,. \tag{7.12}$$

To find the value of E_{xmax} the charge density $\rho(n, |\boldsymbol{E}|)$ has to be integrated along the trajectory. To accomplish this, the spatial coordinate x can be eliminated from (7.1) through substitution of $\mathrm{d}n/\mathrm{d}x$ from (7.2) under the assumption of vanishing J_{nx}, which yields after integration

$$\frac{1}{2} E_x^2(n) = \frac{1}{\varepsilon} \int_{n_1}^{n} \frac{\rho(n', |\boldsymbol{E}|)}{n'} \, \mathrm{d}n' \tag{7.13}$$

or equivalently with the integration path from n to n_3, since E_x equals to zero at n_1 as well as at n_3 (this, in fact, is the content of an equal-areas rule: integral from n_1 to n_3 equals to zero). It can be seen from Fig. 7.11 that the integral along the trajectory from n_2 to n_3 is always greater than the integral from $\overline{n_2}$ to n_3 (i.e. along the line $|\boldsymbol{E}| = E_{co}$), since the function $\rho(n, |\boldsymbol{E}|)$ has no local extreme for $|\boldsymbol{E}| > 0$, $n > 0$. The integral from $\overline{n_2}$ along the constant $|\boldsymbol{E}|$ yields thus the lower estimate to the value of $E_x(n_2)$. Substituting this approximation into (7.12) then leads to the relation

$$E_{co}^2 < E_{th}^2 - \frac{2}{\varepsilon} \int_{\overline{n_2}}^{n_3} \frac{\rho(n', E_{co})}{n'} \, \mathrm{d}n' \,. \tag{7.14}$$

Following the above arguments on the integration path, the second, independent inequality can be found. It has been shown that the integral along the trajectory from n_2 to n_3 is always greater than the integral from $\overline{n_2}$ to n_3 along the $|\boldsymbol{E}| = E_{co}$ line. The integral from n_1 to n_3 along this line is then always negative because the trajectory integral is zero. It means that such a line integral with a constant $|E|$ can equal to zero only for some $|E| = \overline{E}_{co} > E_{co}$. Thus, the following inequality must be obeyed along with (7.14)

$$E_{co} < \overline{E}_{co} , \tag{7.15}$$

where $\overline{E}_{co}$ is given through the equation

$$\int_{n_1}^{n_3} \frac{\rho(n', \overline{E}_{co})}{n'} \, \mathrm{d}\, n' = 0 . \tag{7.16}$$

The integral inequalities (7.14) and (7.15) give the upper estimate for the coexistence field, based on the stationary (i.e. algebraic) solution of the rate equations (7.4). Note that the values of n_1, $\overline{n_2}$ and n_3 can be found as solutions of an algebraic equation (7.8).

One general consequence of the above analysis should be pointed out: the transversal electric field E_x has a strict *upper* limit (see equation (7.12)) in the region of the filament boundary and this is only due to the fact that the E_x-component through $|E|$ influences the rate coefficients of the model. On the contrary, the transversal field component is essentially unbounded (in the framework of the model), if its role in transition processes is neglected. Unfortunately, the influence of the transversal field, especially in a narrow filament boundary, is questionable — both the low number of possible impact events and the confined acceleration path of free electrons make the formulas (7.5) unrealistic in that case. The efficiency of the impact ionization is probably significantly lower in the transversal direction, which leads according to the model to an increase in the local transversal field E_x. Since the approximate width of the interface region is, according to (7.3), inversely proportional to $E_{x max}$, a specific kind of positive feed-back mechanism may not be excluded, which would tend to a further shrinkage of the filament boundary.

7.2.6 Bendable Filament

The filament reconstructions shown in Fig. 7.12 have been obtained with the same sample as in Fig. 7.3, but with a bias just above the impurity breakdown voltage. Besides the expected reduced width, which leads to an overlap of both of the filament boundaries in the image, a new effect can be recognized: a bending of the current filament in the direction of the Lorentz force.

There are several indirect hints that this filamentary structure *qualitatively* differs from the large-area filament of Sect. 7.2.2. In Fig. 7.13 only the central lines of the filament reconstructions are plotted for the range of magnetic field strengths. Whereas the large-area filament in Fig. 7.13a exhibits negligible side shift even for higher magnetic fields, the displacement in the middle of the narrow filament is almost proportional to the magnetic field strength, Fig. 7.13b. Although the width of the narrow filament could agree with the results of the one-dimensional modeling of Sect. 7.2.3, the extent of the lateral displacement runs significantly over the estimate of a side shift due to the formation of the Hall field.

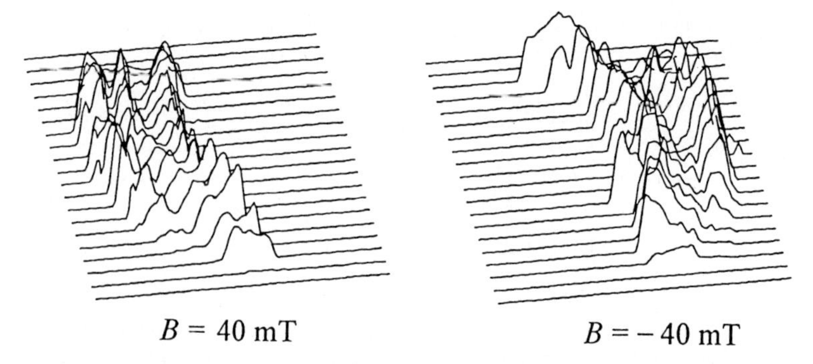

Fig. 7.12. Reconstructions of filaments in perpendicular magnetic fields in immediate post-breakdown regime. The sample was biased via a load resistor of 100 kΩ, stationary current was 100 μA

Let us now discuss the change of the current-voltage characteristic in presence of a perpendicular magnetic field, as is shown in Fig. 7.2. In this nontrivial change the following effect may be distinguished: the occurrence of a new hysteretic loop in the near post-breakdown region. This magnetic field based hysteretic loop separates two new stable stationary states. As qualitatively the same effect has been observed on various n-GaAs samples with point contacts, we suggest this phenomenon to be a generic one.

It seems that the bending filament can be observed only in this new bistability region. Though the filamentary images in both of the stationary states could not have been acquired till now, we suppose that they correspond to the two filament types presented in this paper: to the large-area filament (a narrow one) and to the bendable filament.

Consequently, the bendable filament appears to be formed by another mechanism than that discussed in Sect. 7.2.3. As indicated by some recent simulations, the generation-recombination instability might still be the acceptable one [7.17], combined probably with the effects of enhanced electron mobility in the conducting region.

7.3 Filamentary Structure under Interband Illumination

In Sect. 7.2.1 the principles of the filament imaging method using a scanning laser beam have been outlined, without a detailed explanation of the sample response to the focused illumination. In this section we will analyse the physical mechanism of this reaction. We will point out some aspects which are in our opinion of general importance and which relate the investigated effect to some previously reported phenomena.

7.3.1 Dynamic Behaviour of Illuminated Samples

The setup of the experiment is essentially the same as that for filament reconstruction, Sect. 7.2.1. The sample of Table 7.1 has been biased into a stable

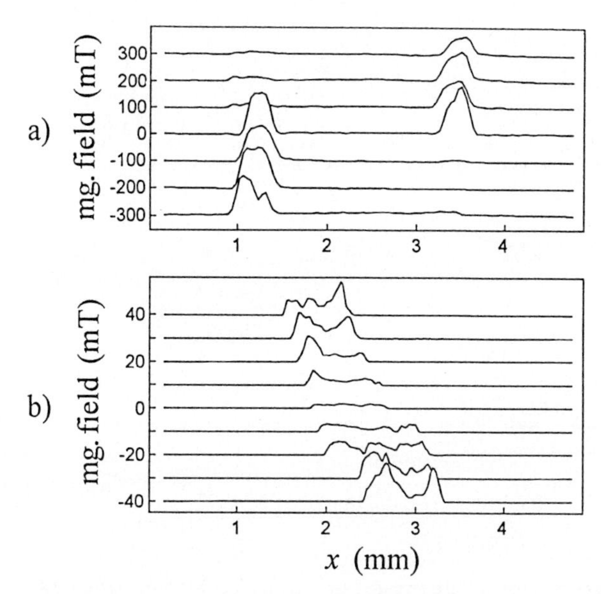

Fig. 7.13. Response of a large-area filament (a) and a bendable filament (b) to a perpendicular magnetic field. Only the central lines of the respective reconstructions have been plotted. The sample was biased via a 100 kΩ load resistor, stationary currents were (a) 1.5 mA and (b) 0.1 mA, respectively

stationary state in the post-breakdown regime. Through the laser beam imaging a stable large-area filament has been proven to exist in this case. In the dark no significant fluctuation of the sample current has been observed. Then the sample has been illuminated by a focused interband light with low intensity and with adjustable focus position and the sample current has been recorded with a transient recorder.

It turns out that upon illumination the previously stable current is broken into repetitive pulses, Fig. 7.14. They are characterized by an initial peak quickly decaying to a discrete, almost constant level of enhanced current. A period of this excitation ends with an abrupt transition to the original current level. Depending on various conditions (e. g. on the stationary current in the dark) pulses of more complex shape may appear. Typically, a second discrete excited level of current can be distinguished during the relaxation process. Quite irregular pulses have then been observed under nonfocused, broad-area illumination. On the other hand, the form of the signal under the focused illumination is quite characteristic and does not qualitatively depend on the focus position. What varies, however, is the height of the pulses and, most apparently, their time scale.

In the first experiment the focus has been shifted along the axis transversal to the current flow (x-axis according to the coordinate system in Fig. 7.5). The correspondingly varying mean pulse length and the interval between pulses have been measured, Fig. 7.15. Both of these dependences across the whole of the sample width are shown in the inset of the figure. It can be seen that they reach

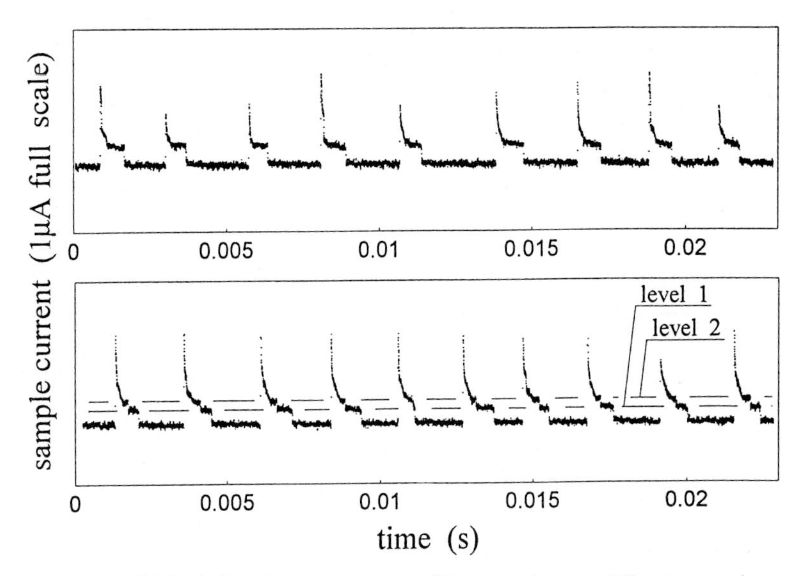

Fig. 7.14. Light induced current pulses. The two pictures differ in sample current and thence in filament width

a sharp minimum at two points which lie symmetrically around the longitudinal current axis. Not surprisingly, these points coincide with the borders of a current filament according to a previous filament imaging. In the figure one of this minima is plotted in detail. Obviously the mean pulse frequency varies over several orders of magnitude.

In a second experiment the height of the pulses has been investigated as shown in Fig. 7.16. Again two symmetric minima can be found, which coincide with the minima of the mean pulse length. Nevertheless, the extremes are flat and reveal locally apparent breaks and discontinuities.

One important conclusion can already be made: the reaction to the illumination, i. e. the occurrence of the pulses, is related to the filament *boundary*, not to the existence of a filament itself. Certain interaction between the filament boundary and the nearest neighbourhood of the filament seems to be essential for the phenomenon. To prove this in another way, a third experiment has been performed. The beam focus has been fixed outside the filament. The bias voltage has been modulated by a triangular signal of small amplitude and low frequency. In that case a filament width is supposed to vary accordingly, which leads to a sweeping of the filament boundaries over the narrow neighbouring areas (less than 20 μm may be estimated from Fig. 7.4). Again the sample voltage has been recorded.

Without any sweeping only the above mentioned current pulses are observed, Fig. 7.17 top (in the figure the sample voltage is plotted, reflecting directly the reversal of the sample current in case without bias modulation). Upon increasing the sweeping voltage amplitude, the pulses begin to merge near the point where the filament reaches its maximal width. Finally, a single pulse is formed

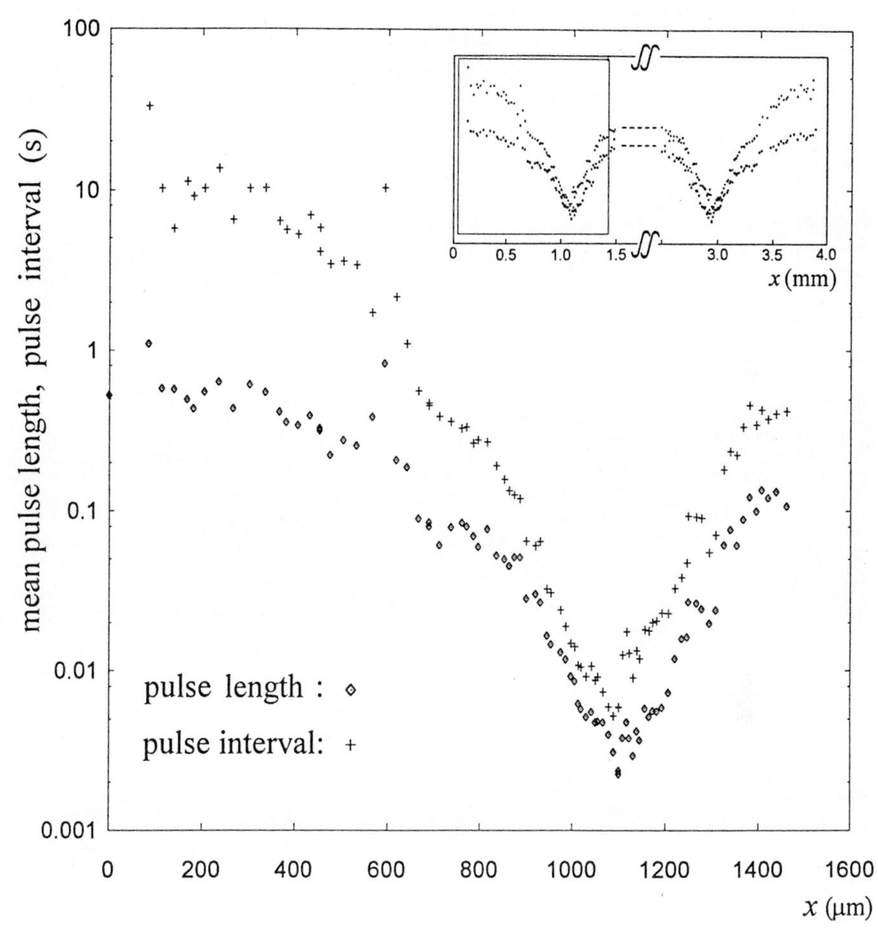

Fig. 7.15. Mean pulse length and interval between pulses under focused illumination. The focus position has been shifted along the middle cross-section of the sample

with a voltage level corresponding to the first level of originally randomly distributed short pulses, Fig. 7.17 bottom. In fact, the event does no more appear to be a pulse: it is an abrupt transition from one to another stationary (or quasi-stationary) state. Moreover, the transition can be induced and controlled externally through the slight shift of the filament boundary. The following conclusion can then be made: the pulses induced through the illumination are to be regarded as a consequence of some *switching* phenomenon rather than of a dynamic process (e. g. local discharge). In the next section we will explain what is meant by switching phenomenon in this case and what we suggest to be the underlying physical mechanism.

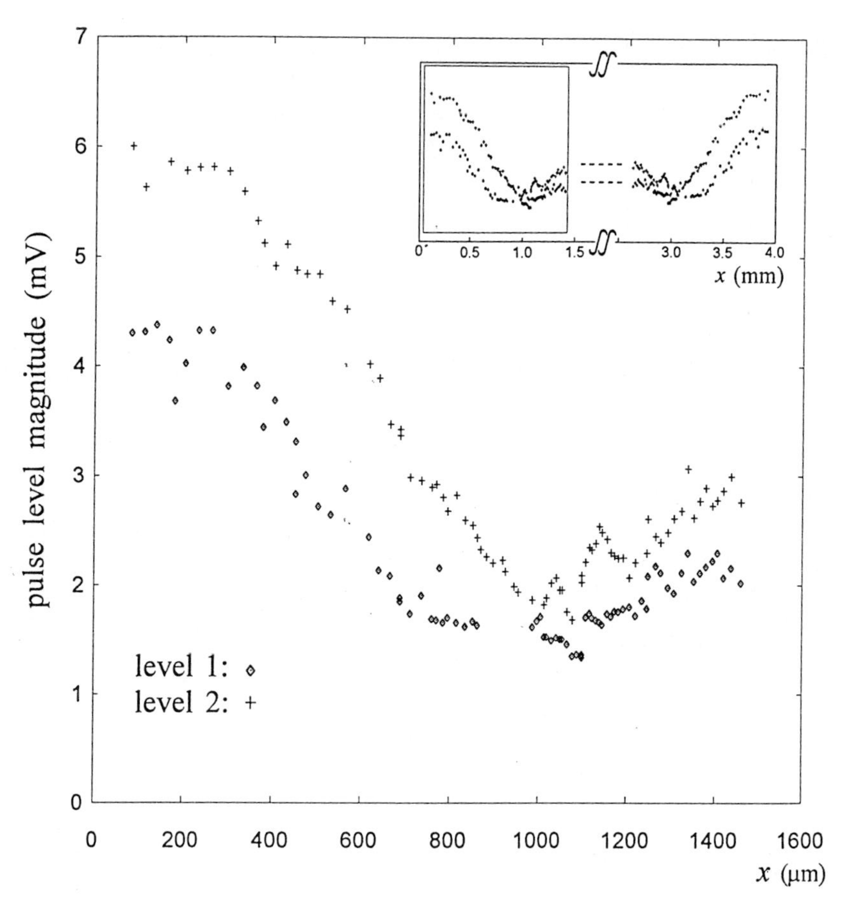

Fig. 7.16. Amplitude of the current pulses under the same conditions as in Fig. 7.15

7.3.2 Hypothesis on the Underlying Mechanism

Focused interband illumination generates locally electron-hole pairs. Outside the filament a long living area of enhanced conductivity is built, Fig. 7.18a, possibly through the mechanism of the persistent photoconductivity according to Queisser *et al.*, [7.21,22], (the generated holes are trapped near the interface between substrate and epitaxial layer, and free electrons remain partially localized through coulombic attraction). The excess conductivity is at first proportional to the local light dose and becomes later saturated. It should be stressed that the mechanism has no threshold with respect to the light dose and that, however small it may be, light of non zero intensity shines on the sample even far from the beam focus.

This area of slowly increasing density of light generated free electrons is separated from the filament by the negative charge layer along the filament boundary; let us note that the existence of the charge double-layer is an intrinsic feature of a stable interface between regions of high and low electron density: it gives

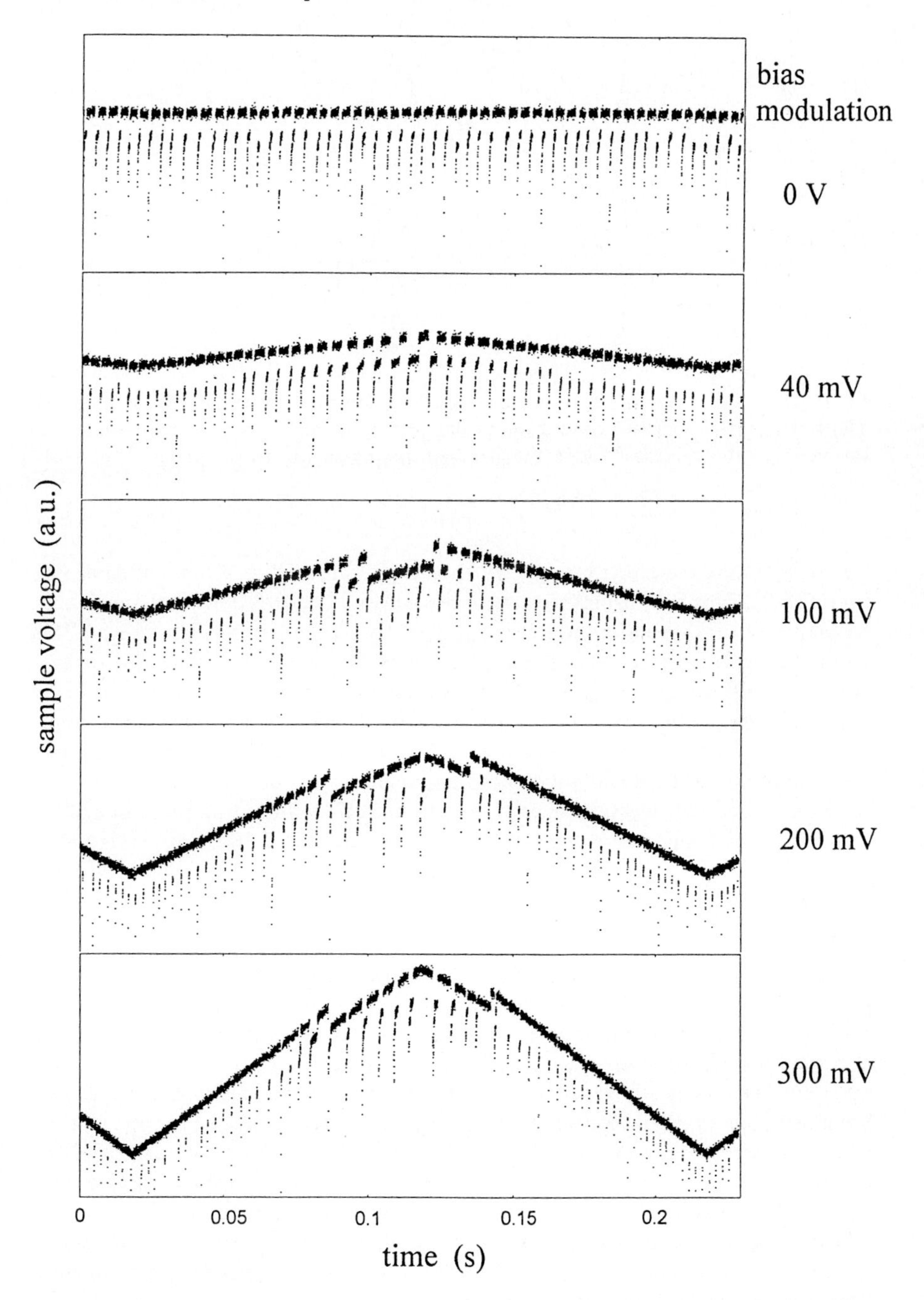

Fig. 7.17. Sample voltage under triangular modulation of the bias voltage. The focus of the illuminating light is fixed near the filament boundary. The amplitude of modulation increases in going from top to bottom

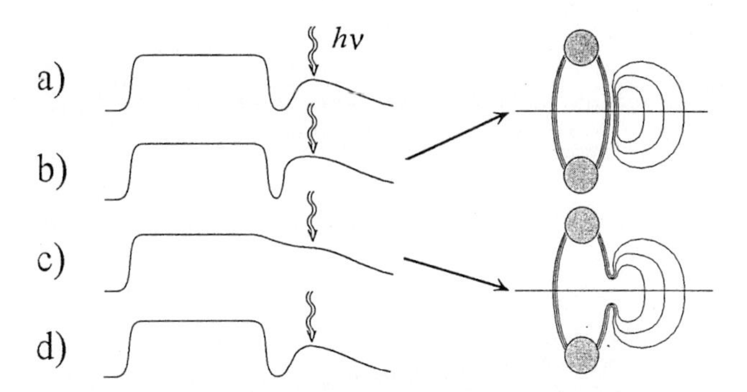

Fig. 7.18. Effect of the induced persistent photoconductivity area outside the filament. The free electron densities are schematically plotted along the central cross-section of the sample. The current flows in the direction perpendicular to the plane of the plot

rise to a transverse electric field, which compensates for electron diffusion (see Sect. 7.2.2). Thus, an insulating layer is formed and the area of the enhanced photoconductivity does not contribute significantly to the current flow, Fig. 7.18b. Upon overcoming a certain critical density of free electrons along the separation layer, the filament and the induced region of persistent photoconductivity suddenly merge, Fig. 7.18c, which leads to the abrupt increase in the sample current. Then the traps are more intensively emptied (especially near the filament boundary) due to the locally enhanced mean energy of conducting electrons. After a period of slow electron density decay the self-organized filament boundary and thence the separation layer are formed again, Fig. 7.18d, and the cycle is repeated.

Let us recall some details of the above experiments which support our explanation.
(i) Excess electron density outside the filament should be proportional to the photon dose supplied by the incoming light. Under the assumption of the exponential-like decay of the light intensity around the focus position, the time period necessary for building up the critical electron density along the separation layer must rise exponentially with the distance between the focus and the filament boundary. This qualitatively agrees with the order-of-magnitude change of time intervals according to Fig. 7.15.
(ii) If the beam is focused far from the filament, the induced conductive area must be large enough to reach the vicinity of the filament boundary. Thence, the relative change of conductivity after the merging should be greater with greater distance between focus position and filament boundary. Experimental confirmation can be seen in Fig. 7.16.
(iii) The shift of the filament boundary towards the induced conductive area should have (temporarily) the same effect as an increase in photon dose has, i. e. a higher probability of merging. Agreement with experiment is straightforward again, as seen from Fig. 7.17.

7.3.3 Large Time Scale Dynamic Behaviour

The periods and lengths of the induced current pulses show apparent fluctuations. In order to investigate the statistical properties of these fluctuations, time intervals between consecutive pulses have been recorded. It turns out that the distribution of these intervals is not Gaussian, Fig. 7.19. Instead, the density of distribution reveals a slowly decaying long time tail, unsymmetry, and even signs of periodicity. The detailed shape of the distribution function changes significantly with a change of illumination intensity, sample bias, focus position, etc. The mean value of the period varies with the focus position according to Fig. 7.15.

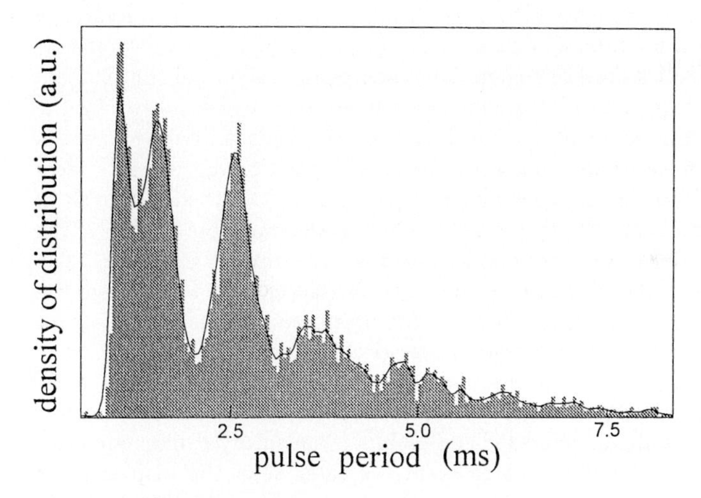

Fig. 7.19. Histogram of time intervals between 8192 consecutive pulses

These observations probably correspond to previously reported results of Aoki *et al.* [7.11, 12], who have observed series of current peaks — firings — in an *n*-GaAs epitaxial layer under very similar conditions. After a frequency

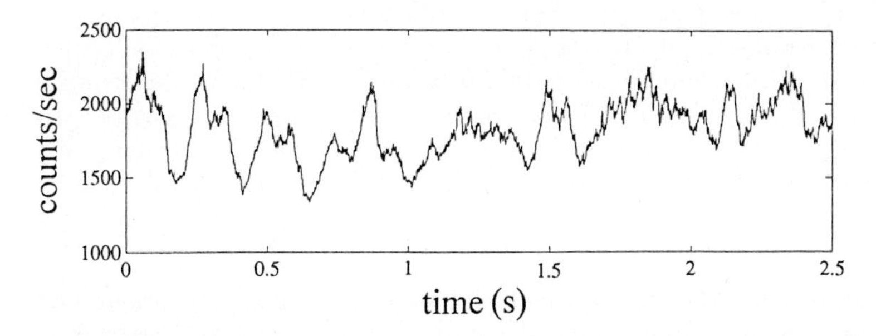

Fig. 7.20. Slowly varying signal, gained from a time series of 8192 consecutive pulse periods through the window-averaging

demodulation the peaks yielded a slowly varying signal — firing density wave — with features typical for nonlinear dynamic system; after a demodulation analogous to that of Aoki, Fig. 7.20, similar wave forms can be found in our case, too. We suggest that in the common background of all of these observations the described mechanism of switching on the boundary of a filamentary structure can be found. Consequently, some of the observed nonlinear dynamics might be attributed to a competition between different, spatially separated switching processes.

7.4 Conclusions

Two types of current filaments have been found by scanning laser microscopy of epitaxial layers of *n*-GaAs. The first one, a large-area filament, exists at currents well above the breakdown region. The macroscopic shape of the filament is essentially determined by the inhomogeneous electric field imposed by the contact geometry. The filament is formed by two independent interfaces, which confine the phase of stable high electron density. The position of these interfaces — filament boundaries — is given by lines of constant local field strength of some specific value — the coexistence field. A perpendicular cross section of the filamentary structure can be described by a simple one-dimensional model. The electric field-imposed filamentary mechanism is efficient enough to suppress the influence of a perpendicular magnetic field on a macroscopic scale. The filament reconstructions show only local changes in regions of filament boundaries.

At lower currents, in the immediate post-breakdown regime, another type of filamentary structure has been observed. Unlike the large-area filament, the low current filament exhibits an apparent bending in a perpendicular magnetic field. The magnitude of the bending is significantly greater than the displacement which is connected with the Hall field formation, and which can be explained in the framework of the one-dimensional model of a large-area filament. Moreover, the current-voltage characteristic in a perpendicular magnetic field shows the existence of two different stationary states in the immediate post-breakdown region. We suggest that they correspond to the two above reported filament types. Accordingly, each of them is assumed to be built by a different physical mechanism.

Well defined pulses have been observed on a previously stable sample current, when the sample in the filamentary regime has been illuminated by focused interband light. Parameters of the pulse series have been found to be strongly correlated with the light intensity in the area of the filament boundary. An experimentally supported mechanism has been proposed to explain this phenomenon. A correspondence with previous results on firing wave instability by Aoki *et al.* is suggested.

Acknowledgments. We wish to thank U. Margull, B. Finger, M. Mayer and A. Bernlochner for providing some experimental data and technical assistance. Also, we thank E. Schöll and R. E. Kunz for helpful discussions. Financial support by the Deutsche Forschungsgemeinschaft is gratefully acknowledged.

References

[7.1] D. Jäger, H. Baumann, R. Symanczyk: Phys. Lett. **A 117**, 141 (1986)
[7.2] F.-J. Niedernostheide, M. Arps, R. Dohmen, H. Willebrand, H.-G. Purwins: phys. stat. sol. (b) **172**, 249 (1992)
[7.3] K.M. Mayer, R. Gross, J. Parisi, J. Peinke, R.P. Huebener: Sol. St. Comm. **63**, 55 (1987)
[7.4] K.M. Mayer, J. Parisi, R.P. Huebener: Z. Phys. B – Condensed Matter **71**, 171 (1988)
[7.5] K. Aoki, U. Rau, J. Peinke, J. Parisi, R.P. Huebener: J. Phys. Soc. Jpn. **59**, 420 (1990)
[7.6] A. Brandl, W. Prettl: In *Festkörperprobleme (Advances in Solid State Physics)*, Vol. 30, ed. by. U. Rössler (Vieweg, Braunschweig 1990) p. 371
[7.7] A. Brandl, M. Völcker, W. Prettl: Appl. Phys. Lett. **55**, 238 (1989)
[7.8] E. Schöll: Z. Phys. B – Condensed Matter **46**, 23 (1982)
[7.9] E. Schöll: Z. Phys. B – Condensed Matter **48**, 153 (1982)
[7.10] E. Schöll: *Nonequilibrium Phase Transitions in Semiconductors* (Springer, Berlin 1987)
[7.11] K. Aoki, K. Yamamoto: Phys. Lett. **98A**, 72 (1983)
[7.12] K. Aoki, O. Ikezawa, N. Mugibayashi, K. Yamamoto: Physica **134B**, 288 (1985)
[7.13] J. Spangler, B. Finger, C. Wimmer, W. Eberle, W. Prettl: Semicond. Sci. Technol. **9**, 373 (1994)
[7.14] V.V. Vladimirov, A.F. Volkov, E.Z. Mejlichov: *Plazma poluprovodnikov* (Atomizdat, Moscow 1979)
[7.15] J.K. Pozhela: *Plasma and Current Instabilities in Semiconductors* (Pergamon, Oxford 1981)
[7.16] G.A. Baraff: Phys. Rev. **128**, 2508 (1962)
[7.17] G. Hüpper, K. Pyragas, E. Schöll: Phys. Rev. **B 47**, 15515 (1993)
[7.18] K. Küpfmüller: *Einführung in die theoretische Elektrotechnik*, 10th ed. (Springer, Berlin 1973)
[7.19] A. Brandl, W. Prettl: Phys. Rev. Lett. **66**, 3044 (1991)
[7.20] B. Kehrer, E. Schöll: private communication
[7.21] H.J. Queisser: Phys. Rev. Lett. **54**, 234 (1985)
[7.22] H.J. Queisser, D.E. Theodorou: Phys. Rev. **B 40**, 4027 (1986)

8 Spatiotemporal Patterns and Generic Bifurcations in a Semiconductor Device

F.-J. Niedernostheide

Universität Münster, Institut für Angewandte Physik, Correnstraße 2/4
D-48149 Münster, Germany

Abstract. A large variety of spatiotemporal current-density patterns can be observed in silicon p^+-n^+-p-n^- diodes by varying only one external parameter. We present experimental and numerical results concerning diverse bifurcation sequences associated with the dynamics of localized dissipative high currrent-density domains, so-called filaments. In particular, we discuss (i) the transition from a spatially uniform to a stable stationary filament, (ii) the bifurcation from a stable stationary to a rocking filament, characterized by a periodic oscillation of the filament around a fixed position in space, (iii) a period-doubling sequence of rocking filaments to chaotic filament motions, (iv) properties of travelling current filaments, and (v) interaction processes of filaments with the semiconductor boundary. It is shown that the different kinds of motion of the localized structures in the devices under consideration are strongly correlated to oscillations of global system variables, e. g., the total current flowing through the device, and, consequently, a determination of the filament motion is possible by investigating only global variables. - The theoretical modelling of the device under consideration is based on a simplified two-layer-model which leads to a two-component reaction-diffusion system of activator-inhibitor type. Numerical calculations reveal that the simple model reproduces all bifurcation sequences observed experimentally.

8.1 Introduction

Spatially non-uniform states, e. g., regions of high electric fields or current density can arise spontaneously in various semiconductors and semiconductor devices (see, e. g., [8.1–8]). Partly detailed experimental and theoretical investigations on dissipative localized current-density structures, subsequently referred to as filaments, have been done, in particular, for semiconductor devices from silicon, which is - even still today - one of the most promising materials for possible applications based on mechanisms of self-organization.

Static multiple filament-states in devices with quasi-onedimensional contact geometries have been observed, e. g., in gold-compensated p-i-n diodes [8.9], in p-n-p-n devices [8.10, 11], or in reverse-biased p-n junctions [8.12], where microplamas arising in a material without defects are explained in terms of large amplitude localized structures [8.13]. The number of existing static filaments is controlled by an external force, the d.c. driving voltage, which provides the

Springer Proceedings in Physics, Vol. 79 **Nonlinear Dynamics and Pattern Formation in Semiconductors and Devices**
Editor: F.-J. Niedernostheide

system continuously with an energy input, thereby compensates the dissipative losses and forces the system in a state far from equilibrium.

Multi-filament patterns between two-dimensional extended electrical contacts have been investigated in semi-isolating silicon [8.14], in reverse-biased α-SiC p-n junctions [8.15], or in high-resistivity zinc-doped silicon [8.16, 17]. High temperature regions, so-called hot spots, appear likewise in n-p-n and p-n-p transistors (e. g., [8.18]) and can be interpreted as dissipative localized structures [8.19]. In conventional thyristors spatially periodic filamentation transverse to the direction of main current flow has been observed [8.20] and can be explained [8.21, 22] in terms of the well-known Turing-instability [8.23]; moreover, even alternately large and small filaments may arise in such thyristors resulting in a spatial period-two arrangement.

Upon variation of an external parameter a static filament may transform spontaneously into a dynamic structure. As basic dynamic structures we consider

- *switching/spiking filaments* defined by large amplitude oscillations that correspond to an alternate ignition and extinction of the filament,
- *breathing/pulsating filaments* characterized by oscillations of the width of the filament,
- *rocking filaments* defined by pendulous motions of the whole filament,
- *travelling filaments* moving with a certain velocity along the sample.

Switching or spiking filaments have been experimentally observed in silicon p-n-p-n [8.10] and p-i-n diodes [8.24, 25]; in the latter case it turned out that complex oscillations of the total current in such devices are caused by a periodic switching of single filaments and in a simple model [8.26, 27] even chaotic spiking is predicted [8.28, 29]. There are experimental results that give indication of breathing filaments in p-germanium [8.30] or n-galliumarsenide [8.31] but they were not observed in silicon devices up to now. Travelling filaments appear in thyristors as transient motion when switching off the device voltage [8.32]. However, in silicon p-n-p-n diodes rocking and travelling filaments can be stabilized as localized structures by the interplay of an autocatalytic and an inhibiting mechanism as reported recently [8.10, 33]. As p-n-p-n diodes show a rich variety of different spatiotemporal structures they are an excellent example in order to study basic mechanisms of nonlinear dynamics and pattern formation in a solid state device.

In this article, recent experimental results together with a simple model are reviewed. First, the applied measuring techniques used to investigate the spatiotemporal behaviour of localized filaments are described. Experimental results are presented in the subsequent sections. In particular, interest is focused on the influence of important semiconductor parameters, e. g., the dc driving voltage, the temperature, or an external magnetic field upon the developing patterns. Furthermore, it is shown that a superposition of a small ac voltage to the dc voltage source leads to interesting effects, e. g., frequency locking if the device is operated in a mode of self-sustained oscillations. In a further section a two-layer model [8.33] based on a division of the device into an active and a passive part is recapitulated. The derived set of equations is a two component set of evolution equations. For a comparison with experimental results, the spatiotemporal current-density and potential distributions in the interface layer between both device parts are calculated as a function of the value of the dc voltage source. Further parameter variations allow predictions about other current-density patterns that may develop in the device under discussion. In a final section, the experimental and numerical results are summed up and some conclusions are drawn.

8.2 Measuring Techniques

Different techniques have been developed to investigate spatially non-uniform current-density structures. For example, it is possible to use the temperature distribution measured on the device surface via a liquid-crystal film whose reflection behaviour for incident light changes with the temperature. Another technique is based on the effect of free charge carrier absorption: The device is illuminated by a laser beam with a wavelength being so large that excitations between the valence band, impurity levels and the conduction band are impossible. Then, the transmitted light signal is only weakened in those regions where many charge carriers are localized. A further powerful tool to visualize spatial structures is provided by the scanning electron microscope. There are essentially two commonly used methods, the potential contrast technique and the EBIC-technique, where the electron beam induced current is used for visualization. Instead of an electron beam one may use likewise a laser beam that is scanned across the device surface. The optical beam induced current can also be used to modulate the video-input of a TV screen.

A large disadvantage of the above-mentioned techniques is the fact that they are either contrast techniques and/or not suitable to record fast temporal changes of the measured physical quantity. This lack can be eliminated by detecting the emitted recombination radiation or by measuring the surface potential with a high-resistance probe. Both methods can be applied successfully to measure the temporal evolution of self-organized spatial structures in p^+-n^+-p-n^- diodes. Besides providing a sufficiently high temporal resolution, these techniques make available information about two important system variables (cf. Sect. 8.5): The infrared electroluminescence signal is a good measure for the current density that will turn out to be a kind of activator and the measured potential distribution can be identified with an inhibiting component the properties of which determine essentially the structures developing in the system. Subsequently, a short description of both applied measuring techniques is given.

Spatially one-dimensional electroluminescence distributions can be measured temporally resolved with a streak camera. The experimentally set-up is shown schematically in Fig. 8.1. For our purposes it is useful to focus a small stripe orientated parallel to the metallic contacts and containing the light being emitted from the p^+-n^+-p part of the device on an infrared sensitive S1-photocathode that converts the incident light into electrons. They are accelerated and deflected vertically when they pass the deflection plates. After being multiplied by a micro channel plate the electrons strike a phosphor screen where they create a visible two-dimensional image. The horizontal direction contains the spatial information whereas the temporal evolution of the light-density distribution is fixed in the vertical direction. Because of the large persistance the image on the phosphor screen can be recorded with a rather slow but sensitive TV camera. With this technique, a temporal resolution up to the picosecond range is possible. In practice, however, the resolution is limited by the extremely weak recombination radiation so that the signal is covered by noise for too fast sweeps of the streak unit.

The areas for the electric contacts of the investigated samples are typically on the order of $\ell_x \times \ell_z = (2\ldots10) \times (0.5\ldots1)\,\mathrm{mm}^2$. The distance of the metallic contacts is $\ell_y = 880\,\mu\mathrm{m}$. From experimental results presented below it turns out that the width of the developing current filaments is on the order of 0.5 mm; therefore, with respect to pattern formation lateral to the direction of the main current flow, the devices can be considered as quasi one-dimensional.

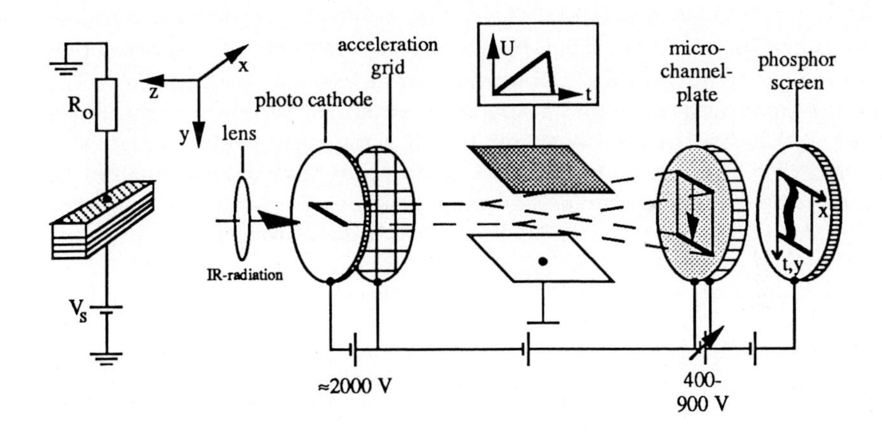

Fig. 8.1. Electrical circuit and basic construction of the streak camera

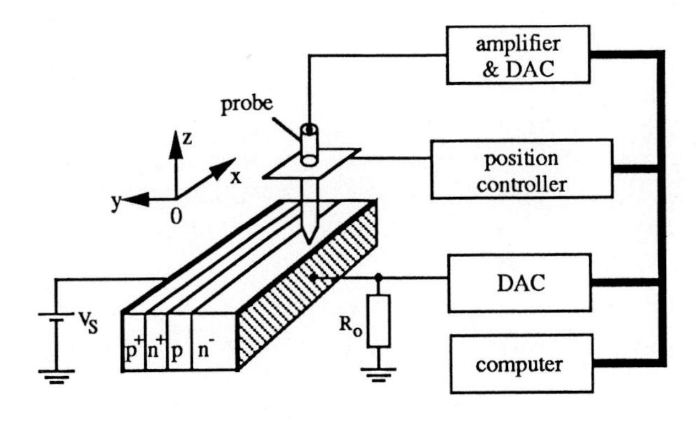

Fig. 8.2. Experimental set-up for measuring the potential distribution on the surface between the metallic contacts

The potential distribution on the device surface between the electrical contacts is measured via a tungsten probe. In order to achieve a large temporal resolution the tungsten probe is connected directly to the gate of a field effect transistor with an input-capacity of only 0.02 pF. Together with the contact resistance between the probe and the semiconductor surface being on the order of 50 MΩ, the typical time resolution is of about 1 μs. By scanning the whole sample surface, a spatiotemporal potential pattern can be reconstructed, if a reference signal, e.g., a self-sustained periodic oscillation of the device voltage is available. Figure 8.2 illustrates the measurement set-up.

8.3 Experimentally Observed Bifurcation Sequences by Increasing the DC Voltage Bias

The spatiotemporal patterns that develop in the p^+-n^+-p-n^- diodes depend on parameters as applied voltage, temperature, or an external magnetic field. The following expositions reveal that nearly all observed structures can be controlled by one parameter, the dc driving voltage, in a wide temperature range with zero magnetic field. On the other hand, it turns out that the patterns can be varied also by the temperature or a magnetic field for a fixed value of the driving voltage and, therefore, the device can be used - at least in principle - as measuring instrument for these quantities.

8.3.1 The Current-Voltage Characteristic

The p^+-n^+-p-n^- devices are connected to a dc driving voltage V_s via a load resistor R_0 as shown in Fig. 8.1. In all experiments presented here, the p^+ layer is biased positive with respect to the n^- layer. Therefore, both outer p-n junctions are forward biased while the middle one is reverse-biased. For low values of the voltage source, merely a small current carried essentially by the thermal saturation current of the blocking junction is allowed to flow. If the voltage is sufficiently large ($\approx$ 40 V), the electric field in the space charge region of this junction is strong enough to cause band-band impact ionization. Thus, an autocatalytic process leading to a charge carrier multiplication takes place. The generated electron-hole pairs are separated in the high field zone; the electrons move towards the p^+ anode, the holes to the n^- cathode. There they may induce additional injection of carriers leading to a positive feedback. A typical measured current-voltage characteristic $I(V)$ of the device is shown in Fig. 8.3.

8.3.2 Transition of a Spatially Uniform State to a Static Localized Filament

Measurements with the streak camera and the potential probe show that the electroluminescence and the surface potential distributions are spatially uniform in the blocking branch of the $I(V)$ characteristic. Both distributions measured at an operating point close to the voltage breakdown are depicted in Fig. 8.4

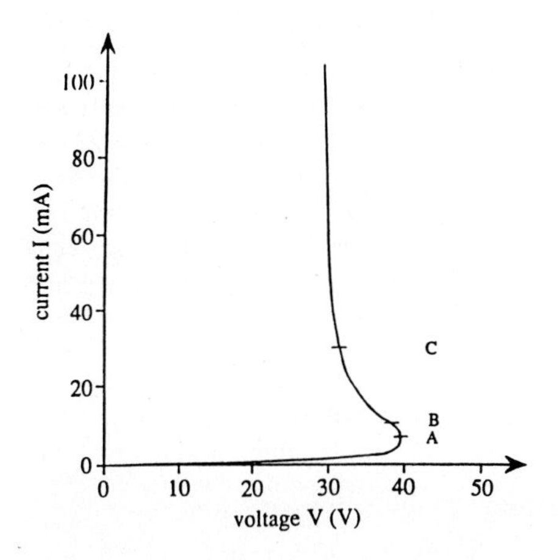

Fig. 8.3. Typical $I(V)$ characteristic of a p^+-n^+-p-n^- diode; the points A-C refer to the current-density and potential distributions in Fig. 8.4 (from [8.34])

(operating point A). The electroluminescence signal has been obtained by evaluating the light intensity near the anode. The potential line has been measured in a distance of about 115 μm from the anode. In the range of the negative differential resistance (Figs. 8.3 and 8.4, operating point B) a contraction of both distributions takes place. When the total current reaches values of about 20 mA the contraction phase is completed and a localized current-density filament has developed near the left sample boundary (Figs. 8.3 and 8.4, operating point C). Proceeding along the $I(V)$ characteristic to larger current values, we observe that the shape of the developed filament remains essentially constant. The absolute value of the filament width and the maximum light intensity are increasing, but the full width at half the maximum (FWHM) depends only weakly on the total current. However, there is still a remarkable difference between the light-density and the potential distribution: the final width of the potential distribution is about five times larger than that of the light density. The reasons and consequences from this will be discussed detailed in Sect. 8.5.

We investigated more than 40 specimens, in most of them the filament is stabilized near one of the two boundaries. However, in a few cases a filament could be observed also in the centre of the sample. An example is shown in Fig. 8.5 where three different filament positions are possible for equal experimental conditions. The three filament positions were observed by switching the sample on and off several times, each time recording the electroluminescence radiation. In each sample, the filament position at the centre of the device is unstable for sufficiently large currents. If a filament is generated in the sample centre and afterwards the current is increased, the filament moves spontaneously to one of the boundaries when a certain current value is reached and becomes again stationary there. This leads to the conclusion that the position in the centre of the device is an unstable position. A stabilization is only possible because of the existence of unavoidable inhomogeneities due to the device preparation. On the other hand, the stable filament positions near the boundaries can be interpreted in terms of nonlinear dynamics as two stable fixed points.

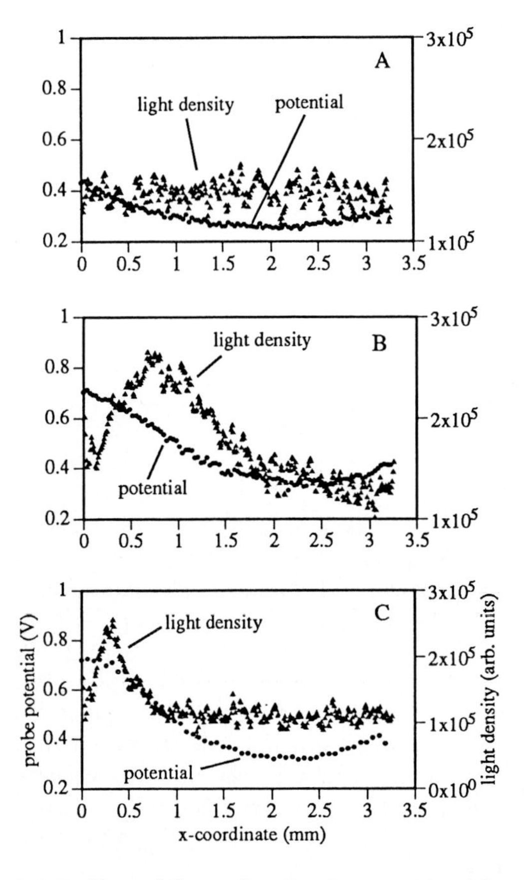

Fig. 8.4. Stationary spatial distributions of the light density and the probe potential for three different values of the driving voltage V_s corresponding to the operating points A - C in Fig. 8.3; parameters are $R_0 = 1.01\,\mathrm{k}\Omega$, $T = 318\,\mathrm{K}$, $(I, V_s) = (6\,\mathrm{mA},\ 45.46\,\mathrm{V})$ (A), $(10\,\mathrm{mA},\ 48.86\,\mathrm{V})$ (B), $(30\,\mathrm{mA},\ 61.8\,\mathrm{V})$ (C), exposure time for the streak image $= 250\,\mathrm{s}$ (A, B), $0.5\,\mathrm{s}$ (C); (from [8.34])

8.3.3 Transition of a Static to a Rocking Filament

When a certain threshold current is exceeded by varying the driving voltage V_s, spontaneous oscillations of the voltage drop across the device appear. A typical time trace of the alternating part $\tilde{V}(t)$ of the device voltage is shown in Fig. 8.6a. The amplitude of the voltage oscillations increases in a square-root like manner with increasing V_s (Fig. 8.8) and is on the order of a few tenths of a percent of the dc voltage part. The frequency is in the range of about 1 to 10 kHz and decreases linearly with V_s. Thus, typical features of a supercritical Hopf-bifurcation are observed. Measurements with the streak camera reveal that the appearance of self-sustained voltage oscillations is correlated with the transition of the static to a rocking filament. In Fig. 8.7a, a streak image for a total current slightly larger than the critical threshold current is shown. As can be seen from this record, the filament oscillates around a fixed position in space with a well-defined frequency that corresponds exactly to the frequency of the global voltage oscillation. The correlation between local filament dynamic and global voltage oscillation is even stronger: the spatial amplitude of the rocking filament increases also in a square-root like manner with the bifurcation parameter V_s.

Further insight into the bifurcation of the static to the rocking filament can

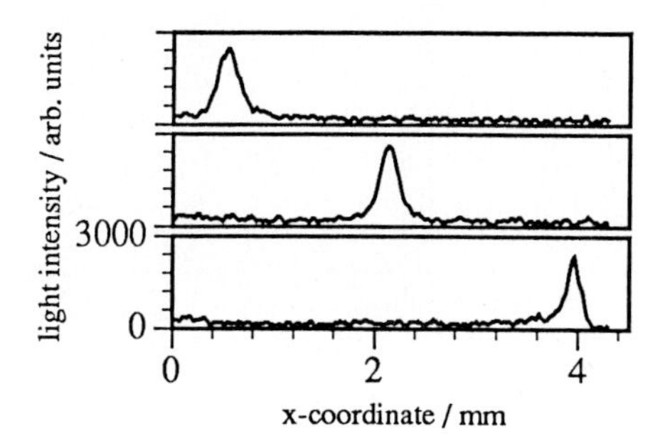

Fig. 8.5. Stationary spatial distributions of the light density indicating different positions of the filament for equal experimental conditions: $R_0 = 209.8\,\Omega$, $T = 318\,\mathrm{K}$, $V_s = 44.4\mathrm{V}$

be obtained by using a scanning electron microscope to induce a perturbation of the static filament. It has been shown in [8.35] that an electron beam striking the static filament near the p^+-n^+-p part of the device generates damped oscillations if the operation point is close to the bifurcation point. These oscillations are correlated with the excitation of a damped rocking oscillation of the filament. Thus, the static filament can be considered as stable focus undergoing a Hopf-bifurcation which leads to the generation of a limit cycle in the form of a rocking filament. With increasing V_s, the damping constant decreases and reaches zero at the transition to the rocking filament corresponding to the self-sustained oscillations. Beyond that, these kinds of perturbation experiments reveal that the linear dependence of the frequency on the bifurcation parameter V_s observed for the rocking filament is also valid for the damped oscillation of the perturbed static filament.

8.3.4 Period-Doubling Cascade and Chaotic Filament Motions

When the total current or respectively the driving voltage V_s is increased further, we observe a sequence of period doublings accumulating at a critical value where the signal becomes chaotic. In Fig. 8.6b-d, a period-two, period-four, and a chaotic time series are shown. Even period-three oscillations have been observed. Thus, the observed period-doubling route is similar to the well-known Feigenbaum-scenario. Figure 8.8 shows a bifurcation diagram where the local voltage maxima $\tilde{V}_m$ of the voltage trace $\tilde{V}(t)$ are plotted versus the control parameter V_s.

Similar to the case of the period-one oscillations these more complex oscillations are strongly correlated to the local dynamic of the rocking filament. In Fig. 8.7b,c streak camera records for a period-two and period-four oscillation are shown. In Fig. 8.7b, e.g., it is obvious that each second elongation to the left is larger than that in between. Each local maximum in the voltage trace $\tilde{V}(t)$ can be associated with a maximum filament elongation to the right, and correspondingly each local minimum of $\tilde{V}(t)$ with a maximum elongation to the left. The larger the voltage extrema the larger the elongations. In the period-four oscillation a similar correspondence between global voltage trace and local filament dynamic is existing.

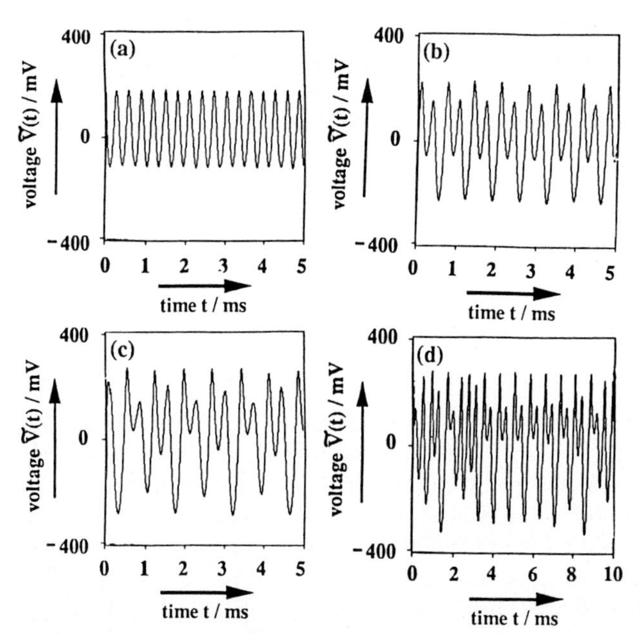

Fig. 8.6. Time traces $\tilde{V}(t)$ observed for increasing $V_s = 168$ V (a), 174 V (b), 180 V (c), 182 V (d); other parameters are $R_0 = 1\,\mathrm{k}\Omega$, T = room temperature (from [8.36])

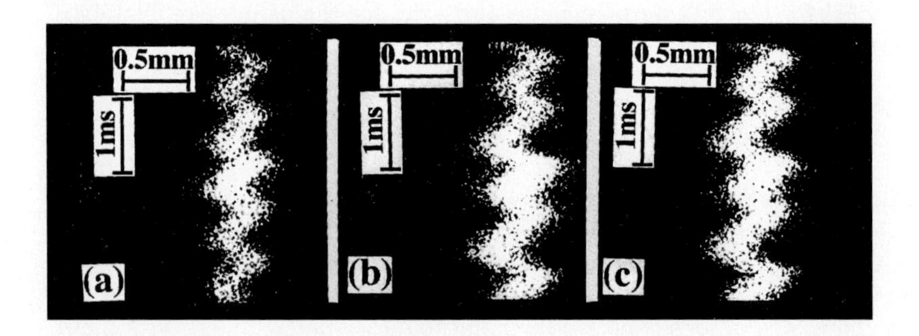

Fig. 8.7. Streak camera records of rocking filaments performing period-one, period-two, and period-four oscillations for $V_s = 95.4$ V (a), 96.5 V (b), 97.1 V (c); other parameters are $R_0 = 606.8\,\Omega$, T = room temperature

Because of the weak recombination radiation it is necessary to integrate some hundreds of periods to obtain images with a sufficiently large signal to noise ratio. For a chaotic filament motion such an integration in not feasible. However, from the strong correlation between the global voltage oscillations and the local filament dynamic in the case of simple and complex periodic oscillations we may conclude that a chaotic time trace $\tilde{V}(t)$ is correlated to a chaotic rocking filament. Note, that all rocking oscillations occur near one of the sample boundaries.

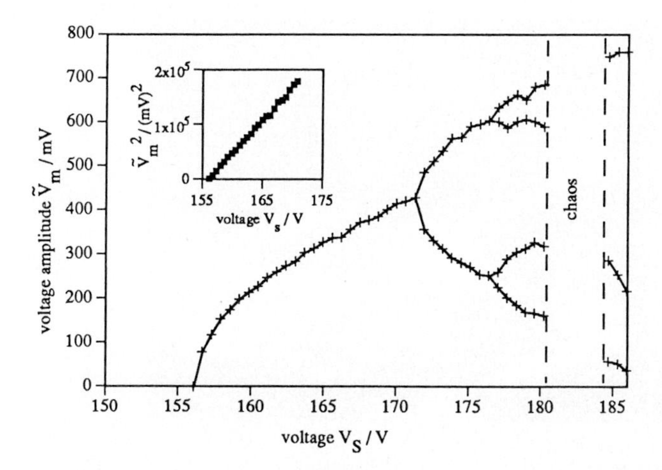

Fig. 8.8. Bifurcation diagram of the voltage maxima $\tilde{V}_m$ vs. the dc driving voltage V_s. The solid curve serves to guide the eye. The inset shows the square of $\tilde{V}_m$ vs. V_s near the threshold where self-sustained oscillations set in

8.3.5 Transition of a Rocking to a Travelling Filament

When the voltage V_s is further increased, beyond a second threshold value the rocking oscillation is superimposed by travelling motions. This can be seen in the streak camera records shown in Fig. 8.9. Near the bifurcation point, the filament is oscillating in the rocking mode close to the sample boundary. After a time varying statistically it travels to the other boundary, continues there the rocking motion, again for a time interval the length of which is not predictable (Fig. 8.9a). The mean time, in which the filament oscillates in the rocking mode decreases linearly with increasing voltage V_s and, finally, a periodic motion composed of well-defined intervals of rocking and travelling oscillations develops (Fig. 8.9b). For a sufficiently large total current/voltage V_s the rocking oscillation vanishes completely and a pure travelling motion of the filament develops, interrupted only by the reflection events at the device boundaries (Fig. 8.9c).

These filament oscillations are again accompanied by oscillations of the total current and the device voltage. In the case of superimposed rocking and travelling oscillations one observes voltage oscillations with a small amplitude emanating from the rocking motion and oscillations with larger amplitudes indicating the travelling mode. In the latter case the filament moves without any interaction with the boundaries and therefore is able to carry a somewhat larger current leading to an increase of the total current and because of the fixed voltage V_s consequently to a decrease of the device voltage. In the case of a pure travelling motion, the trace of the total current contains only the larger peaks indicating the travelling mode, whereas current minima are caused by filament reflections at the boundaries, because the filament suffers there a small contraction. The period for this kind of oscillations is essentially determined by the length of the sample. Typical travelling velocities are in the range of 1 to 10 m/s, so that the

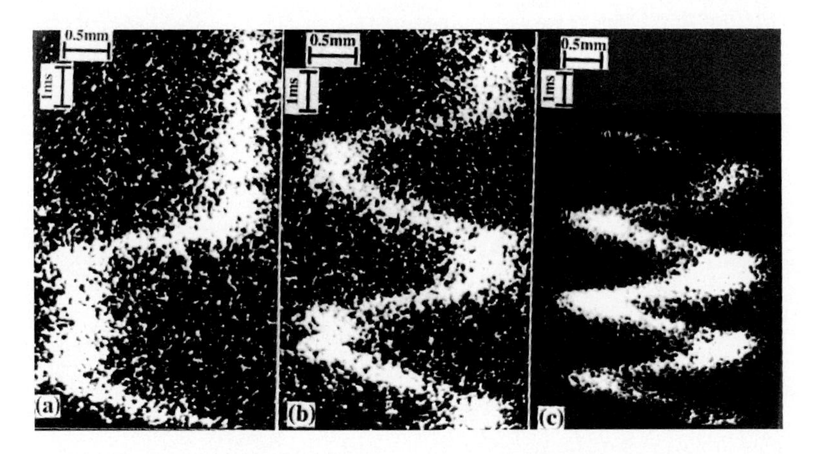

Fig. 8.9. Streak camera records of travelling filaments for $V_s = 91\,\mathrm{V}$ (a), 102.7 V (b), 95.3 V (c); other parameters are $R_0 = 500.8\,\Omega$, $T = 295\,\mathrm{K}$ (a,b) and $R_0 = 163\,\Omega$, $T = 295\,\mathrm{K}$ (c); (from [8.10, 34])

travelling time for a filament in a device with a length $\ell_x = 10\,\mathrm{mm}$ is between 1 to 10 ms. The time needed for a reflection of a filament is usually less than 1 ms.

In order to get insight into the physical mechanisms causing the dynamical behaviour of the filaments, it is useful to consider the current-density represented by the light-density and the potential distributions of a travelling filament during its motion in more detail. In Fig. 8.10 both distributions are shown for six different moments during the travelling motion from the left to the right sample boundary. The potential distributions have been obtained by measuring time traces of the probe potential at 80 equidistantly spaced points on a line situated in the n^- region parallel to the contacts in a distance of about 70 μm from the p-n^- junction. The diagram in Fig. 8.10a presents the situation just before the filament starts to detach from the left boundary. The light-density distribution is spatially more localized than the potential distribution, as in the case of static filaments, and shows a pronounced maximum in a certain distance of the boundary. On the contrary, the potential distribution has its maximum at the boundary. When the filament has detached from the boundary, the potential distribution forms both a precursor in front of the light-density distribution and an extended tail behind (Fig.8.10c,d). With the potential precursor the filament is able to interact with the boundary before the light-density structure reaches the boundary. As outlined in more detail in Sect. 8.5 the potential can be considered as an inhibiting variable, i.e., it counteracts fluctuations of the autocatalytic variable, the current density. Consequently, a filament moving towards a boundary that acts as mirror, e.g., realized when no-flux boundary conditions can be assumed for the respective variables, suffers a damping by its own mirrored potential precursor. By this mechanism the filament is able to control the reflection process: the precursor acts as feeler and retards the filament by its inhibiting property. On the other hand, the potential tail behind the filament governs the "trapping" of the travelling filament when it reaches the boundary: As the potential relaxes relatively slow with respect to the current density, the filament is trapped near the boundary and has to perform there some rocking oscillations until the potential tail is sufficiently declined so that the filament is able to overcome the inhibiting tail of the potential distribution and to start its journey in opposite direction.

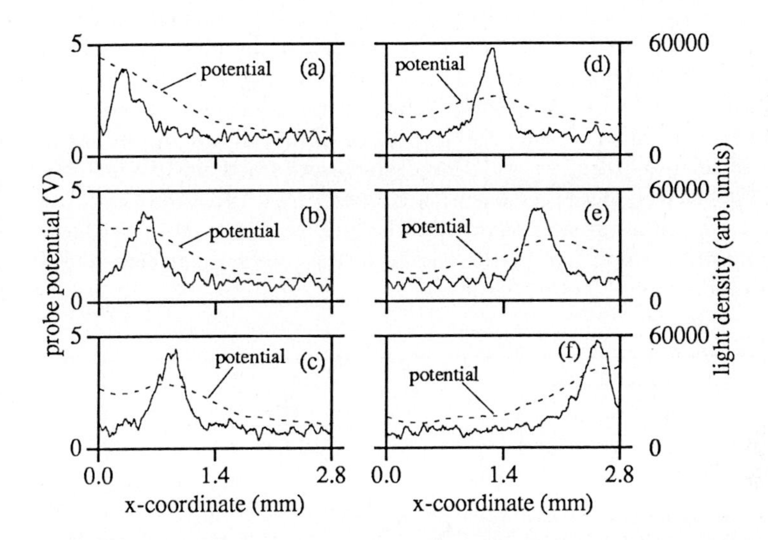

Fig. 8.10. Spatial distributions of the light-density and the potential at six different moments during the travelling motion recorded in time intervals of 100 μs. Parameters are $R_0 = 110/101\ \Omega$, $T = 293/293$ K, $V_s = 59.2/57.6$ V, mean total current = 272/281.7 mA for the light-density and potential measurement, respectively (from [8.37])

8.4 Influence of Other Parameters on the Dynamical Behaviour

8.4.1 Frequency-Locking and Quasiperiodicity

It is well-known that dissipative dynamical systems with two competing frequencies may exhibit phenomena as frequency locking, quasiperiodicity, or transitions to chaotic behaviour; these phenomena have been intensively studied using the circle map (e. g., [8.38, 39]). While in former times main interest was focused to the analysis of integral variables, device voltage or current (see, e. g., [8.40–42]), recently, experimental evidence was reported for frequency-locked and complex spatiotemporal behaviour in ac driven ultrapure germanium [8.43, 44].

In this section we present experimental results obtained when the dc driving voltage V_s supplied to the p^+-n^+-p-n^- diode is superimposed by a sinusoidal ac driving voltage $\tilde{V}_s(t) = \hat{V}_s sin(2\pi f_d t)$, where $\hat{V}_s$ is the drive amplitude and f_d the drive frequency. The results reported here have been obtained by choosing the dc bias V_s such that the device shows pure rocking motions when the amplitude of the ac driving voltage is set to zero ($\hat{V}_s = 0$).

In Fig. 8.11 the winding number $W = f_i/f_d$ vs. the frequency ratio $\Omega = f_0/f_d$ is drawn for fixed values of V_s and $\hat{V}_s$. The fundamental frequency f_0 of the self-sustained oscillation is 4.55 kHz. The fundamental frequency f_i of the system response $\tilde{V}(t)$, the ac part of the voltage drop $V(t)$, has been ascertained by analyzing the power spectrum of $\tilde{V}(t)$ and by evaluating the phase correlation between $\tilde{V}(t)$ and the driving voltage $\tilde{V}_s(t)$. The steps in Fig. 8.11 demonstrate that there are a lot of frequency intervals at which mode-locking appears. The fractions in the figure label distinct locking states. As predicted by the circle map, the locked regions order in a way that regions with frequency ratios $(P+P')/(Q+Q')$ fall between locked regions with P/Q and P'/Q'. The widths of the latter intervals are always greater than that of the locked interval in between. With increasing drive amplitude $\hat{V}_s$ the widths of the intervals increase and a set of Arnold tongues can be observed [8.45].

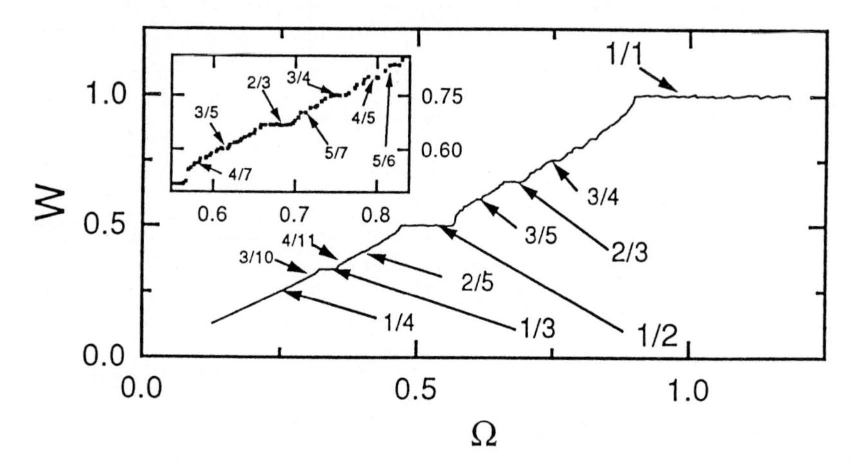

Fig. 8.11. Winding number W vs. frequency ratio Ω; parameters are $V_s = 49.5$ V, $\hat{V}_s = 704$ mV, $R_0 = 480.6\,\Omega$, $T = 314.5$ K

Figure 8.12 shows the time series $\tilde{V}(t)$, the appertaining power spectra p[f], and the Poincaré maps constructed by strobing $\tilde{V}(t)$ at the zeros of $\tilde{V}_s(t)$ with positive slope and plotting $\tilde{V}_{n+1}$ vs. $\tilde{V}_n$ for two different values of the drive amplitude at a fixed drive frequency $f_d = 6.62$ kHz and a frequency ratio $\Omega = 0.695$. For low values of $\hat{V}_s$, a quasiperiodic oscillation develops (Fig. 8.12b). In the appertaining spectrum beyond f_d and f_0, a lot of mixing frequencies are clearly to see. Further evidence for quasiperiodic behaviour is given by the Poincaré map. When the driving amplitude is increased, a 2:3 locking state is observed, as shown in Fig. 8.12a. The two largest peaks in the power spectrum of this signal indicate the drive and the locking frequency. Obviously, all other components are multiples of these frequencies and of their sums and differences. The Poincaré map contains three points, as expected for a 2:3 locking.

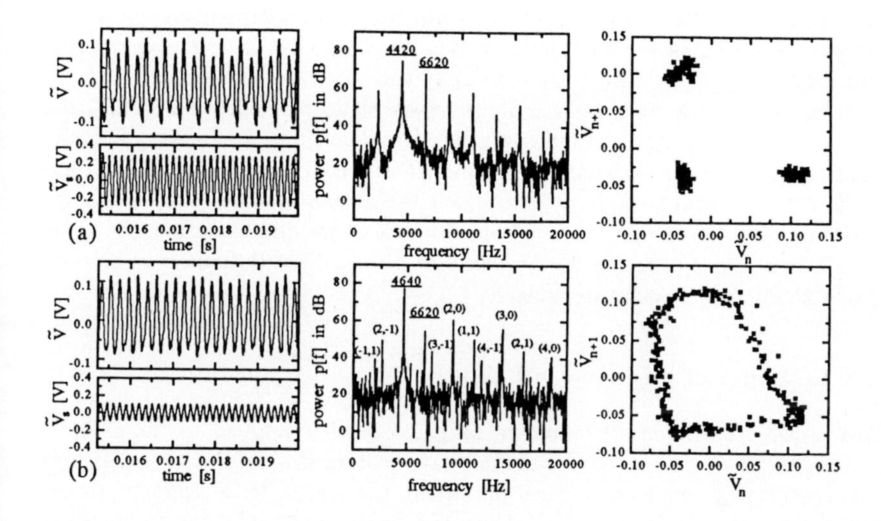

Fig. 8.12. Time series $\tilde{V}(t)$, appertaining power spectra p[f], and Poincaré maps for different drive amplitudes; below $\tilde{V}(t)$ the time trace $\tilde{V}_s(t)$ is shown for comparison; (a) frequency-locked oscillation for $\hat{V}_s = 0.30\,\mathrm{V}$, (b) quasiperiodic oscillation for $\hat{V}_s = 0.10\,\mathrm{V}$, the peaks labelled by (n, m) correspond to frequencies $f = n \cdot 4.64\,\mathrm{kHz} + m \cdot 6.62\,\mathrm{kHz}$; other parameters are $V_s = 49.21\,\mathrm{V}$, $f_0 \approx 4.58\,\mathrm{kHz}$, $f_d = 6.62\,\mathrm{kHz}$, $R_0 = 480.6\,\Omega$, $T = 315.0\,\mathrm{K}$

Because the voltage oscillations are strongly correlated with the filament motions as outlined in detail in Sects. 8.3.3 and 8.3.4, we may conclude that the locked states correspond to a periodic rocking motion the frequency of which is 2/3 of the driving frequency. Accordingly, the quasiperiodic voltage oscillations are connected with quasiperiodic filament motions. We note that these results are based on investigations of only one sample. For a sufficiently large ac driving amplitude one would expect an overlapping of the Arnold tongues and a transition to chaotic behaviour. In the sample investigated it was experimentally not feasible to observe such a transition. Investigations on further samples are in preparation.

8.4.2 Influence of the Temperature

The bifurcations to rocking and travelling filaments can be observed in a wide temperature range. The distinguishing features of the respective bifurcations are experimentally proved to be preserved for device temperatures between 80 K and 350 K.

In the case of a bifurcation of a static to a rocking filament, a temperature increase effects a shifting of the bifurcation point to greater mean total currents and a monotonic decrease of the critical frequency f_c at the bifurcation point. As the $f_c(T)$ dependence is approximately linear for wide temperature ranges, this effect can be used - at least in principle - for measuring temperatures in a very easy way: via a capacitor the device voltage is applied to a frequency-voltage converter the output of which varies linearly with the temperature.

Likewise, we observe a monotonic decrease of both the frequency and the amplitude of the voltage oscillations induced by travelling filaments, when the temperature is increased. The critical current at which travelling filaments appear is shifted towards higher values, i. e., the oscillations are suppressed for sufficiently elevated temperatures. As in the case of rocking filaments, the characteristic variations of frequency and amplitude of the oscillations may be used for easy temperature measurements.

8.4.3 Influence of a Magnetic Field

When a static and spatially uniform magnetic field is applied to the sample, the mobility of a filament is damped or favoured according to the polarity of the magnetic field, which is applied parallel to the z-axis (with reference to the system of coordinates in Fig. 8.1). Thus, the field is perpendicular to both the axis of main current flow and the axis of travelling motions and, consequently, effects a transverse shifting of the filament. Consider a static filament near a boundary. For magnetic fields shifting the filament towards the centre of the sample, the static filament may detach from the boundary, travel to the opposite one and become there static again. Very often it is observed that the filament transforms to a rocking one when reaching the opposite boundary. This can be explained by assuming that the bifurcation points at the two boundaries have slightly different values.

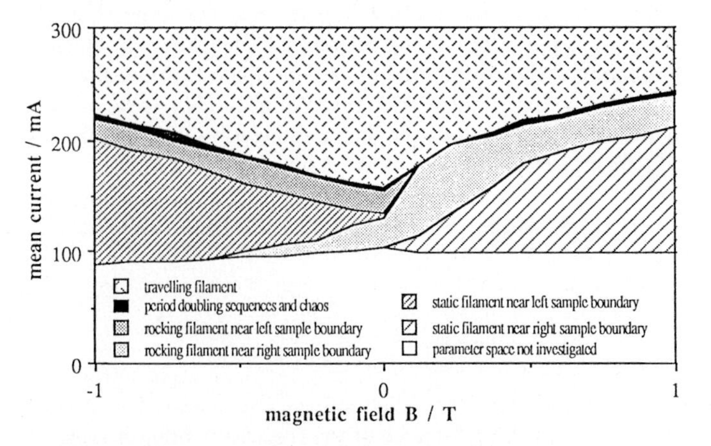

Fig. 8.13. Phase diagram of the filament dynamics in the mean current vs. magnetic field plane; parameters are $R_0 = 100\,\Omega$, $T = 295$ K

Figure 8.13 displays the filament behaviour in the two-dimensional control parameter space spanned by the transverse magnetic field B and the mean total current. The space is classified into regions where static, simple rocking, complex rocking, and travelling filaments appear near the left or the right sample boundary. Starting with a larger negative magnetic field and a mean total current of approximately 200 mA, we observe a bifurcation sequence from a static via a rocking filament with period-doublings to a travelling filament when the magnetic field B is varied from -1.5 T to 0 T, and the reverse bifurcation sequence when B is increased further to +1.5 T.

8.5 Model and Physical Mechanism

8.5.1 Two-Layer Model

It has been shown previously [8.33] that a simplified two-layer approximation of the device under discussion can be used successfully to describe the spatiotemporal structures that may develop in these diodes. Subsequently, the model is described briefly and the final two-component set of evolution equations, that can be derived by using some simplifying assumptions, is presented.

Starting point for modelling the device is a two-layer approximation: The diode is considered to be composed of a p^+-n^+-p transistor and a p-n^- diode. A schematic view is shown in Fig. 8.14. As the device is considered to be quasi-two-dimensional, only the direction of main current flow (y-direction) and one transverse direction (x-direction) are included.

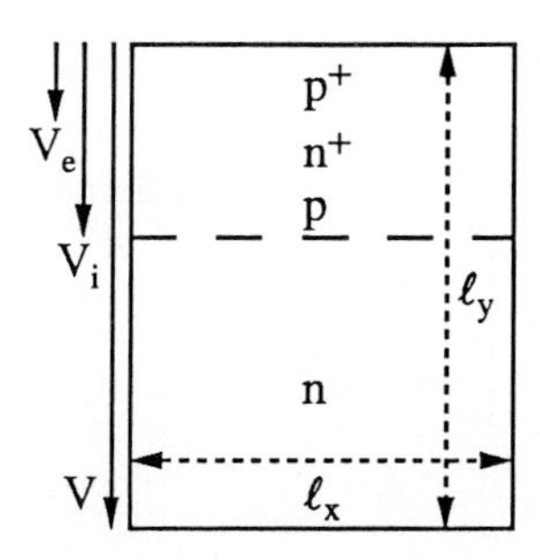

Fig. 8.14. Schematic view of the p^+-n^+-p-n^- diode composed of a p^+-n^+-p transistor and a p-n^- diode; typical dimensions are $\ell_x = 1$-10 mm, $\ell_y = 800\,\mu$m

The following physical mechanisms and assumptions are important for the description of the transistor that works as avalanche transistor and contains the autocatalytic properties:

- avalanche multiplication of charge carriers in the n^+-p junction,
- injection of holes from the p^+ emitter into the n^+ base,
- recombination of electrons and holes in the p^+-n^+ junction,
- a leakage current of the n^+-p junction,
- a spreading of the emitter current in the n^+ base.

The essential assumptions concerning the description of the p-n$^-$ diode are as follows:

- the p-n$^-$ junction is considered as an ideal Shockley-diode,
- injection of holes from the n$^-$ layer into the p layer is neglected,
- the temporal evolution of holes in the n$^-$ layer is dominated by diffusion and recombination to an equilibrium value p_{n0},
- as the main interest is focussed on the current-density distribution transverse to the direction of main-current flow, it is useful to integrate the continuity equation for the hole density in the n$^-$ layer with respect to the y-direction which leads to an equation for the mean hole density $\bar{p}$.

By using these assumptions, the following set of equations can be derived (for details, see [8.33]):

$$C_e \frac{\partial V_e(x,t)}{\partial t} = w\sigma_b \frac{\partial^2 V_e(x,t)}{\partial x^2} - q(V_e, V_i), \tag{8.1}$$

$$\tau \frac{\partial \bar{p}(V(t), V_i(x,t))}{\partial t} = L^2 \frac{\partial^2 \bar{p}(V(t), V_i(x,t))}{\partial x^2} + Q(V_e, V_i, V), \tag{8.2}$$

where V_e, V_i and V are the voltage drops across the p$^+$-n$^+$ junction, the transistor and the whole device, respectively; τ is the lifetime of the mean hole concentration $\bar{p}$ in the n$^-$ layer, C_e is the capacitance of the p$^+$-n$^+$ junction; w and σ_b are the width and the conductivity of the n$^+$ base; they determine the effective diffusion length of V_e. L denotes the effective diffusion length of $\bar{p}$. The mean hole concentration $\bar{p}(V(t), V_i(x,t))$ depends exponentially on the voltage drop $V_p = V - V_i$ across the p-n$^-$ diode:

$$\bar{p}(V(t), V_i(x,t)) = p_{n0} \exp\left(\frac{V_p(x,t)}{V_T}\right) = p_{n0} \exp\left(\frac{V(t) - V_i(x,t)}{V_T}\right) \tag{8.3}$$

with the thermal voltage V_T. The functions $q(V_e, V_i)$ and $Q(V_e, V_i, V)$ are defined by:

$$\begin{aligned} q(V_e, V_i) = {} & (1 - \beta M) j_s \left[exp\left(\frac{V_e}{V_T}\right) - 1 \right] + j_r \left[exp\left(\frac{V_e}{2V_T}\right) - 1 \right] \\ & - M j_{sc} - \frac{V_i - V_e}{\rho_L}, \end{aligned} \tag{8.4}$$

$$\begin{aligned} Q(V_e, V_i, V) = {} & \frac{\tau}{eW} \left\{ M j_{sc} + \beta M j_s \left[exp\left(\frac{V_e}{V_T}\right) - 1 \right] + \frac{V_i - V_e}{\rho_L} \right\} \\ & - (\bar{p}(V, V_i) - p_{n0}), \end{aligned} \tag{8.5}$$

where the multiplication factor M depends on the voltage drop V_c of the n$^+$-p collector. With $V_c = V_i - V_e$, M reads:

$$M(V_e, V_i) = \frac{1}{1 - \left(\frac{V_i - V_e}{V_b}\right)^m}. \tag{8.6}$$

Table 8.1. Set of model parameters and values used in the numerical calculations

Parameter	Parameter meaning	Parameter value
C_e	capacity per unit square of the p$^+$-n$^+$ junction	1×10^{-4} F/cm^2
D_p	diffusion coefficient of holes in the n$^-$ layer	10 cm^2/s
j_r	saturation current density of the recombination current in the p$^+$-n$^+$ junction	3×10^{-7} A/cm^2
j_s	saturation current density of the diffusion current in the p$^+$-n$^+$ junction	1.5×10^{-11} A/cm^2
j_{sc}	saturation current density of the n$^+$-p collector	2×10^{-8} A/cm^2
ℓ_x	sample width in x-direction	0.5 cm
ℓ_z	sample width in z-direction	0.05 cm
L	effective diffusion length of holes in the n$^-$ region, $L = \sqrt{D_p \tau}$	0.02 cm
m	coefficient determining the dependence of the multiplication factor M on the collector voltage V_c	3
p_{n0}	equilibrium concentration of holes in the n$^-$ layer	1×10^7 cm^{-3}
R_0	external load resistor	1 Ω
T	temperature	300 K
V_b	breakdown voltage of the n$^+$-p collector	42 V
V_T	thermal voltage, $V_T = kT/e$	0.026 V
w	thickness of the n$^+$ base	0.0003 cm
W	thickness of the n$^-$ substrate	0.06 cm
β	transport factor of the n$^+$ base	0.6
ρ_L	leakage resistance of the n$^+$-p collector	40 000 Ω cm^2
σ_b	average conductivity of the n$^+$ base	10 (Ω cm)$^{-1}$
τ	effective lifetime of holes in the n$^-$ base	10 μs

For a complete listing of all device parameters see Table 8.1.

Equations (8.1) and (8.2) form a complete set of equations when the device is connected directly to the dc driving voltage V_s, i.e. $V = V_s$. However, in experiment the device is usually driven via a load resistor. Thus, a third equation results from the external electrical circuit:

$$V(t) = V_S - R_0 \ell_z \int_0^{\ell_x} j_c(V_e, V_i)\, dx. \tag{8.7}$$

$j_c(V_e, V_i)$ is the current density at the interface between the transistor and the diode and ℓ_x and ℓ_z denote the lengths of the device in the corresponding directions.

Equations (8.1) and (8.2) have to be completed by appropriate boundary conditions. We have studied the following cases:

- *Homogeneous Neumann boundary conditions* ensuring that there is no lateral current in the n$^+$ base and no lateral diffusion of holes at the sample boundaries:

$$\frac{\partial V_e(x=0,t)}{\partial x}=0, \quad \frac{\partial V_e(x=\ell_x,t)}{\partial x}=0, \tag{8.8}$$

$$\frac{\partial \bar{p}(x=0,t)}{\partial x}=0, \quad \frac{\partial \bar{p}(x=\ell_x,t)}{\partial x}=0. \tag{8.9}$$

- *Mixed boundary conditions*, i. e., Dirichlet boundary conditions for the emitter voltage V_e and homogeneous Neumann boundary conditions for the mean hole density $\bar{p}$:

$$V_e(x=0,t)=V_{e0}, \quad V_e(x=\ell_x,t)=V_{e0}, \tag{8.10}$$

$$\frac{\partial \bar{p}(x=0,t)}{\partial x}=0, \quad \frac{\partial \bar{p}(x=\ell_x,t)}{\partial x}=0. \tag{8.11}$$

  The Dirichlet condition for V_e may be motivated if the emitter voltage is limited or fixed, e. g., by surface recombination.

8.5.2 Activation, Inhibition, Competition

Characteristic properties of the variables V_e and $\bar{p}$. As can be seen by the following considerations, the evolution of the system variables V_e and $\bar{p}$ is based on a competition between both variables. When the device is driven with a bias voltage large enough to cause avalanche multiplication in the n^+-p junction, the generated electrons move to the p^+-n^+ emitter and cause there an additional injection of holes leading to an autocatalytic increase of V_e. This results also in an increase of the emitter current-density j_e that depends on V_e according to

$$j_e(x,t)=C_e\frac{\partial V_e}{\partial t}+j_s\left[exp\left(\frac{V_e}{V_T}\right)-1\right]+j_r\left[exp\left(\frac{V_e}{2V_T}\right)-1\right]. \tag{8.12}$$

For this activating property of the emitter voltage, we call V_e, and loosely, j_e and the collector current-density $j_c=j_e-w\sigma_b\partial^2V_e/\partial x^2$, an activator. An increase of V_e, and consequently an increase of j_e and j_c, causes an injection of holes into the n^- layer which leads to an increase of the voltage drop V_p across the diode part according to (8.3). Therewith, the transistor voltage $V_i=V-V_p$ declines, if the device voltage V is kept constant. That means that the variable $\bar{p}$, or equivalently V_p, counteracts the autocatalytic process in the p^+-n^+-p transistor. The voltage V_p or V_i may therefore be called an inhibitor. As follows from (8.1) and (8.2), the evolution of both variables is determined by reaction and diffusion terms. With regard to localized structures that may develop in such systems, two parameter ratios are of particular importance: the ratio of the relaxation time constants and of the diffusion lengths of both system variables. Roughly speaking, one can expect static localized structures if the relaxation time constant of the inhibitor is smaller and the diffusion coefficient is larger than that of the activator. On the other hand, if the relaxation of the inhibitor is slow with respect to the activator the system tends to form dynamical structures, e. g., travelling structures. If two travelling structures encounter each other, it depends essentially on the ratio of the diffusion lengths of both variables whether a repelling or an annihilation of the structures occurs.

The inhibition caused by the passive diode part is a local inhibition in the sense that it damps out a current-density fluctuation only in a region the size of which is determined by the diffusion length L of the inhibitor. Beyond this local inhibition, the external load resistor effects a global inhibition as a current fluctuation at any position in the device causes an increase of the voltage drop across the load resistor and, consequently, a decrease of the device voltage. Such a global damping mechanism may have an important influence on the generation of spatiotemporal patterns [8.33,46–48].

The null-cline system. The equations $q(V_e, V_i(\bar{p})) = 0$ and $Q(V_e, V_i(\bar{p}); V) = 0$ define the null-cline system of (8.1) and (8.2) for spatially uniform distributions of V_e and $\bar{p}$. In Fig. 8.15, the null-cline system is shown for the case that the device is connected directly to the voltage source, i. e., without a load resistor R_0. From $q = 0$ we obtain a curve $V_i(V_e)$ reflecting the S-shaped current-voltage characteristic of the transistor part. The curves $Q_1 = 0$ and $Q_2 = 0$ belong to two different values of the bias-voltage V_s with $V_{s1} < V_{s2}$ and illustrate that a variation of V_s essentially leads to a vertical shift of the approximately linear decreasing curve defined by $Q = 0$ in the $V_i - V_e$ space.

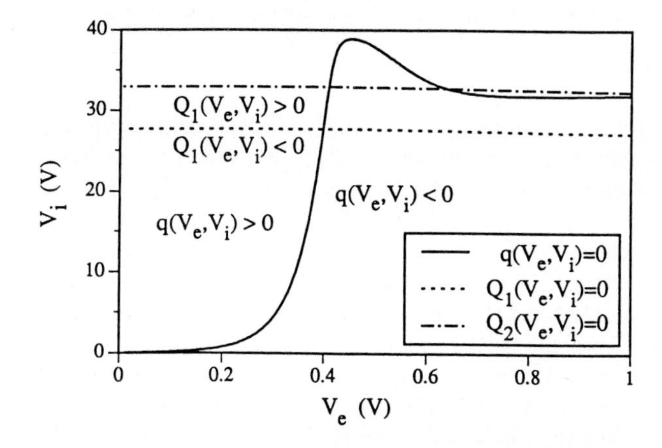

Fig. 8.15. Null-cline system of (8.1) and (8.2); parameters as in Table 8.1, but $R_0 = 0\,\Omega$ and $V_s = 28$ V and 33.2 V for Q_1 and Q_2 (from [8.34])

8.6 Numerical Results

In this section, results of numerical simulations of the set of equations (8.1), (8.2) and (8.7) are discussed and compared with experiments. In particular, the effects of different boundary conditions and, following the experiments, the influence of the dc driving voltage V_s on the evolution of patterns in the device are studied in detail. As the physical mechanisms that lead to the self-organization effects observed in our device can be understood as competing interplay of an autocatalytic and an inhibiting component, the main interest is focussed on the spatiotemporal behaviour of these variables. In order to allow an easy comparison with experiments, the current density j_c at the interface between the active and the passive part of the device and the voltage drop V_p across the passive diode part are chosen to represent the activator and the inhibitor distribution, respectively.

The determination of an appropriate parameter set (see Table 8.1), is based partly on direct measurements of device parameters partly on estimations from the stationary current-voltage characteristic.

8.6.1 Bifurcations to Static, Breathing, Travelling and Chaotically Oscillating Filaments by Using Homogeneous Neumann Boundary Conditions

The computations have been performed on a parallel computer consisting of nine processors of type T800. The finite-difference method has been used for discretization; in particular the Crank-Nicolson technique has been applied for discretization in time. Fluctuations of the system variables are simulated by adding a pseudorandom value of about 0.01 % of the corresponding distributions V_e and $\bar{p}$ in each time step.

Transition of a spatially uniform state to a static localized filament. It is known from the experiments that the current-density and potential distributions are uniform for low driving voltages V_S. Therefore, uniform distributions for V_e and $\bar{p}$ serve as initial distributions and for a given value of V_s, a steady state solution of (8.1), (8.2), and (8.7) is calculated. In the next step, the calculated steady state distributions of V_e and $\bar{p}$ are used as initial distributions, V_s is increased, and the appertaining steady state solution is computed. By repeating this procedure we obtain a current-voltage characteristic $I(V)$ of the device as shown in Fig. 8.16a. Similar to the experiments, the characteristic contains a high-resistance branch, where the current is limited by the reverse-biased n^+-p junction, a branch with a negative differential resistance, and a low-resistance branch for large total currents. In the high-resistance branch, the calculated distributions of V_e and $\bar{p}$, and therefore those of the current densities j_e and j_c and the diode voltage V_p, too, are uniform. Even in the range of negative differential resistance, uniform steady state solutions develop up to total currents of about 0.5 mA, indicated as operating point A in Fig. 8.16a. The distributions j_c and V_p belonging to this operating point are drawn in the uppermost diagram of Fig. 8.16b. For a slightly larger total current, the distribution of j_c becomes non-uniform whereas that of V_p remains still uniform (Fig. 8.16b, B). However, proceeding along the $I(V)$ characteristic to larger current values, we observe that the contraction of the current-density distribution is accompanied by a corresponding deformation of the potential distribution (Fig. 8.16b, C and D). At the operating point E, the contraction phase is completed and the width of both distributions reach a minimum value.

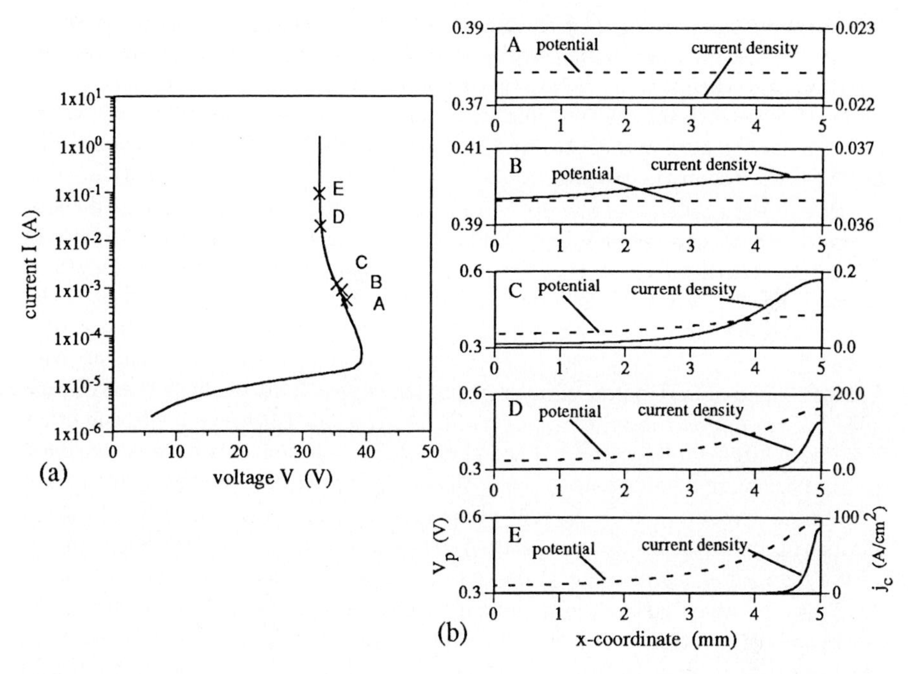

Fig. 8.16. (a) Numerically calculated $I(V)$ characteristic and (b) stationary spatial distributions of V_p and j_c for different values of the driving voltage V_s; the different operating points are labelled by *A* - *E*; parameters as in Table 8.1 but $R_0 = 10\,\mathrm{k}\Omega$, and $V_s = 40.826$ V *(A)*, 45.2 V *(B)*, 47.61 V *(C)*, 229.4 V *(D)*, 952.4 V *(E)* (from [8.34])

The values of the distributions j_c and V_p far away from the centre of the developed structure correspond to solutions of the stationary homogeneous equations, i.e., they solve the equations $q(V_e, V_i) = 0$ and $Q(V_e, V_i; V_s) = 0$; therefore, the generated filament is a localized structure. This is an important difference to nonlocalized filaments that may develop, e.g., in conventional thyristors or many other devices (e.g., [8.4]). The existence of nonlocalized filaments is connected with a positive feedback mechanism. The transistor part of our model, taken as isolated device, would be able to form such a nonlocalized structure. The width of a nonlocalized filament is determined by the length over which the autocatalytically increasing variable, the current density, changes in space. For the evolution of a localized filament in our device, the existence of the p-n$^-$ diode is of special importance; the basic physical mechanism may be illustrated by the following qualitative considerations: Holes are injected from the transistor into the n$^-$ layer. As the injection is most effective in the centre of the filament there is a significant diffusion of holes transversal to the direction of main current flow causing a broadening of the hole distribution in the n$^-$ layer. The voltage V_p across the p-n$^-$ diode increases with increasing hole concentration; therefore, the broadening of the hole distribution is accompanied by a corresponding broadening of the distributions $V_p(x,t)$ and $V_i(x,t) = V(t) - V_p(x,t)$. The width of these distributions is determined by the effective diffusion length L of the holes in the n^- region. By this, in a sufficient distance from the centre of the filament, the distributions reach a value corresponding to the homogeneous state. This would be impossible without the p-n^- diode because in this case the voltage V_i would be fixed along the x-coordinate and its value would be determined exclusively by the processes in the transistor.

Due to the existence of the p-n^- diode, a local increase of the current density leads to a local increase of $V_p(x,t)$ and a corresponding local decrease of $V_i(x,t)$ impeding a further increase of the current density. If the diffusion length L of the inhibitor, the voltage V_p or V_i, is larger than that of the activator, the current-density, the following two consequences are important: At the centre of the filament, V_i declines stronger in comparison with the homogeneous case because there is a netto flow of holes out of the centre. This weakens the damping effect and supports the generation of a high-current-density region. On the other side, in the immediate vicinity of the filament the hole concentration is increased in comparison to a homogeneous case; this supports the damping mechanism and prevents a spreading of the current filament. Thus, the existence of the localized filament proceeds from the competition between the autocatalytic process in one of the device parts and the negative feedback in the other part.

From the distributions shown in Fig. 8.16 it is obvious that the width of the potential distribution is larger than that of the current density. It turns out, that the widths differ by approximately a factor five which was observed also in experiment (cf. Sect. 8.3.2).

Bifurcation of a static to a breathing filament. With increasing total current the width of the filament increases slightly. When a critical width is reached, corresponding to critical values of I and V_s, spontaneous oscillations of the device voltage appear indicating an instability of the static filament. A typical time trace of $V(t)$ is shown in Fig. 8.17a. The oscillation is accompanied by a periodic oscillation of the filament. In Fig. 8.18a, the temporal evolution of the current density j_c belonging to the voltage trace in Fig. 8.17a is shown. Dark areas mark regions where the current density exceeds a certain value, approximately $\frac{2}{3}$ of the maximum current density. Thus, these regions correspond to high-current regions and indicate the position of the filament. From Fig. 8.18a it is obvious that a well-defined oscillation of the filament wall has developed. The static filament has transformed into a breathing one, the width of which widens and narrows periodically. There is a strong correlation between local filament dynamic and global voltage oscillation: The period of both oscillations is the same and, furthermore, a maximum of the filament width means that the total current flowing through the device has a maximum and, consequently, the device voltage has a minimum; accordingly, a maximum device voltage corresponds to a filament state with minimum width. An evaluation of the dependence of the ac part of the device voltage $V(t)$ and the width of the filament on the bifurcation parameter V_s reveals that the square of both quantities increases linearly with V_s. Further, the oscillations set in with a well-defined frequency of about 30 kHz that declines linearly with increasing V_s. Thus, the transition shows typical fea-

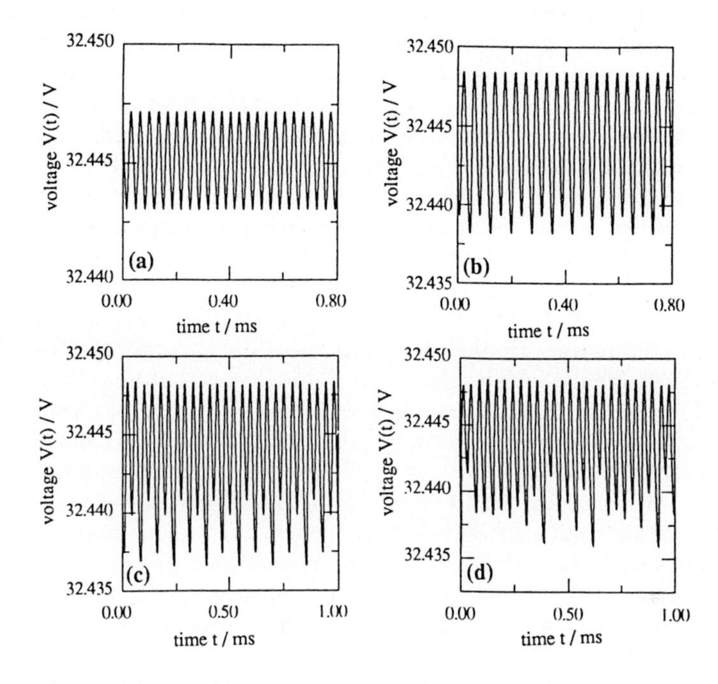

Fig. 8.17. Numerically calculated time traces $V(t)$ for different values of V_s; parameters as in Table 8.1 and $V_s = 33.8900$ V (a), 33.9450 V (b), 33.9495 V (c), 33.9505 V (d)

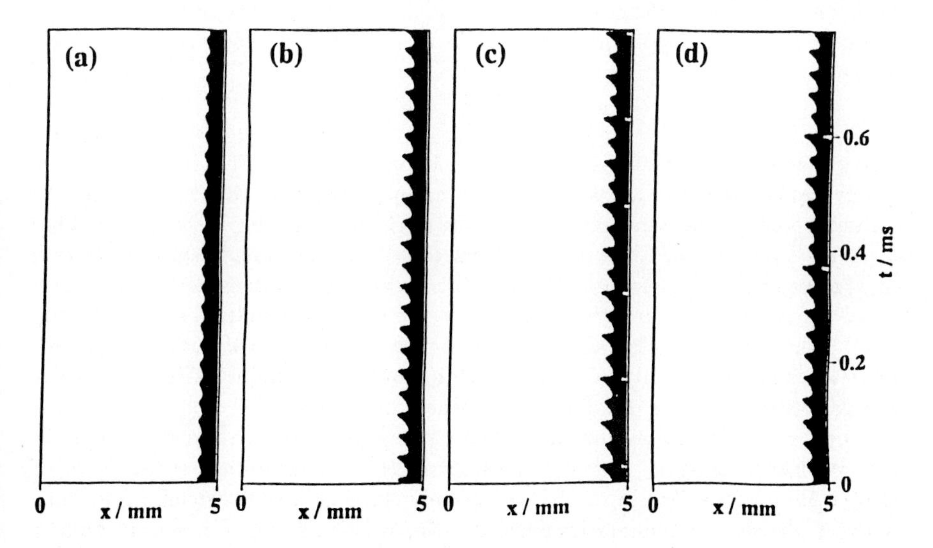

Fig. 8.18. Numerically calculated spatiotemporal evolution of $j_c(x,t)$ for different values of V_s corresponding to the time traces $V(t)$ shown in Fig. 8.17

tures of a supercritical Hopf-bifurcation.

A detailed analysis of the spatial distributions j_c and V_p at different moments during a period of the oscillation reveals that the filament motion is more complex than can be seen from the threshold diagram in Fig. 8.18. The motion is composed of a breathing and an amplitude oscillation. Both oscillations are in anti-phase, i.e., when a maximum width is reached, the amplitude is small and vice versa. This is the reason for the small oscillation of the total current and the device voltage in spite of the relatively large width modulation of the filament.

Period-doubling cascade and chaotic filament motions. A small voltage interval where period doubling routes to chaotic oscillations can be observed adjoin the range of simple periodic oscillations for larger values of V_s. Fig. 8.17b-d displays period-two, period-four and chaotic oscillations of the device voltage $V(t)$ that appear successively with increasing V_s. Fig. 8.18b-d shows the appertaining spatiotemporal evolutions of j_c. Obviously, the period doublings in the global voltage trace are accompanied by period doublings of the wall oscillations. A careful analysis of Fig. 8.18b reveals that each second wall elongation to the left is larger than that in between. The period-four oscillation in Fig. 8.18c is more pronounced: The maximum elongation to the left can be identified clearly by the small detachment of the filament from the boundary repeating in every fourth of the elongations. The filament motion shown in Fig. 8.18d has been followed for some hundreds of fundamental periods. For this period no periodicity was observed. An evaluation of about ten nonperiodic time series of $V(t)$ supplies attractor dimensions between 1.87 and 1.94. However, there is still a correlation between the global voltage signal and the local filament oscillation. For example, each local maxima in the voltage trace corresponds to a local maximum wall elongation.

Bifurcation of a breathing to a travelling filament. The different modes of breathing filaments are stable in a current interval of about some tens of milliamperes. If the device voltage V_s is large enough, the breathing oscillations are superimposed by travelling motions of the filament. As demonstrated in Fig. 8.19a the filament oscillates in the breathing mode near the boundary for a certain period. During this oscillation the width modulation increases and, finally, the filament detaches from the boundary and travels with an approximately constant velocity of about 60 m/s along the system. When it reaches the opposite boundary, the filament is reattached there, and oscillates another few periods in the breathing mode before it travels back to the right boundary. With increasing V_s, the durations of breathing oscillations become smaller (Fig. 8.19b,c) and, finally, vanish completely, so that a pure travelling motion between the sample boundaries develops (Fig. 8.19d). As the filament motion is composed of breathing and travelling motions, the voltage trace $V(t)$ contains characteristic oscillations caused by the breathing and the travelling modes. For details see [8.34].

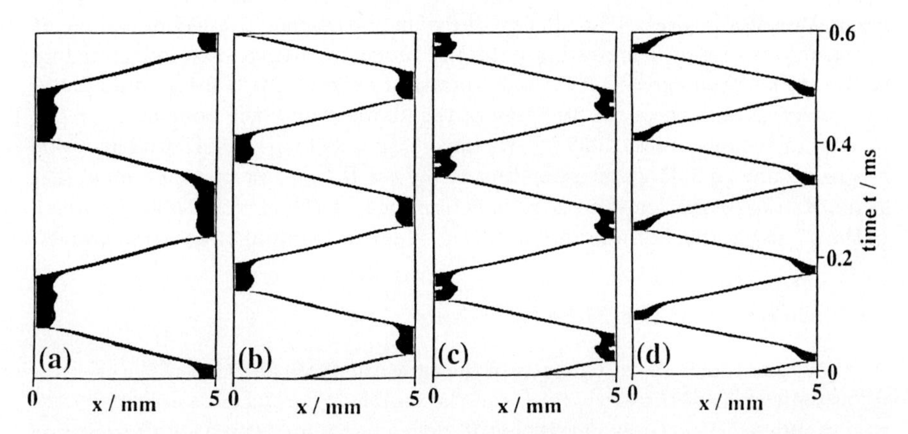

Fig. 8.19. Numerically calculated spatiotemporal evolution of $j_c(x,t)$ for different values of $V_s = 33.89$ V (a), 34.1 V (b), 34.2 V (c), 34.3 V (d), other parameters as in Table 8.1

The spatial distributions of j_c and V_p for different moments during the travelling motion are displayed in Fig. 8.20. The uppermost diagram shows both distributions just before the filament starts to detach from the boundary. Note, that both distributions far away from the filament centre approximate values corresponding to the stationary homogeneous case. In the second diagram, the filament has completely detached from the right boundary and travels towards the left one. In the potential distribution, the precursor and the pronounced tail can clearly be made out confirming the interpretation given for the exper-

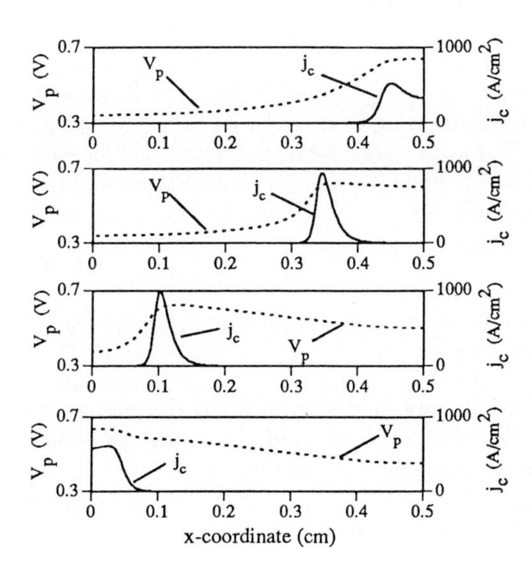

Fig. 8.20. Numerically calculated spatial distributions of j_c and V_p at four different moments during the travelling motion recorded in time intervals of 20 μs; parameters as in Table 8.1, $V_s = 33.964$ V (from [8.34])

imental results in Sect. 8.3.5: Before the filament reaches the left boundary, it interacts with its own inhibiting potential precursor being mirrored according to the applied homogeneous Neumann boundary conditions. On the other hand, the potential tail causes the trapping of the filament near the boundary. As can be seen in the lowermost diagram, we have $V_p(x = 0.4\text{ cm}) = 0.47$ V. This value has to decline to 0.35 V corresponding to $V_p(x = 0.1\text{ cm})$ in the uppermost diagram. As the relaxation time of V_p is determined by the effective hole lifetime τ in the n^- layer, this lifetime dominates the filament trapping near the boundary.

8.6.2 Influence of the Time Constants

As already outlined above, the ratio of the time constants and that of the effective diffusion lengths of the activator ℓ and the inhibitor L determine essentially the spatiotemporal structures developing in nonequilibrium systems with evolution equations of the form (8.1), (8.2). For a constant ratio of the diffusion lengths with $\ell < L$, the structures expected to appear may be classified in a very rough and simplified manner as follows:

- travelling structures, if the time constant of the inhibitor is much larger than that of the activator,
- static, switching, breathing, rocking or travelling filaments, if both time constants are of the same order,
- static structures, if the time constant of the activator is much larger than that of the inhibitor.

Increasing the time constant of the activator by increasing the emitter capacity C_e, indeed leads to a stabilization of static filaments. This is confirmed by the following numerical results: For the parameter set given in Table 8.1 and V_s chosen such that a static filament is stable near the boundary, we choose a filament located in the centre of the system as initial condition. It turns out that this filament moves to one of the boundaries where it becomes stationary. The velocity depends on the ratio of the time constants of activator and inhibitor and can be slowed down by increasing C_e. For $C_e = 10^{-2}\,\text{F/cm}^{-2}$ it is possible to stabilize the filament at any position in the system. If C_e is sufficiently large, it is also possible to stabilize two filaments in the system, because in this case the activator fluctuations located in the filament wall and causing instabilities of a static filament arise very slowly and can be damped out by the inhibitor so that the evolution of dynamical structures is prevented.

8.6.3 Bifurcations to Static, Rocking and Travelling Filaments by Using Mixed Boundary Conditions

The Dirichlet boundary condition for V_e may support or prevent the generation of a filament near the boundary. It depends on the values of $V_e(x = 0, \ell_x)$ whether the boundary exerts a repulsion or an attraction on the filament. With $V_{e,h}$ defined by the solution of $q(V_{e,h}, V_{i,h}) = 0$ and $Q(V_{e,h}, V_{i,h}, V) = 0$, we distinguish the following two cases: If $V_e(x = 0, \ell_x) < V_{e,h}$ the effect of the boundary on a filament is exclusively repulsive because in this case large values of V_e near the boundary are excluded and, consequently, no high current-density region can develop there. As a result, a filament can only be stabilized at the centre of the system. This is confirmed by numerical calculations. In the following passage, we restrict ourselves to the case with $V_e(x = 0, \ell_x) > V_{e,h}$ and the additional condition that $V_e(x = 0, \ell_x)$ is not too large, i.e. smaller than the expected maximum value of the current density in the filament. On these conditions, the V_e-distribution near the boundary cannot decline to any low values; therefore, the boundary has an attractive influence on the filament. On the other hand, a filament suffers a repulsion if the distance to the boundary is too small for the same reason as in the case $V_e(x = 0, \ell_x) < V_{e,h}$. Thus, we can expect that a filament is stabilized at a well-defined distance from the boundary so that the repulsive and attractive influence of the boundary compensate each other.

Evolution of a static localized filament. The $I(V)$ characteristic calculated with mixed boundary conditions has the same shape as in the case of Neumann boundary conditions. In the high-resistance branch, the distributions of j_c and V_p far away from the boundaries are spatially uniform for sufficiently long systems. When the driving voltage is large enough to induce the autocatalytic charge carrier generation, a contraction of the current-density distribution resulting in a filament state occurs. In Fig. 8.21, the stationary spatial distributions of j_c and V_p are shown for an operating point chosen in the low-resistance branch at a total current I of 400 mA. As expected, the filament has developed at a certain distance from the boundary. By decreasing the value of $V_e(x = 0, \ell_x)$ the filament position moves towards the centre of the system.

(a)

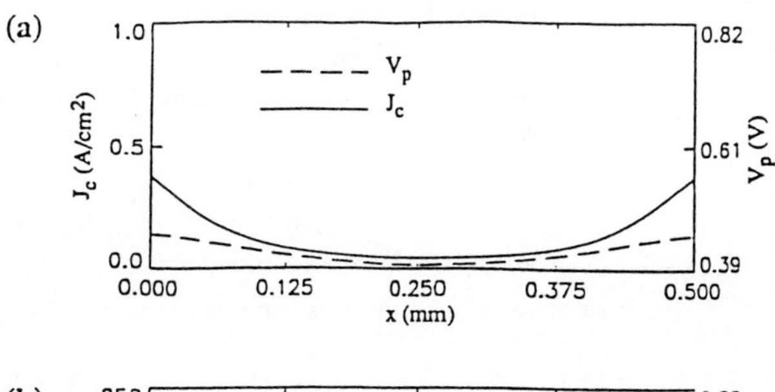

(b)

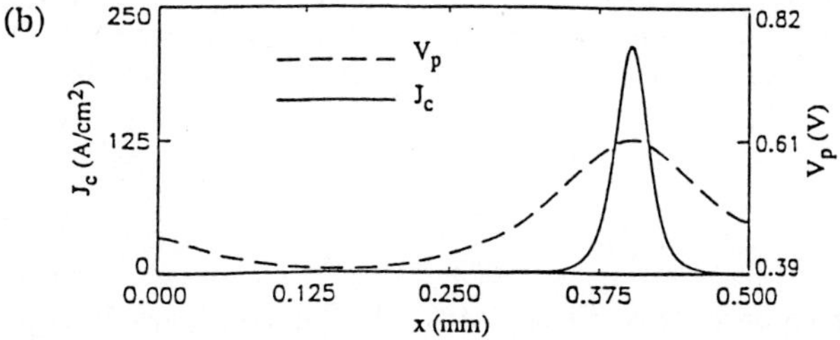

Fig. 8.21. Numerically calculated stationary spatial distributions of j_c and V_p; parameters as in Table 8.1 and $V_s = 32.85$ V, $V_e(x = 0, \ell_x) = 0.63$ V (from [8.34])

Bifurcation of a static to a rocking filament. When a stationary filament is elongated from its equilibrium position, it moves back and performs a damped oscillation around the stationary position as shown in the threshold diagram Fig. 8.22a. The oscillation arises from the delayed response of the inhibitor with respect to the activator. With increasing driving voltage V_s the damping becomes smaller and at a critical value of V_s self-sustained rocking oscillations of the current filament appear (Fig. 8.22b). They are accompanied by oscillations of the device voltage V and the total current I the frequency of which is the same as that of the rocking oscillation. Because the shape of the filament remains nearly constant during the rocking motion, the amplitude of the trace of the total current is less than a percent of the dc part. It turns out that the minima in the current trace are caused by a small filament narrowing when it changes its moving direction. A detailed analysis of the spatial distribution of j_c and V_p representing the activator and the inhibitor, respectively, reveals that the maximum of the V_p distribution catches that of the j_c distribution near the turning points; by this the V_p distribution slows down the motion of the filament and forces it to change its direction. It is worth to note that the spatial amplitude of the rocking motion increases with the square-root of the bifurcation parameter V_s; this is in good agreement with the cxperimental observations.

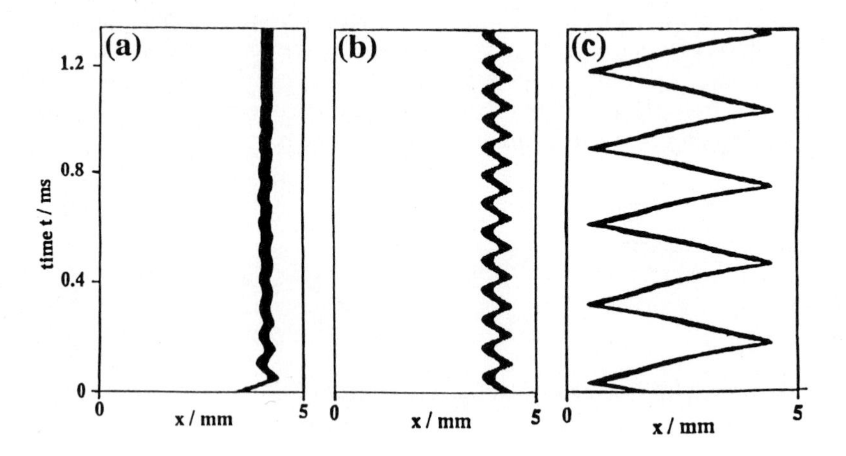

Fig. 8.22. Numerically calculated spatiotemporal evolution of $j_c(x,t)$: (a) transient of a disturbed stationary filament with $V_s = 32.85$ V, (b) rocking filament with $V_s = 32.90$ V, (c) travelling filament with $V_s = 33.05$ V; other parameters as in Table 8.1 and $V_e(x = 0, \ell_x) = 0.63$ V

Bifurcation of a rocking to a travelling filament. For a sufficiently large value of V_s the amplitude of the rocking motion reaches a critical value at which the moving filament is able to surmount the attracting influence of the boundary. This results in a transition from a rocking to a travelling filament. In Fig. 8.22c such a travelling motion is shown. The velocity of the filament in the centre of the system is approximately 30 m/s. When the filament approaches the opposite boundary, the Dirichlet boundary condition for V_e on the one hand impedes the current density to rise near the boundary and, on the other hand promotes an increase of the inhibiting voltage V_p, both leading to the repelling of the filament. The reflection events are accompanied by slight deformations of the filament shape leading to characteristic minima in the time trace of the total current.

8.7 Summary and Conclusions

A survey of stationary and dynamical localized structures developing in an active dissipative medium, a silicon p-n-p-n diode, has been given. The experimental investigations reveal that a sequence of bifurcations connected with transitions between different spatiotemporal current-density patterns occur by varying one single external parameter, the dc driving voltage V_s. The complexity of the observed spatiotemporal patterns increases with increasing V_s. Spatially resolved measurements show that self-generated oscillations of the device voltage indicate the appearance of rocking and/or travelling current-density filaments. Furthermore, the experimental results give evidence for breathing or amplitude oscillations, which superimpose the rocking or travelling motion when the filament reaches the boundary of the device. Using different measurement techniques, we are able to investigate the spatiotemporal behaviour of important device variables the mutual influencing of which is essential for the developing patterns. By measuring the electroluminescence radiation we obtain information about the activating component increasing itself in a certain parameter range. The potential drop across the passive part of the device is related to the inhibiting component that counteracts the activator.

The competition between the activating process (avalanche multiplication) and the inhibiting process (damping due to the p-n^- diode) and the ratios of the characteristic lengths and times of both components determine whether static, rocking, or a travelling current-density filaments develop. This has been illustrated in detail for the case of travelling filaments, where spatially and temporally resolved measurements of both the electroluminescence radiation and the voltage drop across the passive n^- layer have been performed. They illucidate how the delayed response of the longe-range inhibitor with respect to the activator governs the travelling motion, the reflection process and the capture of the filament near the boundary.

The simplified two-layer model based on dividing the device into an active and a passive part leads to a two-component set of evolution equations containing diffusion and reaction terms. Numerical calculations qualitatively reproduce the whole bifurcation scenario observed in experiment. Even period-doublings and chaotic dynamics of filament motions have been found. In particular, the numerical results suggest that the filament dynamic in the chaotic case is controlled by the following modes, representing three possible degrees of freedom for a simplified description of the filament dynamic: amplitude variations, width variations, and variations of the position of the filament centre. Furthermore, we have shown that the applied boundary conditions effect some particular changes of the filament motion; e.g., the rocking motion is preferred when Dirichlet boundary conditions are applied for the emitter voltage. However, the main bifurcation sequence *homogeneous current-density distribution → static filament → oscillating filament with small spatial amplitude → travelling filament* appears independently of the applied boundary conditions.

Acknowledgments. I thank R. Dohmen, A.V. Gorbatyuk, B.S. Kerner, B. Kukuk, M. Or-Guil, and H.-G. Purwins for fruitful discussions, C. Brillert for carrying out the investigations of the ac driven system, and S. Kottemer for his technical assistance. Parts of this work have been supported by the *Deutsche Forschungsgemeinschaft.*

References

[8.1] A.M. Barnett: In *Semiconductors and Semimetals*, Vol. 6, ed. by R.K. Willardson, A.C. Beer (Academic Press, New York 1970) pp. 141 - 200
[8.2] V.L. Bonch-Bruevich, I. Zvyagin, A.G. Mironov: *Domain Electrical Instabilities in Semiconductors* (Consultant Bureau, New York 1975)
[8.3] J. Pozhela: *Plasma and Current Instabilities in Semiconductors* (Pergamon Press, Oxford 1981)
[8.4] E. Schöll: *Nonequilibrium Phase Transitions in Semiconductors* (Springer, Berlin, Heidelberg 1987)
[8.5] Y. Abe (ed.): *Nonlinear and Chaotic Transport in Semiconductors*, Appl. Phys. A, Vol. 48 (Springer, Heidelberg 1989)
[8.6] M.P. Shaw, V.V. Mitin, E. Schöll, H.L. Grubin: *The Physics of Instabilities in Solid State Electron Devices* (Plenum Press, New York 1992)
[8.7] N. Balkan, B.K. Ridley, A.J. Vickers (eds.): *Negative Differential Resistance and Instabilities in 2-D Semiconductors* (Plenum Publishing, New York 1993)
[8.8] B.S. Kerner, V.V. Osipov: *Autosolitons* (Kluwer, Dordrecht 1994)
[8.9] D. Jäger, H. Baumann, R. Symanczyk: Phys. Lett. A **117**, 141 (1986)
[8.10] F.-J. Niedernostheide, M. Arps, R. Dohmen, H. Willebrand, H.-G. Purwins: Phys. Stat. Sol. (b), **172** (1992)
[8.11] F.-J. Niedernostheide, R. Dohmen, H. Willebrand, H.-J. Schulze, H.-G. Purwins: In *Nonlinearity with Disorder*, ed. by F. Abdullaev, A.R. Bishop, S. Pnevmatikos, Springer Proceedings in Physics Vol. 67 (Springer, Berlin, Heidelberg 1992) pp. 282 - 309
[8.12] V.A. Vashchenko, B.S. Kerner, V.V. Osipov, V.F. Sinkevich: Fiz. Tekh. Poluprovodn. **24**, 1705 (1990) [Engl. transl.: Sov. Phys. Semicond. **24**, 1065 (1990)]
[8.13] V.V. Gafiichuk, B.I. Datsko, B.S. Kerner, V.V. Osipov: Fiz. Tekh. Poluprovodn. **24**, 724 (1990) [Engl. transl.: Sov. Phys. Semicond. **24**, 455 (1990)]
[8.14] A.M. Barnett, A.G. Milnes: J. Appl. Phys. **37**, 4215 (1966)
[8.15] B.S. Kerner, B.P. Litvin, V.I. Sankin: Pis'ma Zh. Tekh. Fiz. **13**, 819 (1987) [Engl. Transl.: Sov. Tech. Phys. Lett. **13**, 342 (1987)]
[8.16] Y.A. Astrov: Fiz. Tekh. Poluprovodn. **27**, 1973 (1993) [Engl. transl.: Semiconductors **27**, 1084 (1993)]
[8.17] Y.A. Astrov, S.A. Khorev: Fiz. Tekh. Poluprovodn. **27**, 2027 (1993) [Engl. transl.: Semiconductors **27**, 1113 (1993)]

[8.18] R.M. Scarlett, W. Shockley: IEEE Int. Conv. Rec. **3**, 3 (1963)
[8.19] B.S. Kerner, V.V. Osipov: Mikroélektronika **6**, 337 (1977) [Engl. transl.: Soviet Microelectronics **6**, 256 (1977)]
[8.20] A.V. Gorbatyuk, I.A. Liniichuk, A.V. Svirin: Pis'ma Zh. Tekh. Fiz. **15**, 42 (1989) [Engl. transl.: Sov. Tech. Phys. Lett. **15**, 224 (1989)]
[8.21] A.V. Gorbatyuk, P.B. Rodin: Pis'ma Zh. Tekh. Fiz. **16**, 89 (1990) [Engl. transl.: Sov. Tech. Phys. Lett. **16**, 519 (1990)]
[8.22] A.V. Gorbatyuk, P.B. Rodin: *Proc. 6th Int'l Symposium on Power Semiconductor Devices & IC's* (Davos 1994) paper 6.2
[8.23] A. Turing: Philos. Trans. R. Soc. **237**, 37 (1952)
[8.24] R. Symanczyk, S. Gaelings, D. Jäger: Phys. Lett. A **160**, 397 (1991)
[8.25] R. Symanczyk: this volume, chapter 9
[8.26] A. Wacker, E. Schöll: Semicond. Sci. Technol. **7**, 1456 (1992)
[8.27] A. Wacker, E. Schöll: Z. Phys. B **93**, 431 (1994)
[8.28] S. Bose, A. Wacker, E. Schöll: Phys. Lett. A (1994) in print
[8.29] E. Schöll, A. Wacker: this volume, chapter 2
[8.30] U. Rau, W. Clauß, A. Kittel, M. Lehr, M. Bayerbach, J. Parisi, J. Peinke, R.P. Hübener: Phys. Rev. B **43**, 2255 (1991)
[8.31] A. Brandl, W. Prettl: In *Festkörperprobleme (Advances in Solid State Physics)*, Vol. 30, ed. by U. Rössler (Vieweg, Braunschweig 1990) pp. 371-385
[8.32] K. Penner: Journal de Physique Colloque **C4**, 797 (1988)
[8.33] F.-J. Niedernostheide, B.S. Kerner, H.-G. Purwins: Phys. Rev. B **46**, 7559 (1992)
[8.34] F.-J. Niedernostheide, M. Ardes, M. Or-Guil, H.-G. Purwins: Phys. Rev. B **49**, 7379 (1994)
[8.35] A. Wierschem, F.-J. Niedernostheide, A. Gorbatyuk, H.-G. Purwins: Scanning **17** (1) (1995)
[8.36] F.-J. Niedernostheide, M. Kreimer, H.-J. Schulze, H.-G. Purwins: Phys. Lett. A **180**, 113 (1993)
[8.37] F.-J. Niedernostheide, M. Kreimer, B. Kukuk, H.-J. Schulze, H.-G. Purwins: Phys. Lett. A **191**, 285 (1994)
[8.38] M.H. Jensen, P. Bak, T. Bohr: Phys. Rev. A **30**, 1960 (1984)
[8.39] T. Bohr, P. Bak, M.H. Jensen: Phys. Rev. A **30**, 1970 (1984)
[8.40] E.G. Gwinn, R.M. Westervelt: Phys. Rev. Lett. **57**, 1060 (1986)
[8.41] R.M. Westervelt, S.W. Teitsworth: Physica **23D**, 186 (1986)
[8.42] J. Peinke, J. Parisi, O.E. Rössler, R. Stoop: *Encounter with Chaos* (Springer, Berlin, Heidelberg 1992)
[8.43] A.M. Kahn, D.J. Mar, R.M. Westervelt: Phys. Rev. Lett. **68**, 369 (1992)
[8.44] A.M. Kahn, D.J. Mar, R.M. Westervelt: Phys. Rev. B **46**, 7469 (1992)
[8.45] C. Brillert, F.-J. Niedernostheide, H.-G. Purwins: unpublished (1994)
[8.46] U. Middya, M.D. Graham, D. Luss, M. Sheintuch: J. Chem. Phys. **98**, 2823 (1993)
[8.47] U. Middya, M. Sheintuch, M.D. Graham, D. Luss: Physica D **63**, 393 (1993)
[8.48] F.-J. Niedernostheide, R. Dohmen, H. Willebrand, B.S. Kerner, H.-G. Purwins: Physica D **69**, 425 (1993)

9 Current Filamentation in P-I-N Diodes: Experimental Observations and an Equivalent Circuit Model

R. Symanczyk

Former address:
Gerhard-Mercator-Universität, Fachgebiet Optoelektronik,
Kommandantenstr. 60, D-47057 Duisburg, Fed. Rep. Germany

Present address:
Passauer Str. 48, D-47249 Duisburg, Fed. Rep. Germany

Abstract. Pattern formation and nonlinear dynamics appearing in the way of current filamentation are investigated. The results of extensive experimental observations of stationary structures and spatio-temporal instabilities in Au doped Si, and Cr doped GaAs p-i-n diodes at room temperature are presented. It is demonstrated that a model based on an equivalent circuit can describe the occurrence of current filaments and some of their dynamic properties. A comparison between theoretical predictions and experimental data is carried out showing excellent agreement.

9.1 Introduction

Semiconductor devices are complex dynamic systems which can show a variety of electrical instabilities like switching between different states of conductivity or self-generated oscillations of the current and voltage. For that purpose the device has to be biased in a region of strong nonlinear behaviour which can be achieved technically by high injection rates of charge carriers, large electric fields or intense optical illuminations. In spite of different device structures and operating conditions the observed phenomena often include the formation of spatio-temporal patterns of the current or the electric field in an otherwise homogeneous material [9.1,2].

In recent years there is a growing interest in nonlinear dynamics in semiconductor devices as indicated by a large number of publications, see for example [9.3–6] and references in other parts of this book. Different reasons contribute to this fact: Effects of pattern formation, temporal instabilities and chaotic motions have been observed in nearly all fields of nature [9.7,8]. With improvements in sample preparation and measuring techniques the semiconductors can now act as cheap and easy to prepare model systems for such studies. From the technical point of view, a lot of new devices have been developed making use of strong nonlinear electric characteristics, for example Gunn-diodes, IMPATT-diodes, thyristors, or switches [9.9,10]. The occurrence of pattern formation can be desired or not. The Gunn-diode utilizes the propagation of electric field domains for the generation of microwave power, whereas the other devices need a homogeneous current flow for proper operation. The formation of current density filaments in GTO-thyristors lead to a time delay and possibly irreversible breakdowns [9.11].

Springer Proceedings in Physics, Vol. 79 **Nonlinear Dynamics and Pattern Formation in Semiconductors and Devices**
Editor: F.-J. Niedernostheide

The physical origin of electrical instabilities in semiconductors is traced back to a negative differential conductivity (NDC) of the material [9.12, 13]. Phenomenologically, one can distinguish between two types of current density (j) vs. electric field (E) characteristics. The mechanisms in connection with N-type NDC have been investigated thoroughly in the past. In this field the occurrence of electric field domains is experimentally observed and has led to applications like the Gunn-diode. On the other hand, devices with S-shaped NDC can show a pattern formation perpendicular to the current flow as sketched in Fig. 9.1. Values of j on the descending branch are unstable and a high current density

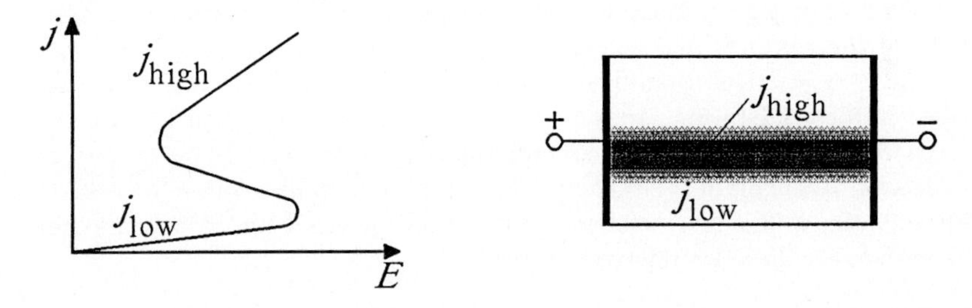

Fig. 9.1. Semiconductor with S-type NDC and sketch of a current density filament

filament is formed embedded in surrounding material with low values of the current density. Although great progress was made in recent years, we are still lacking a complete understanding of these phenomena. Some unanswered questions concern the exact shape of a filament, the effects in the case of temporal instabilities or a theoretical description which can explain all observations.

This contribution exclusively deals with pattern formation and nonlinear dynamics due to current filamentation in p-i-n diodes at room temperature. These devices were choosen for the studies because they exhibit a distinct S-shaped NDC when forward biased. Furthermore, the physical mechanism explaining the electrical character of a p-i-n diode is well known and the fabrication technique uses only conventional semiconductor technology. The devices treated here are rectangular, long and thin diodes with a depth comparable to the contact distance. The material is either silicon or galliumarsenide doped with gold or chromium, respectively. Thus, the high resistivity of the 'i'-layer results from deep trap levels in the band-gap.

The outline of this contribution is as follows. Section 9.2 covers a theoretical treatment which is subdivided into two parts. First, a model explaining the S-shaped NDC of the material is given and numerical calculations are carried out. These results lead to an equivalent circuit model for the extended p-i-n diode presented in the second part. The analysis yields filamentary current density distributions as possible solutions of the derived differential equations. The Sects. 9.3 and 9.4 deal with experimental observations of stationary filaments, and instabilities up to chaotic behaviour, respectively. The technology of the samples and the measuring techniques are briefly surveyed at the beginning. In both sections the data are compared with theoretical predictions. A conclusion following up summarizes the recent research concerning filamentation in p-i-n diodes.

9.2 Theoretical Description

Generally, an NDC of the material is regarded as one precondition for the occurrence of self-generated pattern formation and nonlinear dynamics in semiconductor devices. For that reason a first step in the theoretical description of current filamentation should deal with the j–E characteristic of the homogeneous material. In the case of a p-i-n diode with deep traps the double injection of charge carriers from the contacts and the electric field dependent trapping is responsible for the S-type NDC. Significant electrical properties following from this view taken into consideration, a model for an extended device with transversal inhomogeneities is presented in the second part.

This section will explain the theoretical principles in an illustrative way using only few equations. Readers interested in a more detailed and mathematical analysis with exact data are referred to [9.14–16].

9.2.1 Double Injection with Deep Traps

As mentioned above, the 'i'-layers of the p-i-n diodes treated here are made by compensation. The original semiconductor wafer contains a number of shallow impurities, which are ionized at room temperature and hence produce free charge carriers. By doping with Au in case of Si, and Cr in case of GaAs the free carriers get trapped in levels close to the middle of the band-gap. This results in high values of the resistivity at low values of the applied electric field. Figure 9.2 illustrates the situation by means of the band-structure. In order to be specific, we chose shallow donors as the original impurities led to some negatively charged traps.

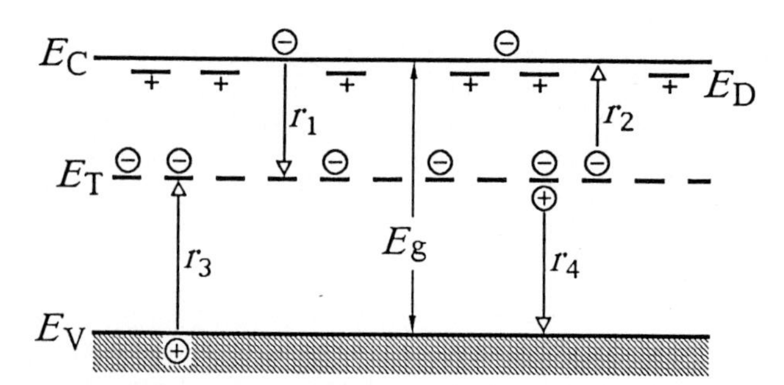

Fig. 9.2. Band-structure of a semiconductor with shallow donors (E_D) and deep traps (E_T). Interactions between the latter and conduction (E_C) and valence band (E_V) are indicated as $r_1 \ldots r_4$

Provided with suitable contacts one gets a p-i-n structure where both types of charge carriers, electrons and holes, can be injected into the 'i'-layer when the device is forward biased. Interaction between deep traps and the conduction band occurs via the capture and reemission of electrons (r_1 and r_2, respectively) as well as the interaction between valence band and negatively charged deep traps via the capture and reemission of holes (r_3 and r_4, respectively). According to the well-known model by SHOCKLEY, READ [9.17] and HALL [9.18] (SRH) the rates $r_1 \ldots r_4$ are proportional to the density of states and the concentration of carriers involved in the process. Since the recombination in such Si and GaAs diodes is dominated by these four interactions $r_1 \ldots r_4$ describe the entire dynamics of the charge carriers.

Proceeding from the SRH mechanism several articles have been published showing analytical and numerical calculations of the current (I) vs. voltage (V) characteristics of double injection p-i-n diodes, see for example [9.19–21]. To some extent they reveal an S-type curvature in excellent agreement with experimental measurements. It is therefore assumed that this model is well suited to describe the electrical properties on a microscopic level. Minding the simple picture in Fig. 9.1 however, the local j–E characteristic is more important than the integral behaviour if engaged in pattern formation. In order to obtain local values of E as a function of j and having regard to specific parameters of the samples we developed a numerical method [9.16]. Starting point are the SRH-model and basic semiconductor equations like Poisson equation and continuity equation. Figure 9.3 demonstrates one of the most decisive results. For both types of diodes j vs. E was calculated at three different distances from the contact. A distinct dependence of the electrical character on the position can be seen: Close to the

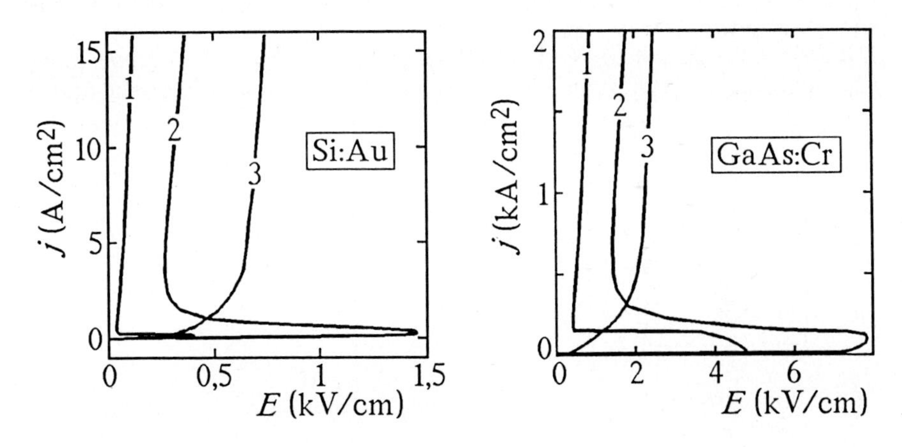

Fig. 9.3. Calculated j–E characteristics at different distances from the contact: 1=close to the p-contact, 2=in the middle, 3=close to the n-contact. The concentrations of donors (N_D) and traps (N_T) are $N_D/N_T = 1.8 \cdot 10^{14}$ cm^{-3}/$2.1 \cdot 10^{14}$ cm^{-3} in case of Au doped Si (left picture) and $N_D/N_T = 1 \cdot 10^{15}$ cm^{-3}/$2 \cdot 10^{15}$ cm^{-3} in case of Cr doped GaAs (right picture)

anode and in the major part of the 'i'-layer the pronounced S-shaped NDC can be found whereas approaching the cathode the NDC disappears and j increases monotonically with increasing E. This inhomogeneous property of compensated material in directon of current flow must be considered when building up an equivalent circuit for an extended p-i-n diode.

9.2.2 Equivalent Circuit Model

The description by means of an equivalent circuit model aims at two basic problems: First, the electrical properties of the material between the contacts have to be reproduced and second, it must enable the formation of an inhomogeneous current density distribution perpendicular to the current flow. To simplify the mathematical treatment we assume a thin p-i-n diode, that means small dimensions and therefore homogeneity in one direction of the contact plane. This leads to a two-dimensional presentation.

The explanations in the preceding section makes it ingenious to subdivide the diode into two regions. It should be noted at this point that a two-layer model presented in an earlier work [9.22] made already use of this fact. But for the lack of analytical solutions and the uncertainty of applicabilty to p-i-n diodes at this time the novel description treated here was proposed. In doing so, the part with an S-shaped NDC is approximated by a nonlinear resistance which reveals the desired characteristic. An ohmic resistor with conductance G on the other hand represents the area with unique $j(E)$ relation close to the n-contact. Furthermore, the dynamic properties of the device have to be considered. BARON and MAYER [9.23] have shown that in case of a p-i-n diode with deep traps an additional capacitance C_d and a series connexion of resistance and inductance L describes the impedance correctly. These reflections yield the equivalent circuit in Fig. 9.4a. To allow an inhomogeneous distribution of the current density

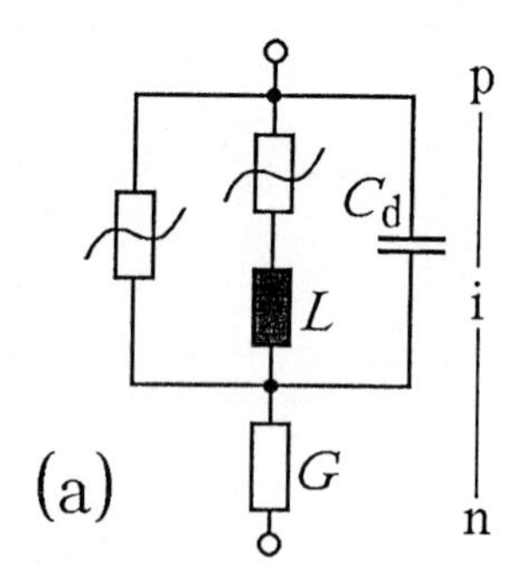

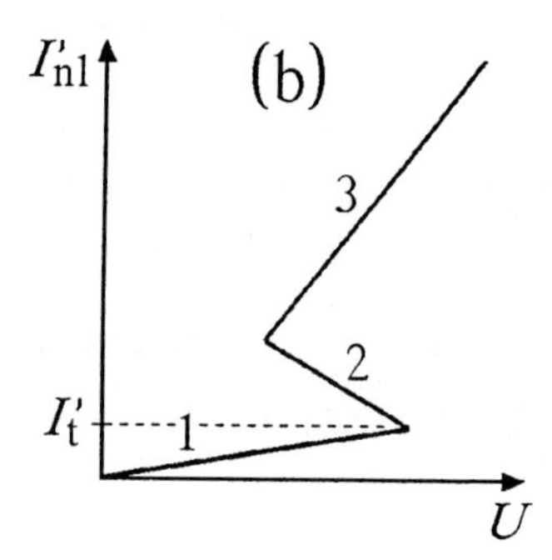

Fig. 9.4. (a) Equivalent circuit of a p-i-n diode with no transversal inhomogeneity and (b) S-type NDC of the nonlinear resistance approximated by three piecewise linear sections

perpendicular to the applied electric field (denoted in the following as the x-direction) the extended device is constructed as a chain of such unit cells. The resistance R introduces a coupling between adjacent cells. Finally, the measured lifetimes of free carriers in our samples are much longer than the dielectric relaxation times. This enables a simplification of the model by omission of the capacitance and equality of both nonlinear resistances [9.23]. Figure 9.5 shows the resulting equivalent circuit for a long and thin p-i-n diode.

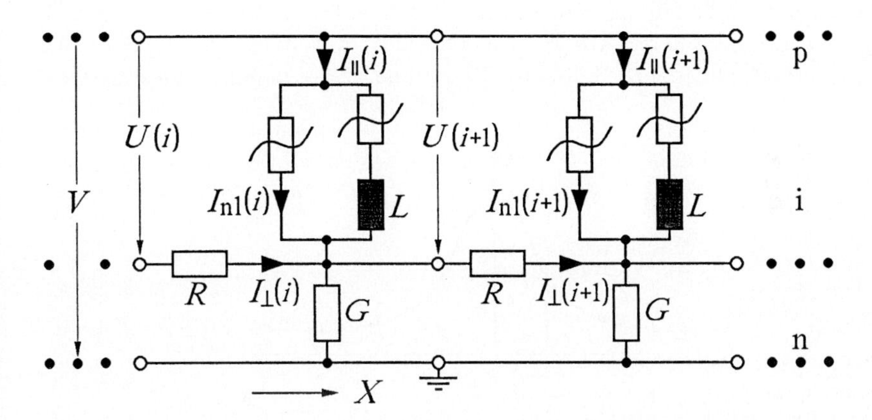

Fig. 9.5. Two unit cells (i and $i+1$) of the equivalent circuit representing the electrical properties of a two-dimensional p-i-n diode

Applying Kirchhoff's laws one easily derives the following equations for the voltage drop U across the nonlinear region, and the transversal current $I_{\perp}$:

$$U(i+1) - U(i) = R \cdot I_{\perp}(i) \quad \text{and} \tag{9.1}$$

$$I_{\perp}(i+1) - I_{\perp}(i) = I_{\|}(i) - G \cdot (V - U(i+1)) \,. \tag{9.2}$$

The transition from discrete to distributed elements implies the introduction of parameters per unit length in x-direction which is denoted by a small hook (') next to the symbol. Combining (9.1) and (9.2) then leads to:

$$\frac{\partial^2 U}{\partial x^2} = R' I'_{\|} - R'G'(V - U) \,. \tag{9.3}$$

This diffusion equation has no unique solution because $I'_{\|}(U)$ is not a mathematical function. In contrast, we can express the voltage drop U as a unique function of the current per unit length through the nonlinear resistance. Thus we obtain a differential equation for I'_{nl}, and an analytical solution can be found if the S-shaped nonlinearity is approximated by piecewise linear sections as shown in Fig. 9.4b. Writing $U = a_i I'_{nl} + b_i$ where $i = 1, 2, 3$ and t denotes the time leads to:

$$2a_i \frac{\partial^2 I'_{nl}}{\partial x^2} = R'(I'_{\|} - G'U) + R'G'f(I'_{\|} - I'_{nl}) + R'G'L' \frac{\partial (I'_{\|} - I'_{nl})}{\partial t} \,. \tag{9.4}$$

Qualitatively, the solutions of (9.4) are the spatial and time-dependent distributions of the current density in a thin p-i-n diode. For constant a_i this equation looks like the well-known FitzHugh-Nagumo equation which gives stationary solutions in form of solitary or kink structures [9.24].

A profound analysis of the stationary solutions has been carried out while studying the phase portrait of the dynamic system represented by (9.4), for details see [9.14, 16, 25]. Current density filaments with a profile as sketched in Fig. 9.6 correspond to trajectories on a homoclinic orbit in the $(I'_{\|}, \partial I'_{\|}/\partial\xi)$-plane where $\xi = \sqrt{R'G'} \cdot x$ is the normalized coordinate. The condition for such a so-

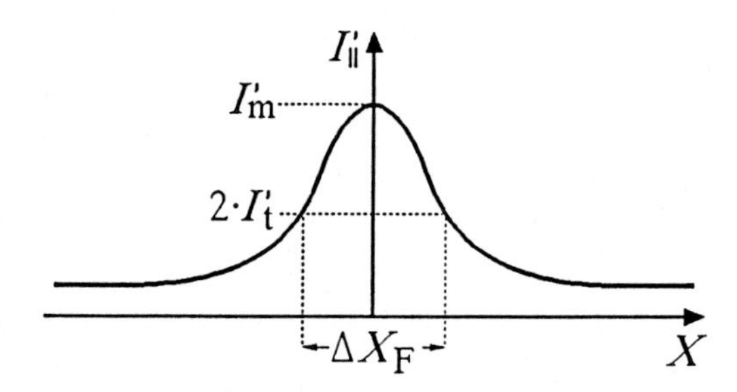

Fig. 9.6. Distribution of the current per unit length transversal to the electric field when following a trajectory on a homoclinic orbit

lution can generally be stated as an equal areas rule. The boundary condition simultaneously determines the maximum current per unit length I'_m in the center. It is derived that filaments exist in a fixed range of the parameters and the applied voltage V. In case of a linearized function $U = a_i I'_{nl} + b_i$ the profile $I'_{\|}(x)$ can be solved explicitely. This enables even the specification of an analytical expression for the width of a filament Δx_F as a function of the resistance elements in the equivalent circuit. For the samples treated in the following sections some approximations hold [9.25] which simplifies to the equation

$$\Delta x_F \approx \frac{2}{\sqrt{R'G'}} \frac{-a_2}{a_1} \propto d\sqrt{f}\,. \tag{9.5}$$

Here f denotes the extension of the region with unique $j(E)$ relation in direction of current flow divided by the contact distance d.

At this point important results of the model come clear. Moreover, they are well suited for a comparison with experimental data. The width of a filament depends not from the applied voltage and therefore likewise not from the total current through the device. On the other hand, Δx_F is proportional to the square root of the contact distance of the device and to the square root of the extension of the monotonic region.

9.3 Stationary Pattern Formation

Extensive experimental observations have been carried out to prove the theoretical predictions. We used Au doped Si, and Cr doped GaAs p-i-n diodes at room temperature. The spatial and time-resolved determination of several physical quantities is necessary to get detailed information about the properties of current density filaments. The next paragraph briefly describes the fabrication of the samples and the experimental set-ups.

9.3.1 Experimental Technique

The devices were made from commercial Cr doped GaAs wafers with resistivities $\varrho = 3 \cdot 10^7 ... 1 \cdot 10^9\ \Omega$cm and n-type Si wafers with $\varrho = 10...200\ \Omega$cm. In case of silicon an Au diffusion process was carried out in order to get compensated, high-resistivity material. The highly doped n- and p-contacts were fabricated on opposite wafer surfaces using spin-on film sources and a further diffusion process. From this a thickness of approximately 1 μm resulted for the layers. Al was evaporated onto both surfaces to ensure a good electrical and thermal contact with the subsequently applied device-mounting. Finally, small samples were cut from the wafer and the newly generated semiconductor surfaces were polished carefully. Figure 9.7a shows the geometry of a finished sandwich p-i-n diode. Typical dimensions are: Length $l = 1...5$ mm, contact distance $d \approx 80\ \mu$m (GaAs), 100...500 μm (Si), and depth $h \approx d$. Due to the fact that the extension in y-direction is comparable with d and small compared with l, the device can be regarded as a long and thin diode as presupposed in the two-dimensional model.

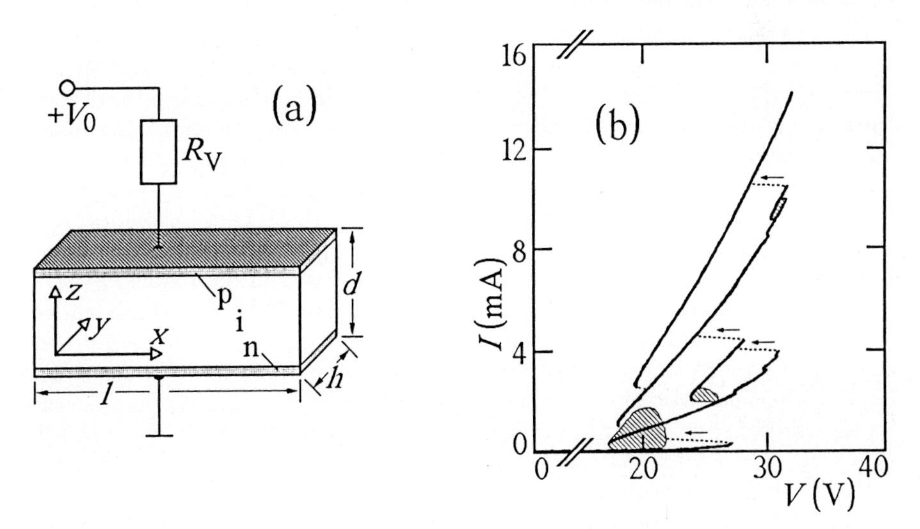

Fig. 9.7. (a) Geometry and electrical connexion of a sample with coordinate-system and (b) measured I–V characteristic of a Si:Au p-i-n diode. The hatched areas mark self-generated oscillations

The I–V characteristic of the device, which differs significantly from the j–E relation, indicates the occurrence of filamentation. Figure 9.7b shows a typical characteristic of a forward biased Si:Au p-i-n diode by applying a dc voltage via a load resistor R_V. Taking GaAs:Cr p-i-n diodes would do just as well since they exhibit the same features [9.25]. One can observe multistability caused by the formation of several current density filaments [9.15, 26]. The different states are seperated by jumps when particular threshold values are reached. Moreover, self-generated oscillations of the current and voltage often occur in the vicinity of the jumps. The measured amplitudes amount to some Volt and mA, respectively. These oscillations are supposed to be connected with instabilities of the filaments and are subject of the fourth section.

The best method in order to observe filamentation would be the spatial and time-resolved detection of the current density. Because up to now a direct measuring technique is not available, we depend on indirect methods. According to Ohm's law and the fact that the mobilities of charge carriers in our devices are supposed to be almost constant, the electric field and the concentrations of free charge carriers account for variations of the current density.

Potential probe measurements using tungsten wires with a tip diamenter of about 1 μm, controlled by a micro-positioning system, reveal an image of the potential distribution on the surface between the contacts [9.26, 27]. A characteristic potential hump indicates the existence of a current filament. Because the measured potential on the surface of the 'i'-layer is nearly equal to the electrostatic potential ϕ [9.23] and according to $E = -\text{grad}(\phi)$, the electric field distribution can be calculated from these data when subtracting the potential values ϕ of adjacent points.

The local concentrations of free electrons n and holes p can be determined by measurements of the infrared absorption as well as the recombination radiation [9.28]. The latter one is especially applicable to GaAs:Cr p-i-n diodes because a semiconductor with direct band-gap causes more radiant recombination processes. A proportionality between the product $n \cdot p$ and the power of the emitted radiation is assumed [9.9]. The second optical method aims at the detection of the free carrier absorption coefficient α in Si:Au p-i-n diodes. According to the Drude model, one gets $\alpha = \sigma \cdot N$ with the concentration N of free electron-hole pairs and the optical absorption cross section σ [9.29]. Experimentally, we measured the transmitted light when the diode was irradiated by a focused laser beam between the contacts. A wavelength of $\lambda = 3.39$ μm was choosen to prevent transitions of carriers between the bands and the traps. To improve the signal to noise ratio we modulated the current through the device by an additional ac voltage in order to use lock-in technique. It was verified by several experiments that the detected optical signals yield a measure for the variation of the free electron-hole pairs N_{eff}.

In the following some experimental results are presented when stable operating points in the I–V characteristic of the devices are adjusted. Emphasis is put upon those permitting a comparison with theoretical predictions.

9.3.2 Observations and Comparison

In the last decade detailed information about stable filament formations in p-i-n diodes have been obtained by utilizing the above described experimental setups [9.15, 26, 28]. A filament produces a potential hump, an enhanced production of Joule's heat, and close to the contacts an increased IR absorption as well as radiation emission. Generally, it turned out that each jump in the I–V characteristic is associated with the generation or disappearance of an individual filament. These filaments have a solitary behaviour, i.e. several filaments in the same device appear likewise with the same profile and this profile is preserved when increasing the external current.

In this passage new quantitative results are discussed and compared with the predictions of Sect. 9.2. As mentioned above, the electric field distribution can be calculated from a spatially resolved potential probe measurement. In doing this, a significant difference in the course parallel to the x-axis becomes evident. Figure 9.8a presents a distribution near the p- and the n-contact of a Si:Au p-i-n diode, respectively. During the potential probe measurement an operating point beyond the first jump in the I–V characteristic was choosen and only the region on the surface of the device showing filamentation was scanned. At z-positions near the p-contact one can observe a double-peaked course. Just this behaviour is expected if the material exhibits an S-shaped NDC and if filamentation means a pulse-shaped current density distribution as illustrated in Fig. 9.8b. On the contrary, the electric field distribution close to the n-contact is single-peaked indicating a monotonic j–E relation in this region. Consequently, the experiments confirm the proposed subdivision of a p-i-n diode into two regions with different electrical character.

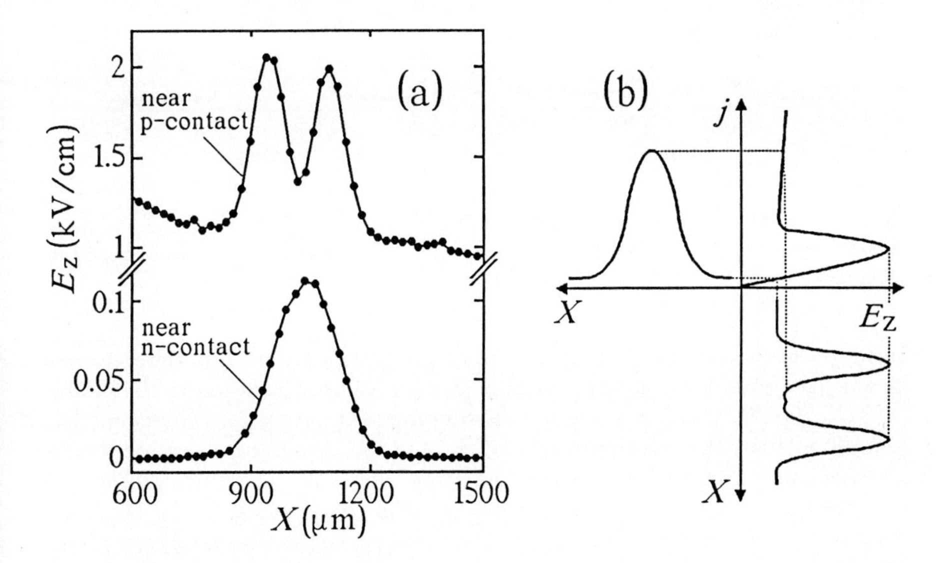

Fig. 9.8. Electric field in the direction of current flow E_z as a function of the x-position in a region with one filament. (a) Calculated from a potential probe measurement and (b) expected course for a material with S-shaped NDC (schematic)

Utilizing the IR absorption technique we determined the concentration of free carriers in the same device when operated at a comparable point, i.e. the current is modulated on a branch beyond the first jump in a way that the dc parts of I and V are unchanged related to the potential measurement. The distribution values taken at identical positions between the contacts are presented in Fig. 9.9. Likewise, we observe a different course. Close to the p-contact the increase is narrower with little foothills at both sides. Combining these data with the electric field data of Fig. 9.8a produces a very important result. Multiplying the values of E_z and N_{eff} at each point and with the electronic charge and the mobilities of the carriers gives the variation of the current density j_{eff} as a function of the x-position, see Fig. 9.9. By this, the filamentary current density distribution

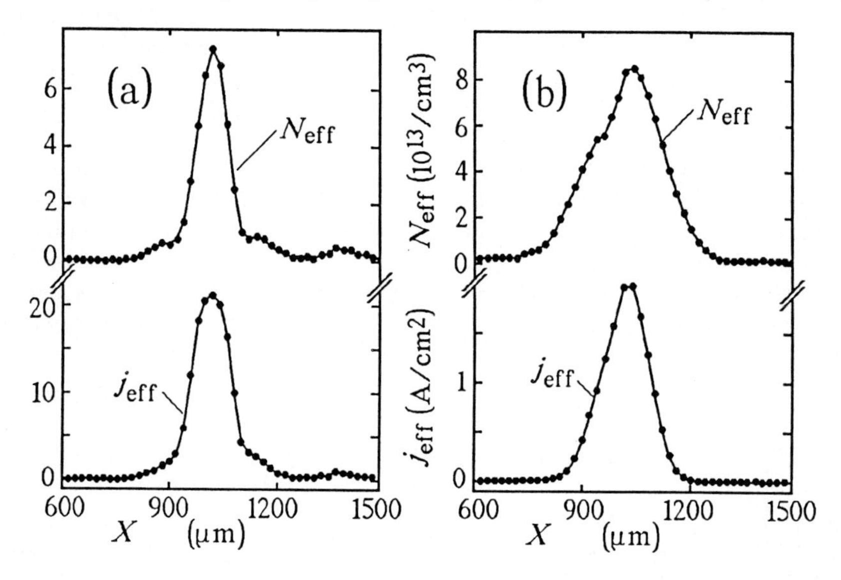

Fig. 9.9. Variation of the concentration of free electron-hole pairs N_{eff} and variation of the current density j_{eff} vs. the x-position. (a) Near the p-contact and (b) near the n-contact

between the contacts has been measured quantitatively for the first time. As can be seen, the course is pulse-shaped at both z-positions and similar to the profile in Fig. 9.6 which sketches a possible distribution in the equivalent circuit model. More than that, the maximum value of j_{eff} in the inhomogeneity close to the n-contact well agrees with an estimated value when calculating the quotient of the external current through the device and the cross section of the filament. Although the extension of the increase close to the p-contact is slightly narrower, the absolute value of j_{eff} seems to be too large. This might be caused by small deviations of the probe positions during the measurements.

For a further comparison between theoretical predictions and experimental observations we have evaluated the width of filaments perpendicular to the current flow. The FWHM values of the measured inhomogeneities show the following properties: First, the extension in x-direction is slightly increasing from p- to n-contact, and different filaments in individual devices have the same size.

Second, the width does not change when the operating point is varied. This fact has already been published [9.25] and it confirms the theoretical statement that the width is independent of the current through the device. To obtain the third result, we compared filaments in different devices and estimated the factor f by means of the potential distribution showing a transition from single-peaked to double-peaked characteristic. Figure 9.10 shows the plotted FWHM values versus the product $d \cdot \sqrt{f}$. The experimental data obey a linear relation indicated by the dashed line. Remembering (9.5) in Sect. 9.2, there is again a very good accordance with a theoretical prediction.

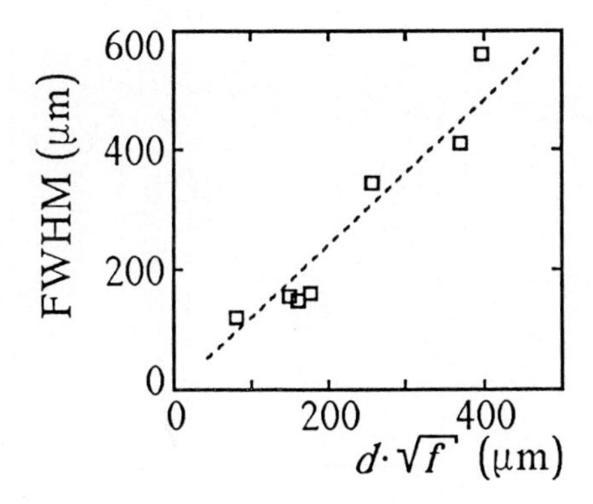

Fig. 9.10. Width of filaments in different Si p-i-n diodes with varying contact distances

In this section a lot of experiments have been presented which yield detailed quantitative information about stable filaments in p-i-n diodes. Wherever a comparison was possible, a good agreement with the statements of the equivalent circuit model followed. In the next section we will investigate temporally unstable patterns.

9.4 Spatio-Temporal Dynamics

On principle, semiconductor devices with descending branches in the electrical characteristics can show self-generated oscillations of I and V if the external circuit and the operating point are choosen suitably [9.12]. These phenomena are often attributed to instabilities of spatial structures as for instance proven true in case of travelling field domains in Gunn-diodes. Even several observations using devices with S-shaped characteristics [9.4,30,31] and p-i-n diodes [9.32,33] have been published. Most of the experiments, however, deal with integral quantities like the current and voltage, giving few insight in the dynamic process of the filament itself. Only some recent work, see the part by NIEDERNOSTHEIDE in this book for example, show spatially and temporally resolved measurements. On the other hand, different theoretical models describe the instabilities as switching, breathing or travelling filaments [9.30,34,35].

Self-generated oscillations of the current and the voltage occur in our Si:Au p-i-n diodes as well as in GaAs:Cr p-i-n diodes, see Fig. 9.7b. The measured frequencies range from some kHz up to MHz, and the shape of the signal versus time varies with the external dc voltage. The current may have a pulse-like or a more sinusoidal course. Intervals with simple periodic oscillations interchange with those showing odd or even numbers of maxima during one period. Likewise, signals with a chaotic behaviour can be observed. Figure 9.11 illustrates a typical scenario close to the first jump in the I–V characteristic of a Si:Au p-i-n diode.

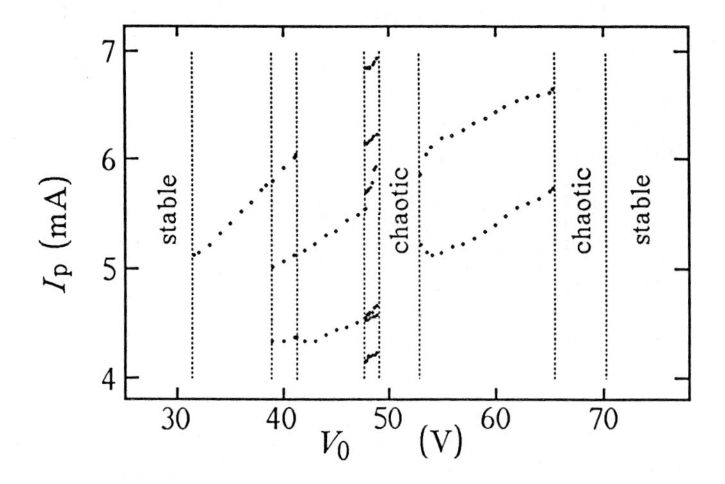

Fig. 9.11. Amplitude of the current pulses I_p vs. the external voltage V_0 across load resistor and device

When utilizing the set-ups described in the preceding section and operating the device in a state with periodic oscillations, we can get information about the underlying mechanism in the semiconductor. In the following those results are presented which give new insight in the dynamics of unstable filaments in p-i-n diodes. The second part of this section briefly describes investigations in connection with chaotic oscillations.

9.4.1 Periodic Oscillations

The potential probe technique enables temporally resolved signal detection if the tungsten wire is directly connected to the gate electrode of a MOS field-effect transistor [9.27]. A contact resistance of about 10 MΩ and an input capacitance < 0.05 pF allows a resolution better than 1 μs. Measurements carried out on the surface of Si:Au p-i-n diodes show spatially limited regions with potential oscillations [9.25, 27]. In case of simple periodic oscillations the potential signals have the same shape and the maxima occur at the same time compared with the external current vs. time signal. Moreover, the extension and location of potential oscillations in x-direction are equal to the width and position of a stable filament appearing when the external current is increased. Consequently, these results can be interpreted as a periodic switching current density filament in an otherwise homogeneous sample.

When operating points with different maxima in the current during one period were adjusted, we always observed several regions with comparable extensions showing oscillations of the potential. The maxima of the potential in different regions either appear temporally locked to the different maxima in the current or the oscillation frequencies of the signals differ from one region to another. These interesting results yield a clear mechanism for the formation of complex integral signals: Period multiplication is caused by a correlated interaction of two or more filaments. Each filament shows a simple periodic switching behaviour but the frequencies or the phase vary across the device.

The same results can be obtained when investigating the GaAs:Cr p-i-n diodes. We used a streak-camera system to detect the spatially and temporally resolved radiation emission. As an example, Fig. 9.12 presents a measured image during simple periodic oscillations of the current. The dark spots indicate regions with intense recombination radiation emission close to one contact due to enhanced concentrations of electron-hole pairs. The temporal distance between the spots amounts to 1.8 μs and they appear in synchronism with the current peaks. Hence, this picture likewise shows the switching of an unstable filament.

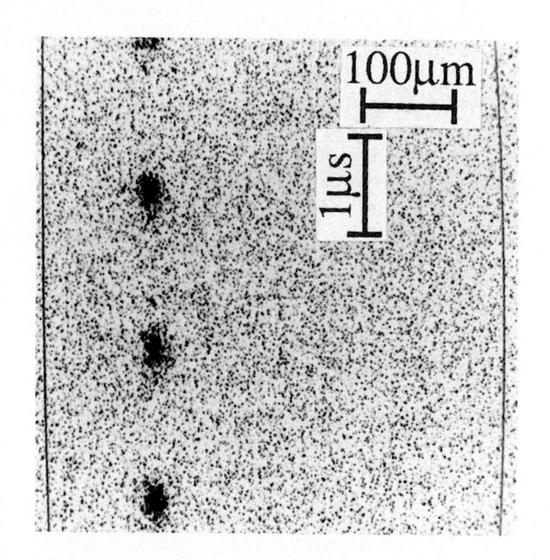

Fig. 9.12. Power of the emitted radiation as a function of the time t and x-position. The lines mark the boundaries of the GaAs:Cr p-i-n diode (length ≈ 0.51 mm). This picture was taken from a photograph

Up to now there are no hints to breathing or travelling filaments in our p-i-n diodes. The fact that the observed periodic oscillations are obviously caused solely by switching processes of current density filaments enables a description of these phenomena by means of the equivalent circuit model. Without loss of generality we can represent a filament by one unit cell, see Fig. 9.5, and neglect the other regions of the sample showing permant high resistance and barely contributing to the current. Concerning the external circuit an extra capacitance C has to be added due to the geometry of the device-mounting and the cable capacitance. These reflections lead to a simplified equivalent circuit in case of simple periodic oscillations as presented in Fig. 9.13. Applying Kirchhoff's

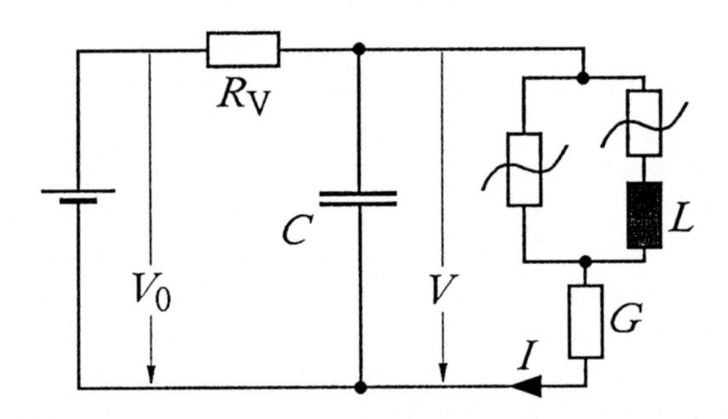

Fig. 9.13. Simplified equivalent circuit for a p-i-n diode with external dc voltage supply, see text

laws one obtains coupled differential equations describing the current $I(t)$ and the voltage $V(t)$ as a function of the time and the other circuit parameters. The equation system can be solved numerically and the results of a calculation are plotted in Fig. 9.14a. For a better comparison, the set of parameters was choosen in a way that they fit to the electric properties of one selected sample. This implies in detail: a_i, b_i and G approximately reproduce the slopes in the stationary I–V characteristic whereas the value of L fits to the measured carrier lifetime of about 1 μs. Figure 9.14b presents measured $I(t)$ and $V(t)$ signals when a dc voltage is applied to a Si:Au p-i-n diode via a load resistor. As can be seen, we get a very good agreement concerning not only the curvature but also the absolute values of the minima and maxima.

It was mentioned above that complex integral signals originate from a correlated interaction of several filaments in a sample. Consequently, we put several unit cells in parallel to represent the experimental observations. This results in a chain of nonlinear oscillators coupled via the external circuit. It is known that such an arrangement can show effects like period doubling, frequency and phase locking and nonperiodic behaviour as a function of the nonlinearity and the coupling strength, see for example [9.36]. Therefore, the equivalent circuit model is even well suited to describe many of the observed self-generated oscillation phenomena in p-i-n diodes.

9.4.2 Nonperiodic Oscillations

Figure 9.11 at the beginning of this section already pointed to the occasional occurrence of self-generated chaotic oscillations. The shape of the nonperiodic

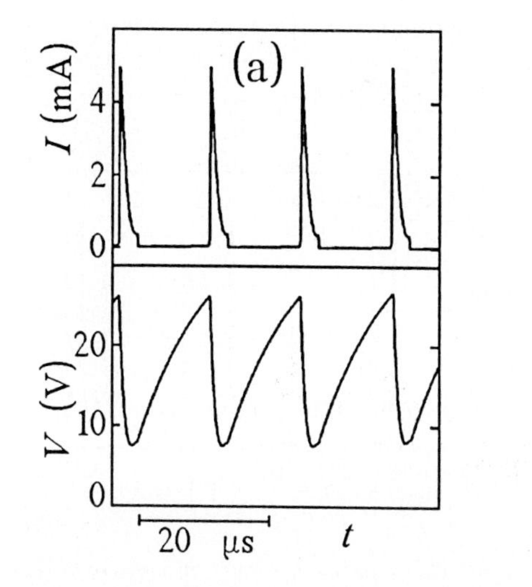

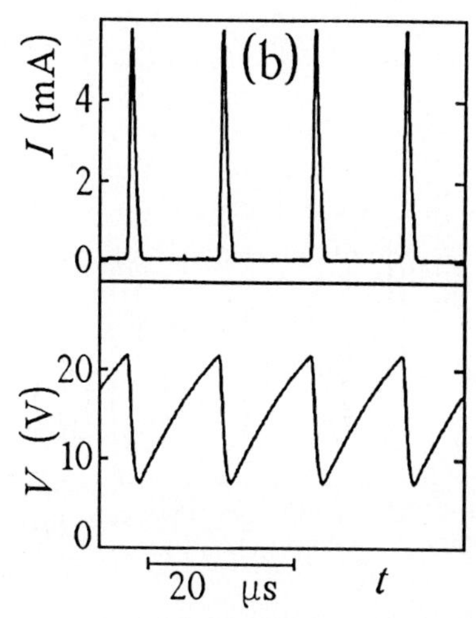

Fig. 9.14. Simple periodic oscillations of the current and voltage obtained by (a) numerical calculations on basis of the equivalent circuit and (b) measurements investigating a Si:Au p-i-n diode. Load resistor $R_V = 60$ kΩ and external dc voltage $V_0 = 40$ V in both cases

current and voltage signals varies with the operating point and differs among the samples. For example, we observed series of spikes with varying temporal distances as well as sequences of pulses showing arbitrary amplitudes. The theory of nonlinear dynamics has proved to be able to characterize these phenomena [9.8,37]. It is assumed that the nonlinear oscillations show deterministic behaviour and obey universal scaling rules. Due to the complexity of observed effects and the number of evolved methods a detailed analysis of the observations would break up the scope of this contribution. Instead, little insight into the chaotic behaviour of the p-i-n diodes is given by means of some examples.

In order to get some information about the processes at different sample positions at the same time, potential probe measurements were carried out using two probes simultaneously. Figure 9.15 demonstrates a result obtained during nonperiodic oscillations of the current. The marked points on the surface of the Si:Au p-i-n diode yield the largest amplitudes of the potential oscillations whereas in the middle between them and at $x > 800$ μm the oscillation vanishes. If the measured signals can be interpreted as spatially limited oscillations of the current density, we have observed two separate regions with chaotic switching filaments. An exact comparison of the moments reveals that at no time filaments in both regions are simultaneously switched on.

Measurements using other samples operated in a state with nonperiodic oscillations of the current show two separate regions on the surface with periodic potential oscillations. But in this case the irrational ratio of the local frequencies causes the integral nonperiodicity. Furthermore, an aperiodic spiking has been observed [9.25]: All maxima in the current vs. time signal have nearly the

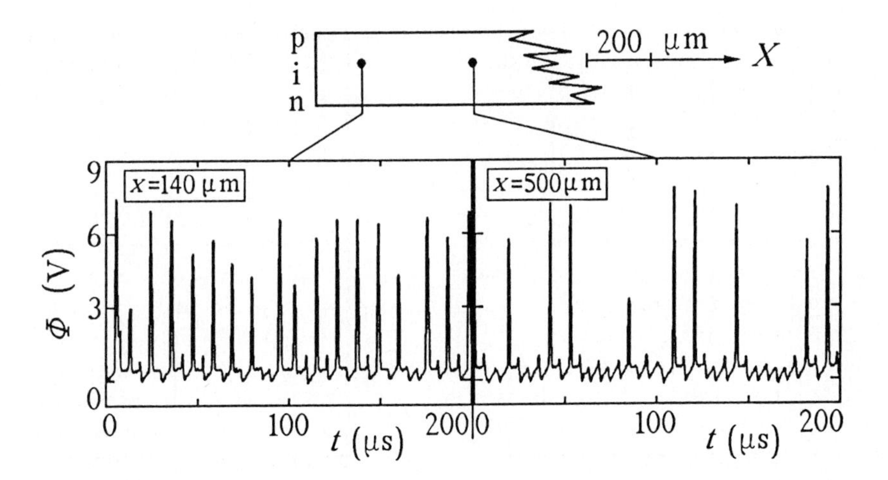

Fig. 9.15. Potential measurements at two differnt surface positions on a Si:Au p-i-n diode during nonperiodic oscillations of the current

same amplitude but the temporal distances differ. The number of spikes per time interval varies with the external dc voltage showing a characteristic scaling behaviour. This seems to be a universal property of some nonlinear systems and it occurs likewise in microwave resonators [9.38].

A numerical method proposed in [9.39] permits the estimation of the Kolmogorov entropy K from discrete time series. This quantity is a measure for the loss of information about the state of a system with time. Applying the method to a measured nonperiodic voltage vs. time signal of a Si:Au p-i-n diode we obtained $K \approx 0.12$. This positive value clearly indicates a deterministic chaotic behaviour.

9.5 Conclusion

The subject of this contribution is the pattern formation due to current density filaments in semiconductor devices, especially p-i-n diodes. It was strongly motivated by the large number of experiments showing filamentation in various and even technically significant devices. Nevertheless, the understanding of these effects is still incomplete.

The samples used for the investigations are Si:Au and GaAs:Cr p-i-n diodes. They exhibit a multistable I–V characteristic due to an S-shaped NDC of the material as one precondition for pattern formation. Numerical calculations on the basis of the SRH-model reveal an inhomogeneous electrical character in direction of current flow which makes a subdivision into two regions ingenious. Considering this result a two-dimensional equivalent circuit model for a long and thin p-i-n diode is proposed. A differential equation describing the spatio-temporal current density distribution perpendicular to the current flow is derived. In the stationary case the analysis shows the possible existence of filamentary solutions and predicts the independence of the filament width from the operating point.

Experimental results are presented utilizing several set-ups and measuring spatially and temporally resolved various physical quantities of the material. Even quantitative values of the filamentary current density distribution between the contacts could be determined. An excellent agreement with theoretical predictions is demonstrated concerning the electrical inhomogeneity in direction of current flow as well as the profile and width of stable filaments. The observed self-generated oscillations are obviously caused by switching processes of unstable filaments. Moreover, the correlated interaction of two or more filaments lead to the occurrence of period multiplication and complex integral current and voltage signals. The equivalent circuit model can also describe the phenomena in connection with periodic oscillations as proved true by means of numerical calculations. A simplified electric circuit exhibits simple periodic oscillations showing the same integral characteristic compared to experimental signals. Finally, measurements during nonperiodic oscillations indicate the participation of at least two interacting and switching filaments even in this case. Several mechanisms may be responsible for the transition into chaotic behaviour. In contrast to the periodic effects, the observed chaos can not be modelled using the equivalent circuit. Applying the methods of nonlinear dynamics, however, we find out universal scaling behaviour and deterministic chaos.

It is assumed, that the results of the equivalent circuit model are transferable to other devices. Any real device needs contacts which cause a differing electrical characteristic compared with the bulk material. This enables a subdivision into two or more regions. The ideas presented in this contribution may possibly lead to novel device concepts utilizing the formation of filaments, or to improvements in those types of devices showing undesired filamentation.

Acknowledgments. The main progress in the experimental and theoretical investigations of current filamentations in p-i-n diodes was achieved during my stay at the Gerhard-Mercator-University. Therefore, I would like to thank PROF. DR. D. JÄGER for helpful discussions and the Stiftung Volkswagenwerk for financial support.

References

[9.1] V.L. Bonch-Bruevich, I.P. Zvyagin, A.G. Mironov: *Domain Electrical Instabilities in Semiconductors*, Studies in Soviet Science (Consultants Bureau, New York 1975)

[9.2] J. Pozhela: *Plasma and Current Instabilities in Semiconductors* (Pergamon Press, Oxford 1981)

[9.3] J. Shah, G.J. Iafrate (eds.): *Hot Carriers in Semiconductors*, Solid-St. Electron. Vol. 31 (Pergamon Press, Oxford 1988)

[9.4] Y. Abe (ed.): *Nonlinear and Chaotic Transport in Semiconductors*, Appl. Phys. A Vol. 48 (Springer-Verlag, Heidelberg 1989)

[9.5] H. Thomas (ed.): *Nonlinear Dynamics in Solids* (Springer-Verlag, Heidelberg 1992)

[9.6] N. Balkan, B.K. Ridley, A.J. Vickers (eds.): *Negative Differential Resistance and Instabilities in 2-D Semiconductors* (Plenum Publishing, New York 1993)

[9.7] G. Nicolis, I. Prigogine: *Self-Organization in Non-Equilibrium Systems* (John Wiley & Sons, New York 1977)

[9.8] H. Haken: *Synergetik, Eine Einführung*, 3rd ed. (Springer-Verlag, Heidelberg 1990)

[9.9] S.M. Sze: *Semiconductor Devices, Physics and Technology* (John Wiley & Sons, New York 1985)

[9.10] G.W. Taylor, J.G. Simmons, R.S. Mand, A.Y. Cho: IEEE Trans. ED **34**, 961 (1987)

[9.11] M. Stoisiek, R. Sittig: In *Festkörperprobleme (Advances in Solid State Physics)*, ed. by P. Grosse, Vol. 26 (Vieweg, Braunschweig 1986) pp. 361-373

[9.12] F. Stöckmann: In *Festkörperprobleme (Advances in Solid State Physics)*, ed. by O. Madelung, Vol. 9 (Vieweg, Braunschweig 1969) pp. 138-171

[9.13] E. Schöll: *Nonequilibrium Phase Transitions in Semiconductors* (Springer-Verlag, Heidelberg 1987)

[9.14] R. Symanczyk, D. Jäger, E. Schöll: Appl. Phys. Lett. **59**, 105 (1991)

[9.15] D. Jäger, R. Symanczyk: In *Nonlinear Dynamics in Solids*, ed. by H. Thomas, (Springer-Verlag, Heidelberg 1992) pp. 68-87

[9.16] R. Symanczyk: *Stromdichtekanäle in pin Dioden: Experimente und Modelle zur Strukturbildung und nichtlinearen Dynamik in Halbleiterbauelementen* (Verlag Tebbert KG, Münster 1993), and Dissertation, Universität - GH - Duisburg (1993)

[9.17] W. Shockley, W.T. Read: Phys. Rev. **87**, 835 (1952)

[9.18] R.N. Hall: Phys. Rev. **87**, 387 (1952)

[9.19] K.L. Ashley, A.G. Milnes: J. Appl. Phys. **35**, 369 (1964)

[9.20] W.H. Weber, G.W. Ford: Solid-St. Electron. **13**, 1333 (1970)

[9.21] I. Dudeck, R. Kassing: J. Appl. Phys. **48**, 4786 (1977)

[9.22] C. Radehaus, K. Kardell, H. Baumann, D. Jäger, H.-G. Purwins: Z. Phys. B - Condensed Matter **65**, 515 (1987)

[9.23] R. Baron, J.W. Mayer: In *Injection Phenomena*, ed. by R.K. Willardson, R.C. Beer, Semiconductors and Semimetals Vol. 6, (Academic Press, New York 1970) pp. 201-313

[9.24] A.C. Scott: Rev. of Mod. Phys. **47**, 487 (1975)

[9.25] R. Symanczyk, S. Knigge, D. Jäger: In *Negative Differential Resistance and Instabilities in 2-D Semiconductors*, ed. by N. Balkan, B.K. Ridley, A.J. Vickers (Plenum Publishing, New York 1993) pp. 269-282

[9.26] D. Jäger, H. Baumann, R. Symanczyk: Phys. Lett. A **117**, 141 (1986)

[9.27] R. Symanczyk, S. Gaelings, D. Jäger: Phys. Lett. A **160**, 397 (1991)

[9.28] R. Symanczyk, E. Pieper, D. Jäger: Phys. Lett. A **143**, 337 (1990)

[9.29] J.P. Woerdman: Philips Res. Rep. Suppl. **7**, 1 (1971)

[9.30] A. Kittel, U. Rau, J. Peinke, W. Clauss, M. Bayerbach, J. Parisi: Phys. Lett. A **147**, 229 (1990)

[9.31] A. Brandl, W. Prettl: Phys. Rev. Lett. **66**, 3044 (1991)

[9.32] I. Dudeck, R. Kassing: Solid-St. Electron. **22**, 361 (1979)

[9.33] S.T. Voronin, A.F. Kravchenko, A.P. Sherstyakov: Sov. Phys. Semicond. **18**, 413 (1984)

[9.34] E. Schöll, D. Drasdo: Z. Phys. B - Condensed Matter **81**, 183 (1990)
[9.35] F.-J. Niedernostheide, M. Arps, H. Willebrand, H.-G. Purwins: Phys. stat. sol. (b) **172**, 249 (1992)
[9.36] T. Hogg, B.A. Hubermann: Phys. Rev. A **29**, 275 (1984)
[9.37] H.G. Schuster: *Deterministic Chaos, An Introduction* (Physik-Verlag, Weinheim 1984)
[9.38] D. Jäger, A. Gasch, H.G. Schuster: Phys. Rev. A **33**, 1451 (1986)
[9.39] P. Grassberger, I. Procaccia: Phys. Rev. A **28**, 2591 (1983)

10 Nonlinear and Chaotic Charge Transport in Semi-Insulating Semiconductors

V.A. Samuilov

Belarus State University, Physics Department, F. Skaryna avenue 4
220080 Minsk, Republic Belarus

Abstract. Recent experimental results on spontaneous periodic and chaotic low frequency current oscillations in semi-insulating crystalline semiconductors are reviewed. Period-doubling bifurcations and intermittent chaotic behavior were observed, depending on the value of the applied voltage. The dependencies of the fundamental frequency of low frequency current oscillations on the applied voltage were found to be sharply nonmonotonical in the samples of GaAs with controlled and different EL2 defect concentrations. Temperature dependencies of frequency modes (f) on an Arrhenius plot of $\log(T^2/2f)$ vs. $1/T$ were found to give information on deep levels in InP.
The experimental results of nonlinear charge transport and low frequency current oscillations in spatially chaotic polycrystalline and nanostructured fractal-like porous Si are also presented. The dependencies of the fundamental frequency on the applied voltage were found to be monotonically increasing in these materials. The transition to chaotic behavior through quasi-periodic route was observed in polycrystalline Si. Semi-insulating polycrystalline Si and porous Si are likely to present a new class of materials for experimental investigations of phenomena concerning nonlinear dynamics.
A model of low frequency current oscillations in semi-insulating crystalline semiconductors with deep levels has been developed. A combination of the approaches of field-enhanced trapping, field-enhanced emission and negative differential mobility of carriers was used. A linear analysis of the model was done. The dispersion relations of the model were found to describe qualitatively the experimentally measured dependencies of the activation energies and of the fundamental frequency on the electric field and the temperature.

10.1 Introduction

In the field of semiconductor physics a large number of reports on nonlinear charge transport and related instabilities was stimulated in the last decade by the introduction of concepts and methods of nonlinear dynamics [10.1,2]. Recent summaries of experimental results and theoretical descriptions of self-organized spatio-temporal structure formation and dynamical chaos in homogeneous crystalline semiconductors and in semiconductor structures and devices were presented in the monographs [10.3–6].

The electronic properties of semiconductors represent a highly productive object for experimental investigations of nonlinear dynamics and deterministic

Springer Proceedings in Physics, Vol. 79 **Nonlinear Dynamics and Pattern Formation in Semiconductors and Devices**
Editor: F.-J. Niedernostheide

chaos [10.2,4]. In semiconductors, as open nonlinear dynamical systems, spatiotemporal structures in the form of various periodic and chaotic oscillations of the electric current, charge density waves, domains of high electric field etc. can spontaneously arise [10.7–9], if an external control parameter, e.g., the applied voltage, is varied.

A number of papers were devoted to the theoretical [10.7, 10–13] and experimental [10.11–20] investigations of current instability phenomena including dynamical chaos in semi-insulating (SI) semiconductors. Low frequency current oscillations (LFO) in SI semiconductors with frequencies of a few Hertz and even less are well suited for experimental observations at temperatures in the vicinity of the room temperature without any additional experimental conditions [10.2,4,12–20].

The experimental analysis of deep levels in SI crystals by investigating the temperature dependence of LFO as trap signatures is well-known. The Arrhenius plots of frequency modes of LFO were analyzed for GaAs [10.10–13, 17] and InP [10.21, 22]. This method is very productive and promising [10.22], because the standard techniques for the determination of deep level parameters, such as transient capacitance or transient current spectroscopies, are not readily applicable to SI semiconductors [10.23, 24]. But still, an analytical model describing experimental dependencies of frequency modes of LFO in SI semiconductors on temperature and the control parameter, the applied electric field, has not been completed yet. The current models of LFO in the most investigated SI GaAs crystals are based on different approaches: field-enhanced trapping [10.10], field-enhanced emission [10.7] and negative differential mobility only [10.13]. Also, the experimentally observed nonmonotonical bifurcation plots (dependencies of frequency modes on electric field) have not been explained yet [10.17, 19].

All the above-mentioned papers were devoted to crystalline semiconductors with homogeneously distributed point defects (shallow or deep energy levels within the forbidden gap) and to structures manufactured from these materials. In the meantime, the study of dynamics in spatially chaotic systems is one of the major tendencies in modern physics [10.25, 26].

In this paper recent experimental results on nonlinear charge transport and related LFO are reviewed not only in SI crystalline GaAs [10.27, 28], InP [10.21, 22] but also in spatially chaotic semiconductors (polycrystalline Si [10.29, 30] and fractal-like porous Si [10.31]). The model of LFO in SI crystalline semiconductors and its linear analysis [10.32, 33] are presented as well.

Our results on memory [10.34] and current instability [10.35] phenomena in recently developed three-terminal heterostructures based on spatially chaotic amorphous Si [10.36] and their possible applications as a new kind of memory cells and reliable and cheap implementations of flexible synaptic connections for artificial neural networks [10.37] are not considered here because of the limited volume of the paper.

10.2 Materials and Experimental Procedure

In our experimental investigations we used the following SI semiconductors: crystalline GaAs with controlled and different EL2 (As antisite related defect) concentration, InP and spatially chaotic semiconductors - polycrystalline Si (poly-Si) and fractal-like porous Si (P-Si).

Semi-insulating wafers of GaAs and InP are the basic material for future high speed VLSI and devices of optical fiber communication for far distances. High resistivity SI substrates are required to attain the desired electrical isolation among devices. A material becomes SI by the presence of high concentration of impurities or defects with deep levels within the forbidden gap. SI GaAs wafers are usually Cr-doped or nominally undoped [10.38] and InP wafers are usually Fe-doped or nominally undoped [10.39]. SI behavior ($\rho \geq 10^7\ \Omega$cm) in undoped GaAs comes from the presence of the deep donor EL2 [10.38]. The EL2 concentration in SI GaAs could be varied by arsenic pressure controlled annealing [10.40]. The correct determination of energy position and concentration of defects with deep energy levels in SI semiconductors is an important problem. For this purpose, the investigation of peculiarities of nonlinear dynamic charge transport and LFO in the samples of SI GaAs with different and controllable concentrations of defects with deep levels is promising [10.14, 15, 20].

10.2.1 GaAs

Undoped liquid encapsulated Czochralski GaAs single crystals were subjected to two-zone furnace annealing under 1000 °C for 5 hours in the sample zone. The As_2 pressure was controlled in the range $P(As_2) = (0 - 5)$ atm through the arsenic zone temperature. The concentration of the deep centers EL2 in the annealed samples was found to be changed nonmonotonically with the maximum at about 1 atm [10.40]. The investigated samples were slabs ($8 \times 4 \times 1\ \mathrm{mm}^3$) with planar configuration of the contacts placed 3 mm apart.

Figure 10.1 shows the dependencies of the concentration of EL2 centers [10.40] and the resistivity of the samples on the pressure $P(As_2)$. It should be pointed out that both of the dependencies present maximum at about $P(As_2) = 1$ atm.

The current-voltage (I-V) characteristic and time-domain techniques with fast Fourier transform (FFT) method of data sets analysis were used. The samples were biased with a low noise power supply. An external illumination was excluded. Current oscillations with amplitudes greater than 10^{-11} A were measured by a precise current-voltage amplifier. The output was passed through a low-pass filter to avoid aliasing effects, which could introduce false peaks above the Nyquist frequency, to an ADC converter with sampling rates of 200 Hz or less. The digitized $I(t)$ signal was then analyzed using the FFT method to extract the dependencies of the power spectra and the frequency modes on the applied voltage. The experiments on SI GaAs were performed at room temperature.

10.2.2 InP

The investigated samples of semi-insulating Sumitomo Inc. liquid- encapsulated Czochralsky (100) InP:Fe consist of a 10 mm by 10 mm rectangle, 450 μm thick. A sandwich structure of the sample was chosen because it produces a more uniform electric field in comparison with the planar structures used in [10.14, 15]. A miniature Joule-Thompson refrigerator was interfaced to an IBM-PC.

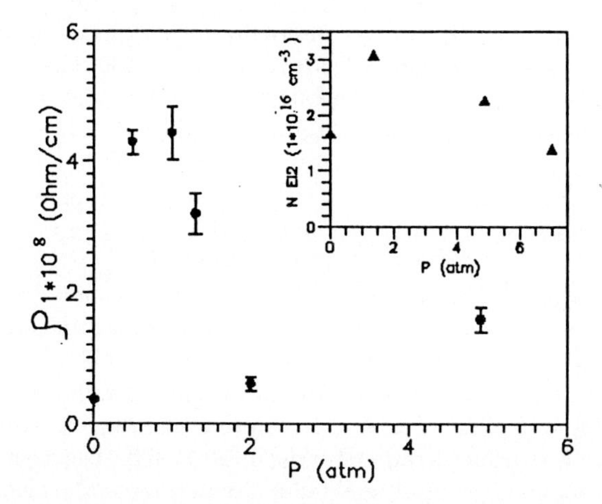

Fig. 10.1. Resistivity ρ of the samples of undoped SI GaAs and concentration N of EL2 centers (in the inset) vs. the pressure $P(As_2)$

The specimen chamber was evacuated (< 1 mTorr) using a sorption pump. A constant voltage (150 V) was applied across the sample giving an average field of $4.3 \cdot 10^3$ V/cm. The current was measured using a current-to-voltage amplifier. The output from the amplifier was passed through a low-pass filter ($\tau = 10$ ms) to an ADC/DAC board. Direct memory access (DMA) was used with a sampling rate of 200 Hz, for 32,768 samples. The temperature was reduced from 295 to 250 K in steps of one degree [10.21, 22].

10.2.3 Polycrystalline Si

Poly-Si layers are widely used in microelectronics [10.41–43]. Poly-Si is composed of small crystallites with grain boundaries between them. Defects caused by atomic dangling bonds and disordered material result in deep trapping states at grain boundaries that reduce the number of free carriers and create space charge regions in the crystallites and potential barriers, that impedes carrier motion [10.44, 45]. For a given average grain size there exists a critical doping concentration N^* for which the grains are completely depleted and the poly-Si layer is semi-insulating, if N is smaller than $N^* = Q_t/L$, where Q_t is an effective trapping state density and L an average grain size. Otherwise, the grains are only partially depleted [10.44, 45].

Poly-Si layers with a thickness of 0.4 μm grown by low pressure vapor chemical decomposition of monosilane on oxidized single-crystal silicon substrates were investigated. These layers were doped by ion implantation of boron with an acceleration energy of 30 keV and a surface density of $1.3 \cdot 10^{12}$ - $1.9 \cdot 10^{15}$ cm^{-2}. Annealing at 1000 °C for 30 min ensured a uniform distribution of the impurity atoms across the layer thickness. Photolithography was used to form planar structures with Al contacts deposited on p$^+$-type regions which had been annealed before. In most cases the size of the resistive body was 150×40 μm^2. The average grain size was about 110 nm as shown by investigations with a transmission electron microscope. Silicon grains with a size smaller than 120 nm are completely depleted if the doping concentration is less than about $1{\cdot}10^{17}\,cm^{-3}$ and possess SI properties [10.45]. Thus, lightly doped poly-Si layers are semi-insulating.

10.2.4 Porous Si

P-Si has gained a great interest since the discovery of its visible luminescence properties [10.46] due to silicon nanostructures [10.47, 48], exhibiting quantum confinement effects [10.49, 50]. A lot of attention has been paid to this material in the attempt to produce light emitting devices that are completely compatible to silicon technology and a number of different electroluminescence structures based on porous Si have been already developed [10.51–57]. The possible cheap implementation of monolithically integrated solid state opto-electronic devices could be very promising for high speed information transfer in future computer systems.

A detailed analysis of *I-V* curves measured at room temperature indicates a large value of the series resistance [10.56, 57] in sandwich structures with junctions. A very large value of the applied bias drops across the series resistance, not across the junction. The first reason of the relatively large series resistance is in the SI properties of porous silicon layers because of the completely depleted silicon nanostructures [10.56,58]. The second one is in a large density of interface states on the surface of the silicon nanostructures [10.57, 58]. The presence of electrically active defects, which can act as luminescence killers, must be carefully analyzed [10.58]. Also, the fractal-like structure of porous Si appears to influence the nonequilibrium charge relaxation in porous Si [10.59, 60].

The investigated structure was similar to that described in [10.61] originally designed to measure quantum transport phenomena in the transverse electrical conduction process in P-Si, but without the p-n junctions under the contacts. P-Si layers with a thickness of 1.5 μm were fabricated using a standard electrochemical dissolution of p-Si with 10 Ω cm resistivity (111) Si in a 48% HF electrolyte. The p-type Si with low doping level was used, because porous structures formed in it are characterized by an interconnected network of nanometer sized silicon ligaments with porosities on the order of 40 - 60% and pore dimensions of less than 10 nm [10.48]. The anodized wafers were rinsed in deionized water for 30 minutes and dried in nitrogen atmosphere. A thin aluminium film was deposited on the p^+-type boron-doped back side of the wafers and annealed at 450 °C for 15 min to form an ohmic contact in order to provide a uniform current distribution during anodization. The porosity of P-Si layers (40 - 55%) was varied with a forming current density change in the region of 5 - 80 mA/cm^2. A half of each wafer was treated by dipping it in 5 wt.% aqueous HF solution for about 5 min before the deposition of Al contacts. An Al layer of about 0.5 μm was deposited and was patterned using a standard rf sputtering, photolithography and etching steps, in order to fabricate contacts to porous silicon. The contact dimensions were approximately $1 \times 1\,mm^2$; the distances between the contacts were about 1 mm. Some of the samples were then annealed at 450 °C for 15 min in nitrogen. Current-voltage characteristics of P-Si layers were investigated in high electric fields and current instabilities were measured using a time-domain technique [10.31].

10.3 Results and Discussion

10.3.1 GaAs

Some experimental observations concerning the influence of the technological treatment (as a result of different concentrations of EL2 defects) of SI GaAs on peculiarities of nonlinear dynamics during charge transport are presented here.

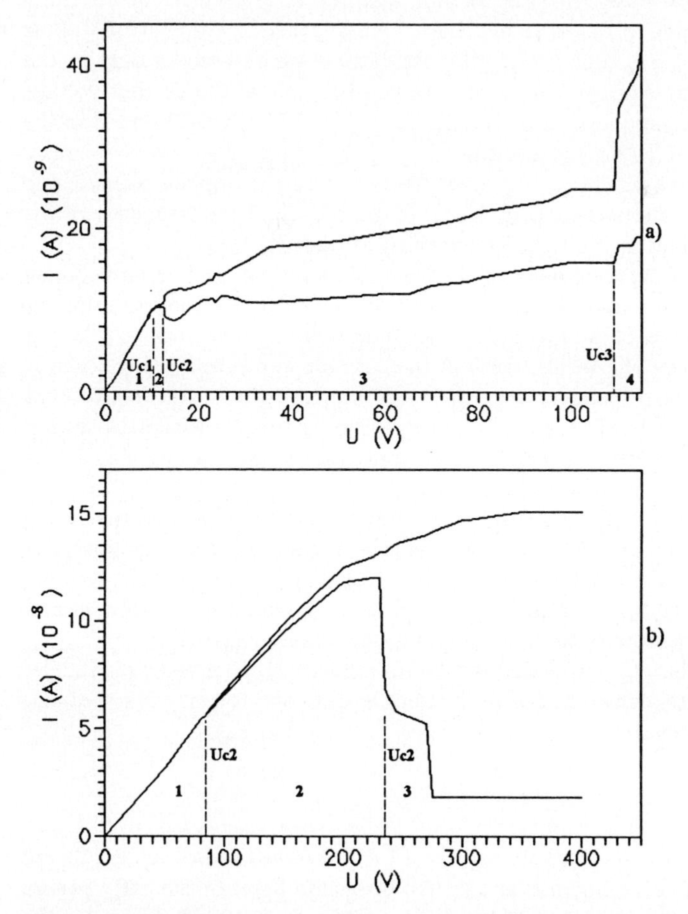

Fig. 10.2. Typical current-voltage characteristics of a) a sample annealed at $P = 0.0\,\mathrm{atm}$; b) a sample annealed at $P = 4.9\,\mathrm{atm}$; 1 - linear and sublinear I-V dependencies; 2 - smoothly growing amplitude, small oscillations; 3 - high amplitude quasiperiodic or relaxation pulse-like oscillations; 4 - chaotic oscillations

All investigated samples showed I-V characteristics very similar to those reported in [10.62] and shown in Fig. 10.2.

The following distinct regimes could be recognized: 1) a smooth and slightly sublinear characteristic at low applied voltage, 2) small regular oscillations before the saturation, 3) high amplitude quasiperiodic oscillations for the sample annealed at $P(As_2) = 0.0$ atm and relaxation pulse-like oscillations for the other samples, 4) chaotic oscillations for the sample annealed at $P(As_2) = 0.0$ atm. Figure 10.3 shows typical current oscillations and power spectra at different applied voltages for the sample annealed at $P(As_2) = 0.0$ atm.

The applied voltage appears to be the bifurcation parameter of the system. At low applied voltages regular oscillations were observed (Fig. 10.3a). The period-doubling bifurcations [10.19] were found for intermediate voltage values (Fig. 10.3b). At applied voltages greater than about 100 V intermittent chaotic behavior [10.63] was observed (Fig. 10.3c). The period doubling and intermittent chaotic behavior in SI GaAs had been found earlier in [10.17, 19, 63]. It is interesting to point out that the chaotic oscillations were observed only in the samples annealed at $P(As_2) = 0.0$ atm. The third regime of the current voltage characteristics with regular puls-like oscillations (Fig. 10.2b) was observed in the samples annealed at $P(As_2) \neq 0.0$ atm.

The dependencies of the fundamental frequency on the applied bias voltage was found to be nonmonotonic (Fig. 10.4). A sharp drop of the frequency of the oscillations was found in the samples annealed at high $P(As_2)$.

According to [10.62] the difference between the regimes 2) and 3) corresponds to the transition from sinoidal current oscillations (weak, fluctuating domains forming at some points of the cathode border) into pulse like oscillations (distinct domains forming over the whole width of the cathode and propagating towards the anode, where they collapse and the cycle starts again). We believe that the sharp drop of the frequency could be explained by this transition. A similar frequency drop of oscillations linked with spatio-temporal oscillations of a current density filament had been found in [10.64].

Real time recording of the spatial and temporal behavior in semi-insulating semiconductors had been done by using spatially resolved photoconductivity measurements and the electron-beam-induced change in the electric conductivity [10.65], the voltage contrast spectroscopy in a scanning electron microscope [10.17], and electro-optic light modulation technique [10.62].

In order to explain the presented results it seems useful to investigate the dependencies of the frequency modes on the temperature and to carry out spatially resolved measurements.

10.3.2 InP

The peculiarities of the method of analysis of LFO observed in InP [10.21,22] are very important for obtaining new results and presented here. In the early papers [10.14,15] LFO were observed with frequencies varying from 10^{-2} to 10^2 Hz as the temperature varies from 250 K to 320 K. To resolve the fundamental frequency the sampling time was varied. A set of 1024 samples was then obtained at each temperature and treated by FFT. In contrast to [10.14,15] the signal was sampled

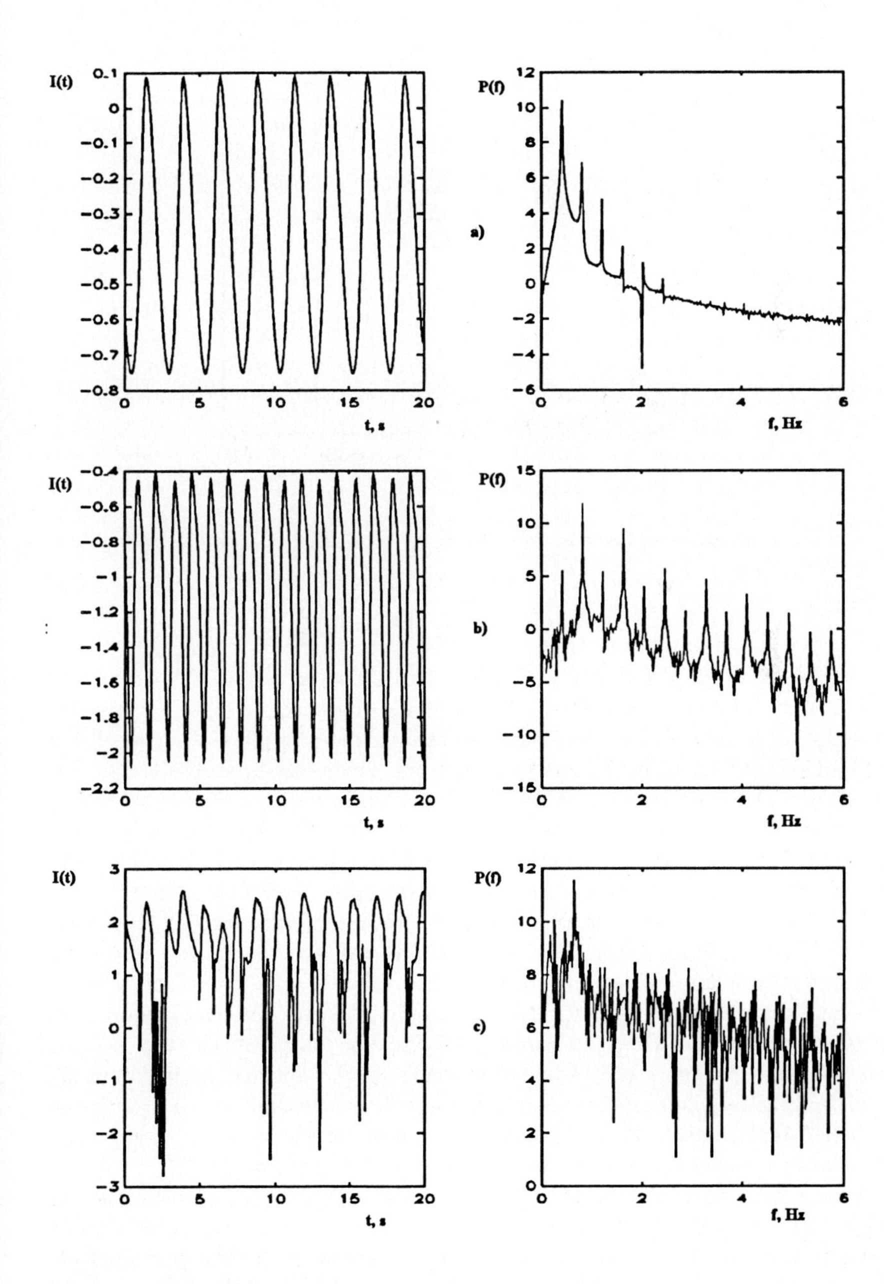

Fig. 10.3. Current vs. time and power spectra, log(power) vs. frequency. Applied voltage: a) 15 V, b) 35 V, c) 100 V

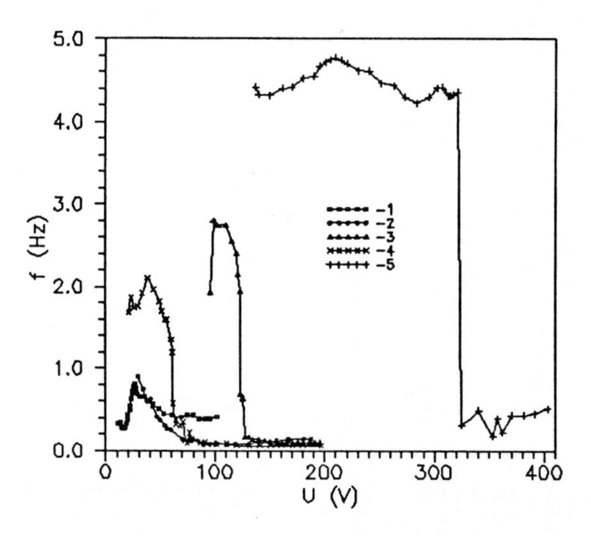

Fig. 10.4. Fundamental frequency vs. applied voltage. $P(As_2)$ (atm.): 1-0.0; 2-1.0; 3-1.3; 4-2.0; 5-4.9

at a fixed rate of 200 Hz [10.21, 22]. Figure 10.5 shows examples of two seconds of data from various temperatures. To avoid aliasing effects [10.66], frequencies above the Nyquist frequency were removed using a low-pass filter. Aliasing effects could introduce false peaks to the power spectra. From the data obtained in this way, slower sampling rates were obtained by simulating in software the effect of another low-pass filter. The low frequency oscillations were detectable with average applied fields as low as 800 V/cm, but as the field was increased more spectral peaks appeared - hence the choice of a rather large field, $4.3 \cdot 10^3$ V/cm [10.21]. The power spectra were obtained by means of the maximum entropy method which, as Press et al. [10.66] state, is preferable to the Fourier transform technique when fitting sharp spectral features. The Fourier transform spectra contained sharp peaks which increased in amplitude from the background level typically by a factor of about 10 over about 1 Hz. The maximum entropy method with 160 poles produces the same sharp peaks as the Fourier method but the smoother spectra produced by the maximum entropy method are more suitable for the automated extraction of peaks, so this method was used for all analyses. The Fourier transform method was used only for comparison [10.22].

Also, a procedure of overlapping of each set with its neighbors by half its length was done [10.22] to reduce the spectral variance per data point [10.66]. Figure 10.6 shows the oscillations appearing to switch mode: the dominant oscillation at 1.5 s was about 6 Hz, but at 5 s had shifted to 4 Hz. This mode of switching did not preserve the phase, i.e. when the dominant oscillation was again 6 Hz it was not in phase with the original 6 Hz wave train. The broadening and weakening of the spectral peaks introduced by the lack of coherence was reduced by computing spectra from sets of 400 measurements, each set being less

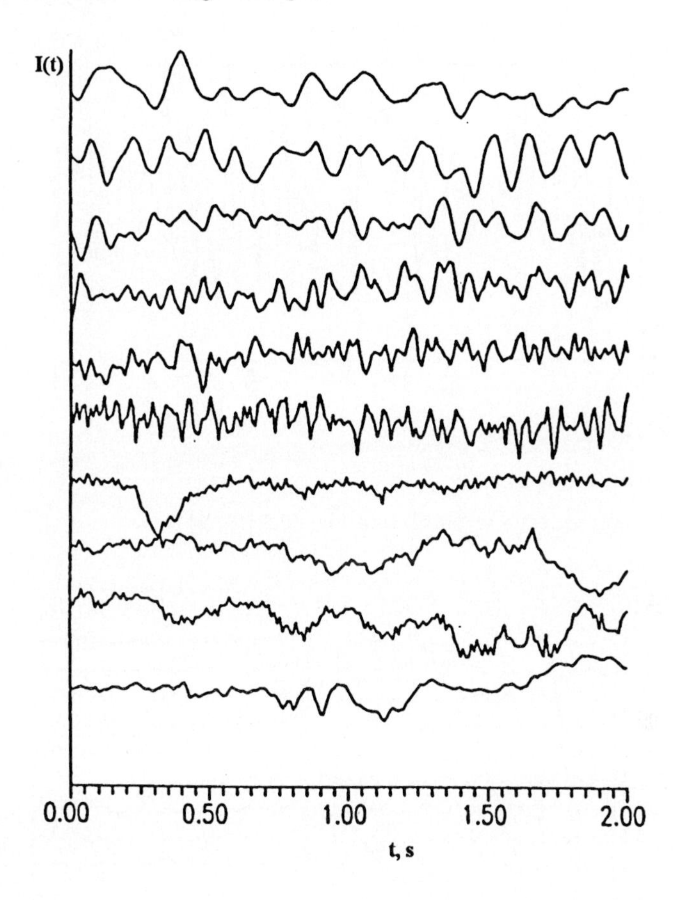

Fig. 10.5. Current vs. time at intervals of 5 K from 295 K at the bottom to 250 K at the top. An average field of $4.3 \cdot 10^3$ V/cm was applied. Since the magnitude of the fluctuations changed significantly with temperature, the vertical scale was determined individually for each temperature [10.21, 22]

likely to contain phase-shifts instead of using all 32768 measurements at once. The spectra calculated were averaged to reduce the amplitudes of random fluctuations. With the same total number of points used, the spectra indeed showed sharper peaks when the set size was lower [10.22].

The averaged spectra consisted of a number of peaks superimposed on a smoothly varying $1/f^{3/2}$ background as found in [10.13]. In order to identify weaker peaks an estimate of the background was removed by subtracting from each point the median value determined within 5 Hz of that point. The median was used since it was less affected by the presence of nearby peaks than an average. The analysis program extracted the local maxima and sorted them in order of amplitude, discarding those with amplitudes less than twice the standard

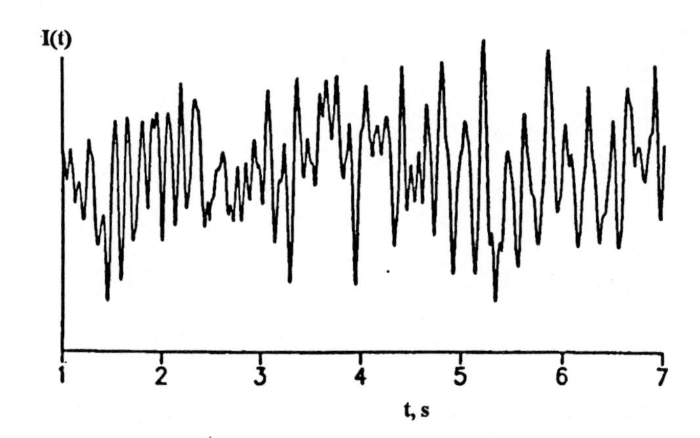

Fig. 10.6. Current vs. time at 255 K as in Fig. 10.5 [10.22]

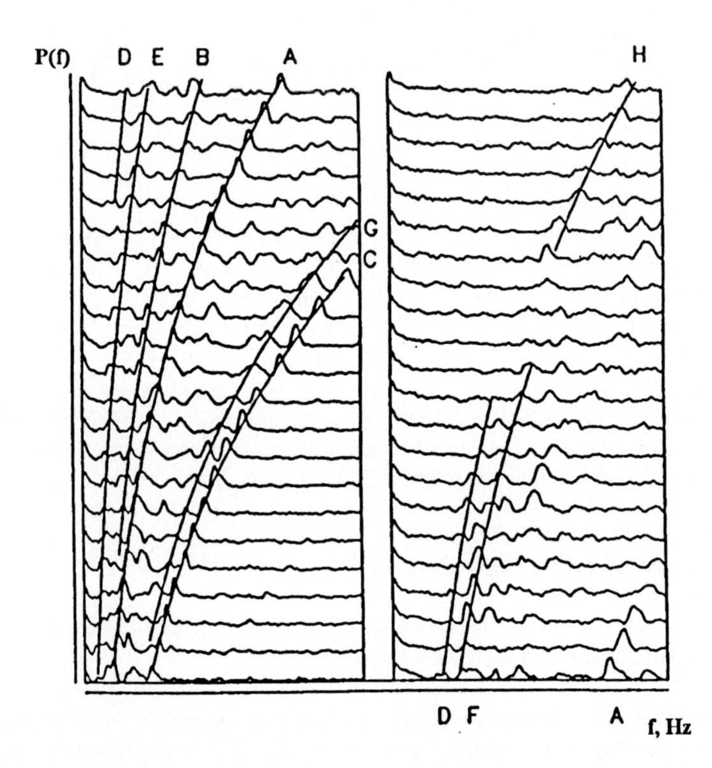

Fig. 10.7. Power spectra, log(power) vs. frequency. Upon the spectra are plotted the lines described by the peak frequencies as a function of temperature as calculated from the trap parameters shown in Fig. 10.8, with single-letter labels according to Table 10.1 [10.22]

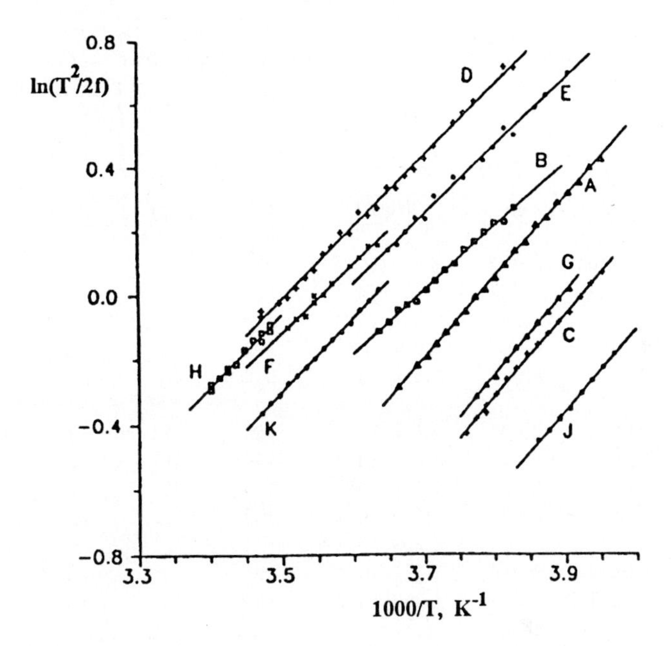

Fig. 10.8. Arrhenius plot of $ln(T^2/2f)$ vs. $10^3/T$ for the main oscillations. Labelled as in Table 10.1 [10.22]

deviation in the background within 5 Hz of that frequency [10.22]. The calculated oscillation frequencies with baseline removed were plotted on the power spectra (Fig. 10.7). The largest 10 of the remaining maxima were plotted on an Arrhenius plot of $ln(T^2/2f)$ vs. $1/T$ (Fig. 10.8). The 14 detected spectral peaks were listed in Table 10.1.

Although the observed original method of analysis of power spectra is very productive, it did not propose that all the 14 peaks correspond to 14 different levels. Apart from one lone peak with the activation energy of 0.3 eV, they could be grouped around 0.39, 0.41, 0.44 and 0.49 eV where in each group the differences in activation energies were not statistically significant [10.22]. Two observed peaks in SI GaAs with very close activation energies (0.69 and 0.70 eV) were observed in [10.14,15] and attributed to distinct traps identified as EL0 and EL2. Two previously known levels in InP with Fe content were identified in [10.22] as 0.494 eV (0.47 eV [10.68]) and 0.299 eV (0.29 eV [10.69]).

Also, it was mentioned in [10.22] that domains could be triggered at more than one position in the specimen, and therefore had different transit times. Thus one type of a trap could launch domains with different frequencies. So, the question was if the peaks in each group were produced by the same level, each giving a fundamental frequency, and harmonics are governed by the same activation energy [10.22]. A more acceptable model [10.67] suggests that the passage of a high field domain could modify the occupancies of the traps. The relaxation of the occupancy to the steady-state could involve the thermally activated emission

Table 10.1. Activation energies, capture cross sections and their uncertainties for the peaks shown in Fig. 10.7, arranged in 5 groups, each containing peaks with similar activation energies. Sample frequencies are given and the ratio, R, of these to the first frequency is given

Label	W (eV)	$\pm$	σ (cm^2)	$\pm$%	f (Hz)	Ratio
Group 1						
P	0.480	0.019	4.7×10^{-15}	217	56	1.0
A	0.494	0.003	1.1×10^{-14}	114	72	1.3
G	0.512	0.009	5.0×10^{-14}	149	153	2.7
C	0.507	0.007	4.7×10^{-14}	138	178	3.2
J	0.502	0.027	7.5×10^{-14}	346	351	6.3
Group 2						
D	0.438	0.006	2.4×10^{-16}	128	17	1.0
H	0.444	0.021	3.5×10^{-16}	236	19	1.1
K	0.450	0.004	7.5×10^{-16}	119	32	1.9
N	0.434	0.008	4.1×10^{-16}	143	35	2.0
Group 3						
F	0.414	0.016	1.1×10^{-16}	193	22	1.0
E	0.417	0.011	1.4×10^{-16}	160	25	1.1
Group 4						
L	0.384	0.007	5.3×10^{-17}	135	37	1.0
B	0.389	0.007	7.7×10^{-17}	137	44	1.2
Group 5						
M	0.299	0.012	1.9×10^{-19}	162	5	1.0

of carriers with corresponding time constants, with accompanying modulation of the conductivity. The Fourier transform of the exponential decay would produce a peak at a frequency which would be governed by the activation energy of the carrier emission.

On the other hand, it has been pointed out in [10.62] that the temporal evolution of a complete domain cycle can be described by two characteristic times: the domain decay time at the anode and the transit time of the domain. If the former time is much less than the latter, the two states do not interfere. Otherwise one may find a range of these times that give rise to period doubling or nonlinear coupling of the two states, eventually leading to chaotic behavior [10.62].

For nonlinear dynamics and its possible applications for material research of SI InP as well as other crystalline SI semiconductors then the question arises, if a precise experimental analysis like that done in [10.22] could extract a number of independent frequency modes and be useful as a method of obtaining information on deep levels in semi-insulating crystals by analyzing the temperature dependence of LFO as trap signatures due to the formation and transit of high field domain, or if these oscillations are chaotic. In other words, if the transition to turbulence of the current in SI InP is as an indefinite cascade of instabilities (Landau [10.70]) or chaotic behavior appears already with few interacting modes (Ruelle and Takens [10.71]). In contrast to the former "linear" interpretation, the dimension analysis [10.72–74] gives a totally different picture of low-dimensional chaotic dynamics of a few degrees of freedom only [10.4].

10.3.3 Polycrystalline Si

Experimental *I-V* characteristics of poly-Si had a hyperbolic-sine behavior as in [10.45]. When the current-voltage curves were recorded under static conditions in poly-Si layers with a boron concentration N less than $2{\cdot}10^{17}\,\mathrm{cm}^{-3}$ ultralow frequency current oscillations in a frequency range 10^{-2} - $2{\cdot}10^{-1}$ Hz were observed in high electric fields ($F > 10^4\,\mathrm{V/cm}$) at room temperature [10.29,30]. The amplitude of the oscillations reached a few tens of percent of the steady-state value. The time series of the current $I(t)$ were treated by FFT and the power spectra $P(f)$ of the signal $I(t)$ were analyzed. Figure 10.9 presents current waveforms, their power spectra and phase plots at different applied voltages as an external parameter. In the case of the periodic oscillations (Fig. 10.9a) the current power spectrum consists of a fundamental frequency and a number of its harmonics.

To identify the nature of the oscillations in poly-Si layers the signals from side potential contacts separated along a sample (these signals were applied to the X and Y inputs of a potentiometer via electrometric amplifiers) were also measured in addition to $I(t)$. These measurements are believed to produce a two dimensional projection of phase patterns. If the oscillations had a form close to sinoidal, the phase pattern looked like an ellipse, providing a direct proof of the motion of charge density inhomogeneities in the form of charge density waves [10.75]. LFO had been observed earlier in compensated single crystals Ge:Au and interpreted using a model of spatial trap-recharging waves [10.75,76]. However, additional investigations of the dependencies of the fundamental frequency on the applied voltage and the intensity of illumination of a sample revealed wave-like charge motion processes which were incompatible with the model [10.75]. The fundamental frequency of current oscillations in poly-Si samples increases when the applied voltage is increased and decreases when the illumination intensity is increased. For these dependencies the very reverse is true in the case of spatial trap-recharging waves in compensated single crystals Ge:Au [10.76]. An experimental determination of the phase velocity of the waves in poly-Si layers from phase plots gave $2{\cdot}10^{-3}\,\mathrm{cm/s}$. Taking into account the average grain size $L = 110\,\mathrm{nm}$, the effective time of single barrier recharging could be estimated as $5{\cdot}10^{-3}\,\mathrm{s}$. This is in a reasonable agreement with the capture rate of a hole in a lightly doped (nearly intrinsic) poly-Si with an average hole concentration $p = 10^{11}\,\mathrm{cm}^{-3}$ [10.77] and the capture cross section of grain boundary states for trapping a hole $\sigma_p = 10^{-16}\,\mathrm{cm}^{-3}$ [10.78].

Summing up, a qualitatively new type of wave charge transport, namely the barrier recharging waves in polycrystalline Si layers, was observed [10.29,30]. The physical mechanism of the oscillations has not been satisfactory explained yet, but it is believed to be based on barrier height variation controlled by injected carriers [10.79,80] and their nonequilibrium capture at the grain boundary trap states and reemission in high electric fields.

Quasiperiodic oscillations of the current (Fig. 10.9b) were observed as well

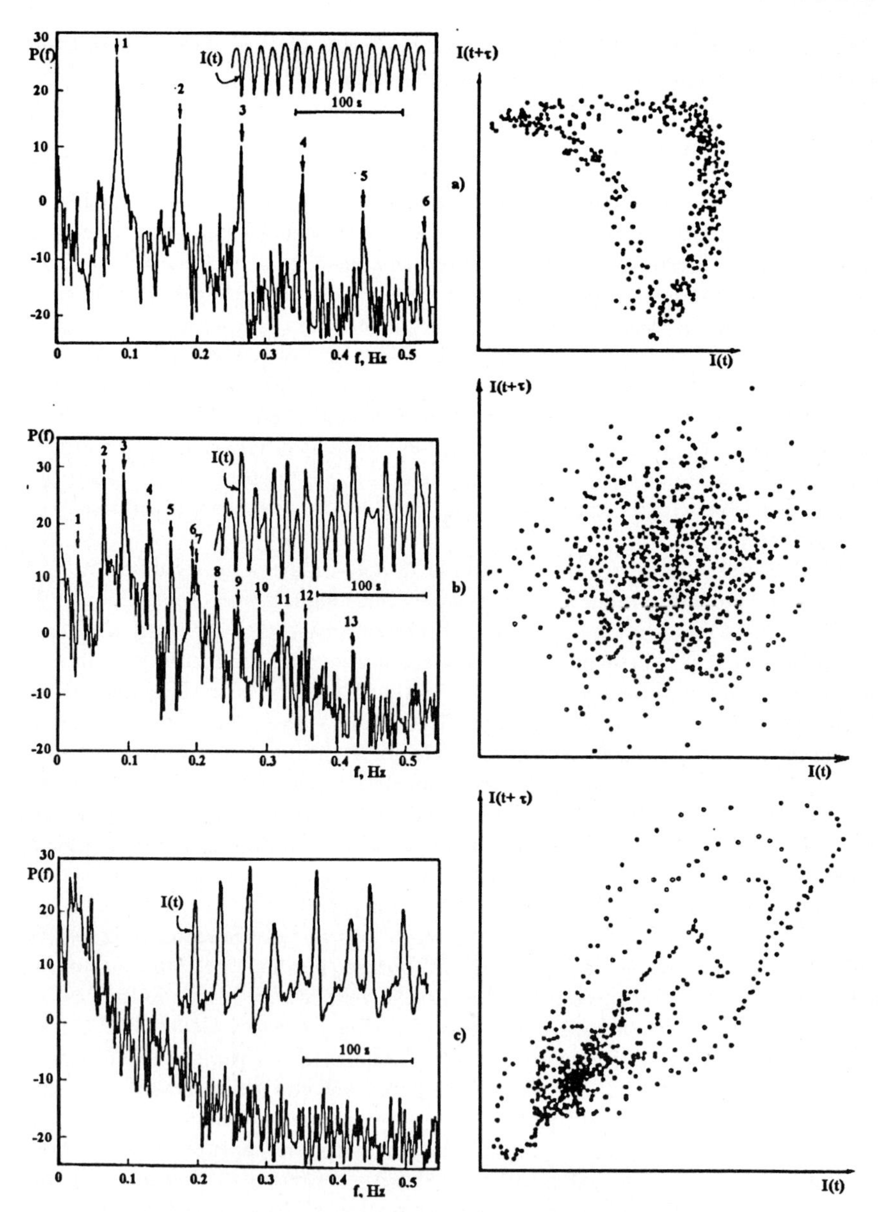

Fig. 10.9. Current waveforms $I(t)$, power spectra $P(f)$ and phase plots $I(t+\tau)$ vs. $I(t)$ of ultra low frequency oscillations in poly-Si layers obtained at different values of the electric field F: a) F_1=1.05·10^4 V/cm, b) F_2=1.15·10^4 V/cm, c) F_3=1.25·10^4 V/cm. a1) f_1; a2) $2f_1$; a3) $3f_1$; a4) $4f_1$; a5) $5f_1$; a6) $6f_1$; b1) f_1-f_2; b2) f_2; b3) f_1; b4) $2f_2$; b5) f_1+f_2; b6) $2f_1$; b7) $3f_2$; b8) $2f_2+f_1$; b9) $2f_1+f_2$; b10) $3f_1$; b11) $2f_1+2f_2$; b12) $3f_1+f_2$; b13) $3f_1+2f_2$

as periodic oscillations when the applied voltage was increased. The power spectrum of the quasiperiodic oscillation consists of two main peaks, indicating the frequencies f_1 and f_2, and their linear combinations. Further increasing of the applied voltage results in chaotic oscillations of the current (Fig. 10.9c).

The electric field appears to be the bifurcation parameter of the system. The transition to the chaotic behavior is believed to follow the Ruelle-Takens-Newhouse route [10.81]. The spectrum of the chaotic oscillations in poly-Si shows the $1/f$ features. Since the $1/f$ noise may derive from collective modes [10.82,83], previously observed $1/f$ noise in poly-Si resistors [10.84] may be explained as a result of chaotic oscillations there.

10.3.4 Porous Si

I-V characteristics of P-Si layers were investigated in high electric fields. Current instabilities were observed when the *I-V* curves were recorded under static conditions. The treatment of porous Si in 5 wt.% aqueous HF solution before Al contacts deposition was found to change the *I-V* curves dramatically (Fig. 10.10).

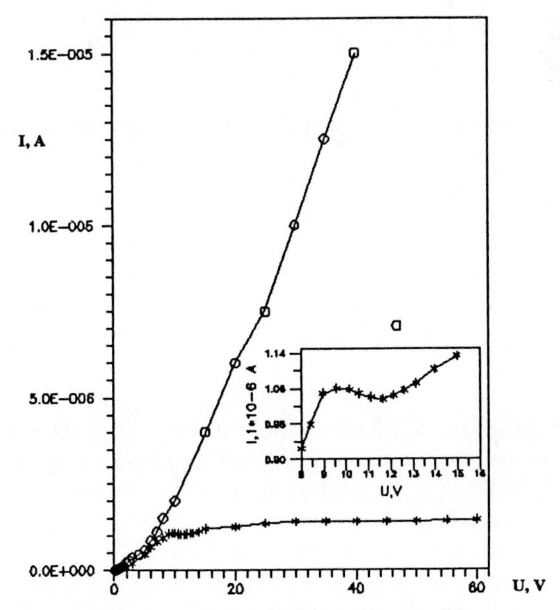

Fig. 10.10. Typical current-voltage characteristics of porous Si structures with Al contacts. * - without HF treatment; o - with HF treatment before Al contacts deposition. Inset a: the negative differential conductivity region of the *I-V* curve

Without this treatment of the samples the *I-V* characteristics were sublinear and even *N*-type curves with a negative differential resistance region were observed. The sublinear regime at high bias appears to be the reason for the majority carrier depletion in thin semi-insulating (SI) P-Si layers because of the interface defects acting as active carrier traps [10.58]. But after the HF treatment, the *I-V* characteristics were found to be superlinear. The hysteresis of *I-V* curves was found to appear in the structures without previous treatment (Fig. 10.11). The hysteresis effect could be caused by a disorder-induced gap

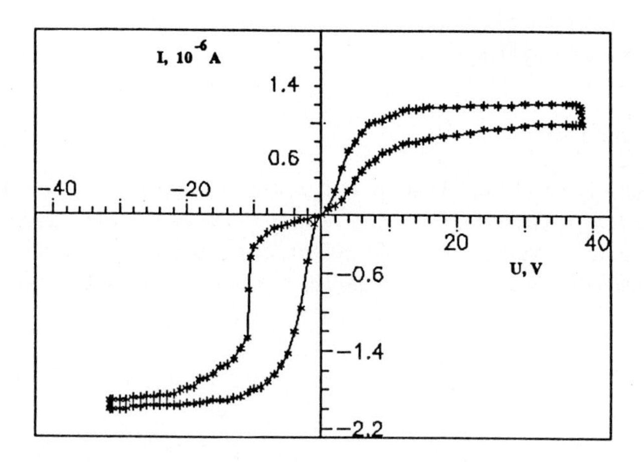

Fig. 10.11. The hysteresis of *I-V* curves of porous Si structures with Al contacts

state continuum which exhibits a continuous distribution of density of states both in energy and in space at the P-Si-SiO_2 interface [10.58], resulting in an extremely wide range of time constants [10.85].

Using the time domain technique, LFO ($f = 0.1 - 10\,\mathrm{Hz}$) have been observed for the first time in porous Si (Fig. 10.12a) [10.31]. The FFT of LFO was also done (Fig. 10.12b).

As mentioned in [10.58], some instabilities were also observed in Al-P-Si-Si heterostructures, when two successive time transients were recorded without having carefully controlled the bias between these two records. The dependence of the fundamental frequency on the applied voltage in P-Si (Fig. 10.13) was found to be monotonically increasing.

As it follows from Fig. 10.12b additional frequencies appear when the applied voltage is increased. The LFOs in P-Si layers are expected to follow the Ruelle-Takens-Newhouse route to chaos [10.81]. The process of LFO in SI P-Si appears to correspond to the following mechanism. Injected carriers are trapped by defects of the interface SiO_2 - P-Si and a space charge field builds up. When this field reaches a certain threshold, the impact ionization of the trapped carriers sets in, increasing the free carrier density in semi-insulating P-Si and resulting in an enhanced dielectric relaxation. Subsequently, the field drops below the impact ionization threshold, trapping of carriers prevails and completes the cy-

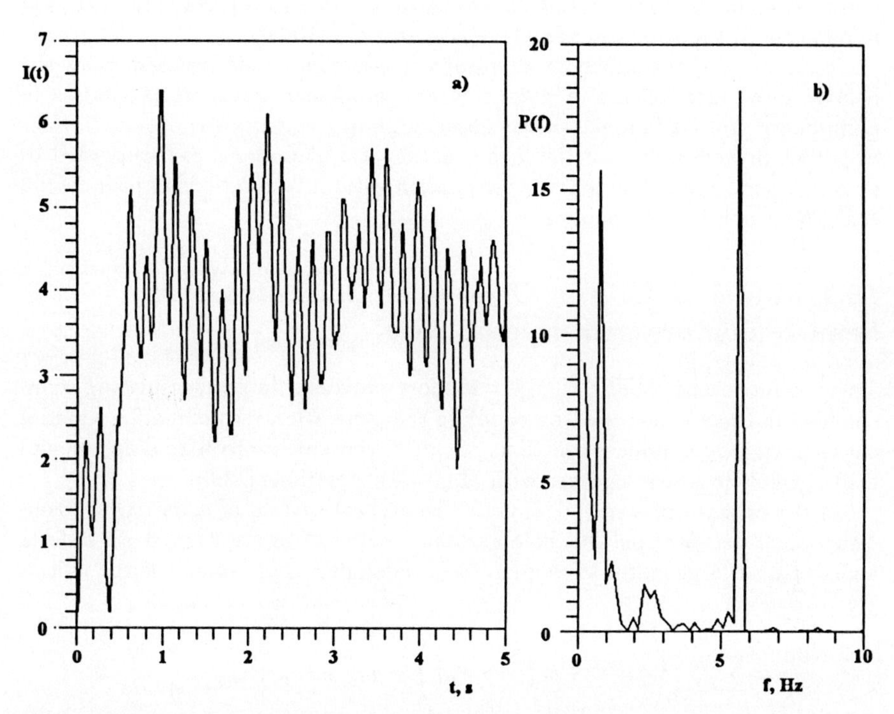

Fig. 10.12. a) Low frequency current oscillations in porous Si structures with Al contacts; b) Power spectrum of low frequency current oscillations in porous Si structures

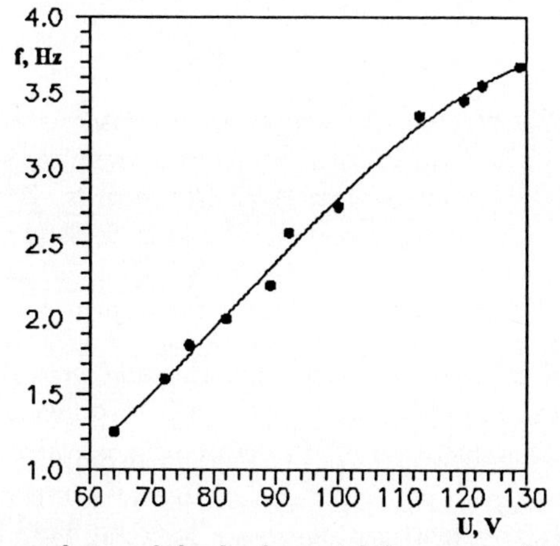

Fig. 10.13. Dependence of the fundamental frequency of low frequency current oscillations in porous Si on the applied voltage

cle [10.9]. From the point of view of material science, it is important that low frequency current oscillations were found in porous Si. This self-organizing phenomenon seems to be promising as a diagnostic tool for deep levels in porous-Si in addition to the time transient behavior analysis [10.58].

Summing up, the influence of standard microelectronic processing on the density of surface defects in porous Si and nonlinear transport properties of aluminium - porous Si junction was observed. Low frequency current oscillations in porous Si were found for the first time. Fractal-like porous Si is supposed to be a new semiconductor material for experimental investigations of phenomena concerning nonlinear dynamics.

10.4 Model of LFO in Crystalline Semi-Insulating Semiconductors with Deep Levels

The nonlinear and chaotic charge transport processes in semi-insulating semiconductors can be described by coupling the generation-recombination kinetics characterized by nonlinear functions of carrier concentrations with diffusion and drift current densities together with Maxwell's equations [10.8].

In the absence of a magnetic field, the general system of equations on condition that the electron and hole exchange between multi-charged ith defects being in jth charge states takes place only through c- and v-bands is the following [10.5, 85]:

$$\frac{\partial n}{\partial t} = \sum_i \sum_j \left(g_{n i}^{j-1} + e_{n i}^{j-1} - r_{n i}^{j} \right) + g_{np} + e_{np} + r_{np} + \frac{1}{q}\frac{\partial J_n}{\partial x};$$

$$\frac{\partial p}{\partial t} = \sum_i \sum_j \left(g_{p i}^{j+1} + e_{p i}^{j+1} - r_{p i}^{j} \right) + g_{np} - e_{np} - r_{np} - \frac{1}{q}\frac{\partial J_p}{\partial x};$$

$$\begin{aligned}\frac{\partial N_i^j}{\partial t} &= g_{n i}^{j-1} + e_{n i}^{j-1} + r_{p i}^{j-1} - g_{n i}^{j} - e_{n i}^{j} - r_{n i}^{j} - \\ &\quad g_{p i}^{j} - e_{p i}^{j} - r_{p i}^{j} + g_{p i}^{j+1} + e_{p i}^{j+1} + r_{n i}^{j+1};\end{aligned}$$

$$J_n = q\left(n\mu_n F + D_n \frac{\partial n}{\partial x}\right); \; J_p = q\left(p\mu_p F - D_p \frac{\partial p}{\partial x}\right);$$

$$\frac{\partial F}{\partial x} = \frac{q}{\epsilon\epsilon_o}\left(p - n + \sum_i \sum_j j N_i^j\right);$$

$$\sum_j N_i^j = N_i = const, \quad (10.1)$$

where q is the charge of an electron, n, p - electron and hole concentrations, μ_n, μ_p, D_n, D_p - their drift mobilities and coefficients of diffusion; F - electric field strength; N_i^j - the concentration of ith defect (donors, acceptors, or deep traps) in jth charge state; $g_{ni}^j, g_{pi}^j, e_{ni}^j, e_{pi}^j, r_{ni}^j, r_{pi}^j$, - light, thermal generation and recombination rates of (i, j)-defect; g_{np}, e_{np}, r_{np} - band to band light, thermal generation and recombination rates; $i = 1, 2, \cdots, l$ - number of kinds of defects, $j(i) = +u(i), u-1, \cdots 0, \cdots, -z+1, -z(i)$ - charge states of ith defect.

The system (10.1) can be analyzed for a number of experimentally important situations.

For a unipolar conduction with one type of carriers (electrons), in undoped SI GaAs [10.38] without photoexcitation and by neglecting the minority carriers (holes), the set of equations (10.1) can be rewritten:

$$\begin{aligned}
\frac{\partial n}{\partial t} &= \sum_i \sum_j (e_{ni}^{j-1} - r_{ni}^j) + \frac{1}{q}\frac{\partial J_n}{\partial x},\\
\frac{\partial N_i^j}{\partial t} &= e_{ni}^{j-1} - e_{ni}^j - r_{ni}^j + r_{ni}^{j+1},\\
J_n &= q\left(n\mu_n F + D_n \frac{\partial n}{\partial x}\right),\\
\frac{\partial F}{\partial x} &= \frac{q}{\epsilon\epsilon_o}\left(\sum_i \sum_j jN_i^j - n\right),\\
\sum_j N_i^j &= N_i.
\end{aligned} \qquad (10.2)$$

For an arbitrary number of deep levels in the energy band gap the set of equations (10.2) can be reduced using the main concepts of synergetics: the slaving principle and the order parameters [10.86]. The order parameters describe the macroscopic behavior of the system. According to the slaving principle, once the macroscopic observables are given, the behavior of the slaved subsystems is determined. In this way a reduction of the degrees of freedom is achieved. The reduction of the degrees of freedom and, consequently, a reduction of the number of equations in the system (10.2) can be done rigorously by the adiabatic approximation, if the relaxation constants Γ_i can be represented in a hierarchical structure [10.86]:

$$\Gamma_1 \ll \Gamma_2 \ll \cdots \ll \Gamma_i \ll \cdots \ll \Gamma_l. \qquad (10.3)$$

This is possible because the emission rate of the ith level depends exponentially on the ionization energy E_{ti} [10.14]:

$$\Gamma_i \approx e_i^{-1} = \theta\sigma_i T^2 exp(E_{ti}/k_b T), \qquad (10.4)$$

where σ_i is the capture cross section of the ith level, θ is a constant for a particular material and k_b is the Boltzmann constant.

For simplicity, consider a single deep donor trap level in the energy band gap. Then the model of periodic LFO [10.10, 13] can be rewritten as follows:

$$\begin{aligned}
\frac{\partial n}{\partial t} &= -C_n(F)(N_t - n_t)n + X_n(F)n_t + \frac{1}{q}\frac{\partial J_n}{\partial x},\\
\frac{\partial n_t}{\partial t} &= C_n(F)(N_t - n_t)n - X_n(F)n_t,\\
J_n &= q\left(n\mu_n F + D_n \frac{\partial n}{\partial x}\right),\\
\frac{\partial F}{\partial x} &= \frac{q}{\epsilon\epsilon_o}(N_A - n_t - n),
\end{aligned} \qquad (10.5)$$

where N_A is the density of compensated, fully ionized acceptors, n - the concentration of electrons in the c-band, n_t - the concentration of electrons trapped in deep donor trap levels like EL2 with the density N_t, and $C_n(F), X_n(F)$ - the field-dependent capture and ionization coefficients, respectively.

The set of equations (10.5) can describe the main features of the dependencies of the fundamental frequency on an electric field, temperature, equilibrium concentration of electrons, compensating acceptors and donor traps, their capture cross section and ionization energy.

In the absence of an external electric field the capture and emission rates are equal to each other due to the principle of detailed balance [10.7]. As a result at or close to the thermodynamic equilibrium, a unique steady state exists, and spontaneous oscillations cannot occur even though the capture-emission rates are strongly nonlinear functions of the carrier densities [10.8].

We have improved the simulation method of [10.13] for linear analysis of the set of equations (10.5). The appearance of oscillations in the model [10.13] is based on the presence of negative differential mobility only, but the capture and emission coefficients are equal to each other due to the principle of detailed balance and do not depend on F.

It is widely known that negative differential mobility can exist in the case of a nonmonotonic field dependence of the drift velocity. For undoped GaAs negative differential conductivity may appear for $F > 3.2 \cdot 10^3\,\mathrm{Vcm}^{-1}$ [10.10]. Thus, according to the model [10.13] the oscillations can not occur at $F < 3.2 \cdot 10^3\,\mathrm{Vcm}^{-1}$.

It is a subtle point that for an instability at low electric fields the capture and emission coefficients must meet additional conditions driving them far from their equilibrium values [10.8]. As mentioned in [10.12], it is a matter of great significance to determine the capture and emission rates correctly.

There are two kinds of models for additional conditions driving the coefficients far from their equilibrium values: 1) field enhanced trapping [10.10,12] and 2) field enhanced emission [10.7]. The model 1) is based on empirical relations for the capture coefficient that is an increasing function of the electric field and the nonmonotonic dependence of the drift velocity on the electric field [10.10]. The model 2) is based on thermal emission given by detailed balance and field-enhanced emission (Poole-Frenkel effect) due to the reduction of the ionization energy and on the nonmonotonic dependence of the drift velocity on the electric field as well [10.7].

The nonmonotonic dependence of the electron drift velocity v on F was used in a widely used form [10.9]:

$$v(F) = v_0 \frac{2F/F_0}{1 + (F/F_0)^2}, \tag{10.6}$$

where v_0 is the maximum of the drift velocity at $F = F_0$. According to (10.6) the low field drift mobility is given by $\mu(F \to 0) = 2v_0/F_0$.

The dependence of the capture coefficient on the electric field was used in the form [10.5]:

$$C_n(F) = C_n(0)\left(1 + \beta_1 \frac{(F/F_1)^{\beta_0}}{1+(F/F_1)^{\beta_0}}\right) \qquad (10.7)$$

where $C_n(0) = \sigma v_{th}, \sigma$ is the capture cross section of deep donor trap levels like EL2 for electrons at zero electric field, $v_{th} = \left(3k_bT/m^*\right)^{1/2}$ - the thermal velocity of electrons, k_b - the Boltzmann constant, T - the temperature, m^* - the effective mass of electrons in a lower valley (Γ). We did not include the capture of electrons from the upper (L) valley, since the model for EL2 electron field-enhanced capture from the L-valley did not fit the experimental data [10.12]. The simple empirical relation (10.7) based on currently accepted EL2 parameters [10.12] was used without further improvements, just to describe qualitative behavior of frequency modes of LFO.

We also included into the model the field enhanced emission due to the reduction of the ionization energy. The field-dependence of the emission coefficient was used from [10.7]:

$$X_n(F) = C_n(0)N_c(T)exp\left(-\frac{E_t - \beta E_{PF}(F)}{k_bT}\right), \qquad (10.8)$$

where $N_c(T) = 2(2\pi m^* k_bT/h^2)^{3/2}$ is the effective density of states in the conduction band, β - a dimensionless geometric factor and $E_{PF} = (Fe^3/\epsilon\epsilon_0)^{1/2}$ is the reduction of the ionization energy.

The linearization of equations (10.5) taking into account (10.6) - (10.8) was done with the substitutions:

$$n = n_s + \delta n, n_t = n_{ts} + \delta n_t, F = F_s + \delta F, \rho = \delta\rho, \qquad (10.9)$$

$$C_n = C_{ns} + \delta C_n, \quad \text{where } \delta C_n = \left(\frac{\partial C_n}{\partial F}\right)_{F=F_s} \delta F,$$

$$X_n = X_{ns} + \delta X_n, \quad \text{where } \delta X_n = \left(\frac{\partial X_n}{\partial F}\right)_{F=F_s} \delta F,$$

$$v = v_s + \delta v, \quad \text{where } \delta v = \left(\frac{\partial v}{\partial F}\right)_{F=F_s} \delta F. \qquad (10.10)$$

By neglecting the terms of the order greater than one, the equations (10.5) can be rewritten in the following form:

$$\frac{\partial \delta n_t}{\partial t} = \left((N_t - n_{ts})n_s\left(\frac{\partial C_n}{\partial F}\right)_{F=F_s} - n_{ts}\left(\frac{\partial X_n}{\partial F}\right)_{F=F_s}\right) -$$

$$- (C_{ns}n_s + X_{ns})\delta n_t + C_{ns}(N_t - n_{ts})\delta n,$$

$$\frac{\partial \delta n}{\partial t} = -\frac{\partial \delta n_t}{\partial t} + v_s\frac{\partial \delta n}{\partial x} + n_s\left(\frac{\partial v}{\partial F}\right)_{F=F_s}\frac{\partial \delta F}{\partial x} + D\frac{\partial^2 \delta n}{\partial x^2},$$

$$\frac{\partial \delta F}{\partial x} = -\frac{q}{\epsilon\epsilon_0}(\delta n + \delta n_t). \qquad (10.11)$$

For the verification of the stability of the stationary solution n_s, n_{ts}, F_s, small perturbations were chosen in the form of plane waves [10.13, 87]:

$$\begin{aligned} \delta n &= \delta n_k exp[ikx + i\omega(k)t - \tau^{-1}(k)t], \\ \delta n_t &= \delta n_{tk} exp[ikx + i\omega(k)t - \tau^{-1}(k)t], \\ \delta F &= \delta F_k exp[ikx + i\omega(k)t - \tau^{-1}(k)t], \end{aligned} \tag{10.12}$$

where k is a real wave number, $\tau^{-1}(k)$ and $w(k)$ - Re and Im parts of the complex frequency.

As a result we obtain a set of dispersion relations:

$$-\frac{qn_s}{\epsilon\epsilon_0}\left(\frac{\partial v}{\partial F}\right)_{F=F_s}(1+B_1)+(1+B_1)\tau^{-1}-B_2\omega-Dk^2=0,$$

$$-\frac{qn_s}{\epsilon\epsilon_0}\left(\frac{\partial v}{\partial F}\right)_{F=F_s}B_2+(1+B_1)\omega+B_2\tau^{-1}+v_s k=0,$$

where

$$B = B_1 + B_2 = \frac{C_{ns}(N_t - n_{ts}) + i \cdot A}{C_{ns} \cdot n_s + X_{ns} - \tau^{-1} - i \cdot (\omega + A)},$$

and

$$A = \frac{q}{\epsilon\epsilon_0 k}\left((N_t - n_t s)n_s\left(\frac{\partial C_n}{\partial F}\right)_{F=F_s} - n_{ts}\left(\frac{\partial X_n}{\partial F}\right)_{F=F_s}\right). \tag{10.13}$$

It is interesting to point out that the equations (10.13) can be transformed to the identical form of the dispersion relations (9) from [10.13], if $A = 0$ (C_n and X_n do not depend on F) and $X_{ns} = C_{ns}N_c exp[-E_t/k_b T]$ according to the principle of detailed balance.

The set of equations (10.13) can be solved numerically to yield real (τ^{-1}) and imaginary (ω) parts of a complex frequency in (10.11) as functions of not only the wave vector (k) as in [10.13], but also of the electric field (F) and the temperature (T).

The parameters used in the calculations of the dispersion relations, the dependencies of the frequency modes on the electric field and the temperature are listed in Table 10.2.

In Figs. 10.14a and 10.14b the results of dispersion relations (ω and τ^{-1} as function of k at fixed $F = F_s$ and $T = T_s$) are shown.

The form of these dependencies is analogous to the dispersion relations from Fig. 1 in [10.13]. Nonuniform oscillations occur at $k > 0$, but sufficiently large oscillating distributions of the electric field in the sample (domain structure formation) are expected if $k > 2\pi/L$, where L is the length of the sample.

By fixing the wave number $k = k_s$ and the temperature $T = T_s$ we can get ω and τ^{-1} as functions of the electric field. These dependencies are shown in Fig. 10.14b and Fig. 10.15b.

The period of the observed LFO consists of the domain formation and annihilation times and the domain transit time, determined by the domain velocity v_{dom} [10.17]. If the phase velocity ($v_{ph} = \omega/k$) [10.13] is equal to v_{dom}, for large samples ($L \gg 150\mu$m) [10.17] the observed frequency of the LFO's mode is

Table 10.2. Parameters used in the modeling of low frequency oscillations in undoped GaAs

Parameter	Values	References
N_t	$5.0 \cdot 10^{21}$ m^{-3}	[10.13]
n_{ts}	$2.5 \cdot 10^{21}$ m^{-3}	[10.13]
n_s	$1.0 \cdot 10^{11}$ m^{-3}	[10.13] $n_{ts}/n_s = 10^{10}$
μ	$0.6\,\mathrm{m^2V^{-1}s^{-1}}$	[10.13]
v_0	$0.96 \cdot 10^5\,\mathrm{ms^{-1}}$	
$v_0 = \mu F_0/2$		
F_0	$3.2 \cdot 10^5\,\mathrm{Vm^{-1}}$	[10.10]
F_1	$2.1 \cdot 10^5\,\mathrm{Vm^{-1}}$	[10.10]
β_0	4.0	[10.10]
β_1	9.0	[10.10]
σ	$9.71 \cdot 10^{-22}\mathrm{m^2}$	[10.10]
$C_n(0)$	$4.35 \cdot 10^{-16}\mathrm{m^3s^{-1}}$	[10.10]
m^*	$0.068\,m_e$	
β	2.0	
ϵ	12.5	[10.17]
E_t	$1 - 0.75; 2 - 0.70; 3 - 0.68; 4 - 0.66;$ $5 - 0.64; 6 - 0.62; 7 - 0.59$ (eV)	1 - [10.17], 2 - [10.14, 15] 7 - [10.20]
F_s	$2.5 \cdot 10^4\,\mathrm{Vm^{-1}}$ (low field), $5.0 \cdot 10^5\,\mathrm{Vm^{-1}}$ (high field)	
T_s	300 K	
k_s	$2.0 \cdot 10^3\mathrm{m^{-1}}$	If $k_s = 2\pi/L$: $L \approx 3 \cdot 10^{-3}$ m

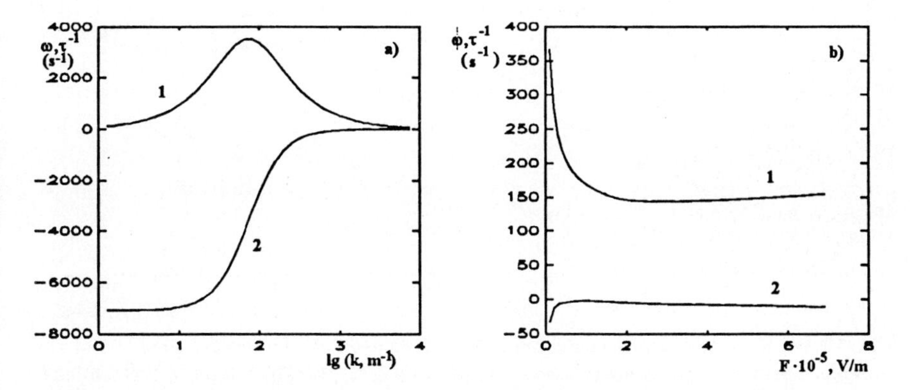

Fig. 10.14. a) The solution of the dispersion relation (10.13) for the LFO (1 - ω, 2 - τ^{-1}): $E_t = 0.59\,\mathrm{eV}$, $F_s = 2.5 \cdot 10^4\,\mathrm{Vm^{-1}}$. b) Electric field dependencies of the real (τ^{-1}) and the imaginary (ω) part of the complex frequency; $E_t = 0.59\,\mathrm{eV}$, $k = 2 \cdot 10^3\,\mathrm{m^{-1}}$, $T = 300\,\mathrm{K}$

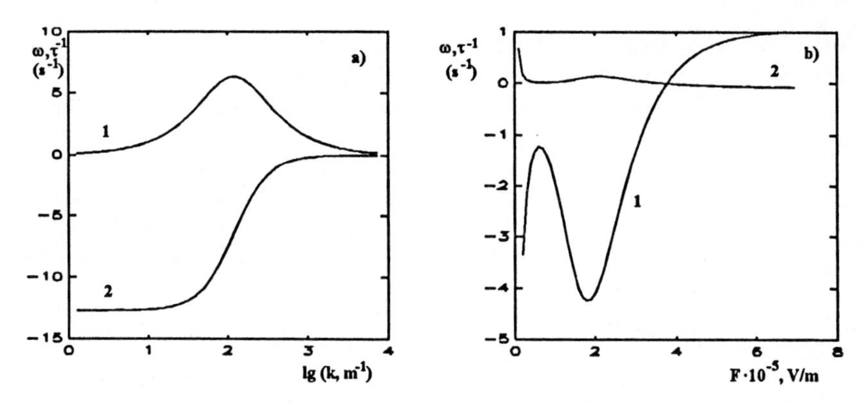

Fig. 10.15. a) The solution of the dispersion relations (10.13) for the LFO (1 - ω, 2 - τ^{-1}): $E_t = 0.75\,\mathrm{eV}$, $F_s = 5 \cdot 10^5\,\mathrm{Vm}^{-1}$. b) Electric field dependencies of the real (τ^{-1}) and the imaginary (ω) part of complex frequency; $E_t = 0.75\,\mathrm{eV}$, $k = 2 \cdot 10^3\,\mathrm{m}^{-1}$

$f = \omega/kL$. Thus, the observed frequency modes (f) are less than the calculated values of ω. Taking into account that oscillations are possible only for negative values of τ^{-1} (according to (10.12)), we find that in the case of the lower ionization energy ($E_t = 0.59$ eV) oscillations exist for all investigated values of the electric field. However, for the higher ionization energies ($E_t = 0.75\,\mathrm{eV}$), τ^{-1} changes the sign at $F \approx 3.8 \cdot 10^5\,\mathrm{V/m}$. So these very low frequency oscillations can exist only at a high electric field. This is more clearly shown on Fig. 10.16: the oscillations exist for values $\omega > 0$, $\tau^{-1} < 0$.

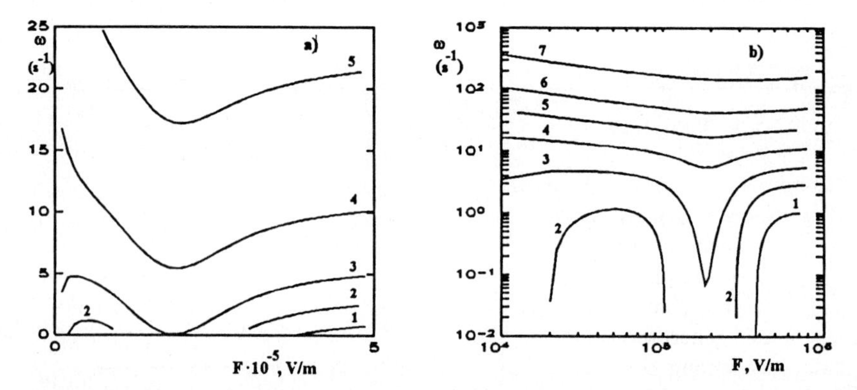

Fig. 10.16. a),b) Identical electric field dependencies of the imaginary part (ω) of the complex frequency in different scales. E_t (eV): 1-0.75, 2-0.70, 3-0.68, 4-0.66, 5-0.64, 6-0.62, 7-0.59; $k = 2 \cdot 10^3\,\mathrm{m}^{-1}$

It is also interesting to point out that the dependencies of the calculated frequency modes on the electric field are nonmonotonic for a different E_t. This corresponds to the nonmonotonic electric field dependencies of frequency modes in the experimental bifurcation plots (Fig. 3b from [10.17] and Fig. 3f from [10.19]).

The calculated temperature dependencies of frequency modes for different E_t are plotted in Fig. 10.17. The slope of the plots corresponds to E_t at low electric

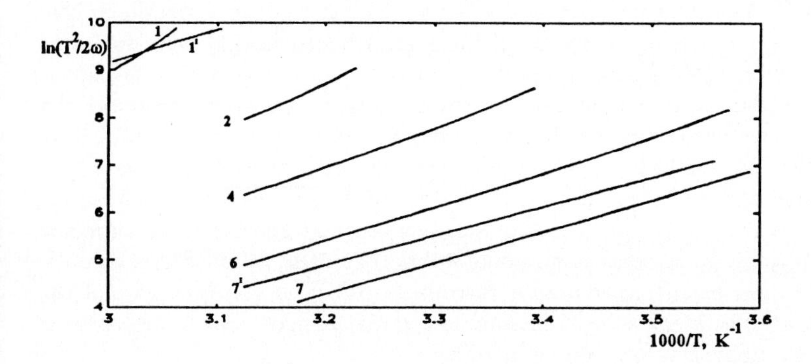

Fig. 10.17. Arrhenius plots of $ln(T^2/2\omega)$ vs. $10^3/T$. $k = 2 \cdot 10^3\,\mathrm{m}^{-1}$; E_t (eV): 1,1'-0.75, 2-0.70, 4-0.66, 6-0.62, 7,7'-0.59; $F\,(\mathrm{Vm}^{-1})$: 1,2,4,6,7 - $2.5{\cdot}10^4$; 1',7' - $5{\cdot}10^5$

fields. But comparing the plots 1 with $1'$ and 7 with $7'$ it is necessary to point out that with increasing electric field the activation energy decreases (the slope of the plot). This corresponds to the experimental results from [10.20].

The model describes qualitatively the nonmonotonic dependencies of frequency modes on the electric field and the influence of the electric field on the activation energies.

10.5 Conclusions

Nonlinear and chaotic charge transport was observed in a number of semi-insulating semiconductors: crystalline GaAs with controlled and different EL2 concentrations, InP and spatially chaotic semiconductors (polycrystalline Si and fractal-like porous Si). All of them were investigated at room temperature. This was possible, because deep levels in the energy band gap of the semiconductors under consideration play the main role in the generation-recombination processes of nonequilibrium charge carrier transport in high electric fields. The frequencies of the self-generated oscillations are on the order of 1 Hz. We believe that this reflects a strong correlation between the concentration of trapped carriers (fast variable) and the electric field (slow variable for semi-insulating semiconductors) [10.9]. This appears to display the "slaving" principle [10.86] in an experiment, when the fast variable is slaved by the slow variable and low frequency current oscillations are observed only. The frequency range and the fact that the oscillations appear at room temperature frequency range are "convenient" for experimental investigations.

The results obtained with the system independent characterization methods emphasizing the underlying nonlinear dynamics may give some further insight into the physics of semiconductors [10.4]. Period-doubling bifurcations and intermittent chaotic behavior were observed in dependence on the applied voltage. The dependencies of the fundamental frequency of low frequency current oscillations on the applied voltage were found to be sharply nonmonotonic in the samples of GaAs with controlled and different EL2 defect concentrations. The observed temperature dependencies of the main frequencies (f) were found to give information on deep levels in InP.

From the point of view of material science, it is important that self-organized phenomena have been used as a diagnostic tool for deep levels in semiconductors on the basis of LFO [10.14–16]. This method is very useful and informative [10.16] because the standard techniques for the determination of deep level concentrations, such as transient capacitance or transient current spectroscopies, are not readily applicable to SI semiconductors [10.17]. The method of LFO is complementary to deep level transient spectroscopy (DLTS); no modifications of the equipment are necessary when the sample-type is changed and information is more easily obtained in SI semiconductors.

The experimental results on nonlinear charge transport and low frequency current oscillations in spatially chaotic polycrystalline and nanostructured fractal-like porous Si are also briefly presented. In these materials the fundamental frequency increases monotonically with the applied voltage. The transition to chaotic behavior through a quasi-periodic route was observed in polycrystalline Si. Semi-insulating polycrystalline Si and porous Si are likely to present a new class of materials for experimental investigations of phenomena concerning nonlinear dynamics.

A model of low frequency current oscillations in semi-insulating crystalline semiconductors with deep levels has been developed. The combination of the approaches of field-enhanced trapping, field-enhanced emission and negative differential mobility of carriers was used. The dispersion relations of the model were found to describe qualitatively the experimentally measured dependencies of the activation energies and the frequency modes on the electric field and the temperature.

Acknowledgments. This work was supported in part by the Fundamental Research Programs of the Ministry of Education of the Republic Belarus: "Intelligent Materials" and "Physics of Information Processing" and Belarus State University "Synergetics". I am greatly thankful to all my colleagues and coauthors for providing the samples, the assistance in the experimental investigations, and the fruitful discussions of the results. I am grateful to Dr. C. Backhouse and Prof. L. Young for their kind permission to present Figs. 10.5 - 10.8 and Table 10.1 of their paper [10.22].

References

[10.1] Y. Abe (ed.): Appl. Phys. A **48**, 93 (1989)
[10.2] R.P. Huebener, J. Peinke, J. Parisi: Appl. Phys. A **48**, 107 (1989)
[10.3] E. Schöll: *Nonequilibrium Phase Transitions in Semiconductors* (Springer-Verlag, Berlin 1987)

[10.4] J. Peinke, J. Parisi, O.E. Rössler, R. Stoop: *Encounter with Chaos* (Springer-Verlag, Berlin 1992)
[10.5] E. Schöll: In *Handbook on Semiconductors*, Vol. 1, 2nd ed., ed. by P.T. Landsberg (Elsevier, Amsterdem 1992) pp. 419-447
[10.6] M.P. Shaw, V.V. Mitin, E. Schöll, H.L. Grubin: *The Physics of Instabilities in Solid State Electron Devices* (Plenum Press, New York 1992)
[10.7] E. Schöll: Physica Scripta **T29**, 152 (1989)
[10.8] E. Schöll: Appl. Phys. A **48**, 95 (1989)
[10.9] E. Schöll: J. Phys. Chem. Solids **49**, 651 (1989)
[10.10] H.K. Sacks, A.G. Milnes: Int. J. Elect. **28**, 565 (1970)
[10.11] K. Pyragas, Yu. Pozhela, A. Tamashyavchyus, Yu. Ul'bikas: Sov. Phys. Semicond. **21**, 335 (1987)
[10.12] D.A. Johnson, R.A. Puechner, G.N. Maracas: J. Appl. Phys. **67**, 300 (1990)
[10.13] S.P. McAlister, Z.-M. Li, D.J. Day: Canadian J. Phys. **69**, 207 (1991)
[10.14] H. Goronkin, G.N. Maracas: In *Proc. IEDM Conf.* (1984) pp. 182-185.
[10.15] G.N. Maracas, D.A. Johnson, H. Goronkin: Appl. Phys. Lett. **46**, 305 (1985)
[10.16] D.J. Miller, M. Bujatti: IEEE Trans. ED-**34**, 1239 (1987)
[10.17] G.N. Maracas, D.A. Johnson, R.A. Puechner, J.L. Edwards, S. Myhailenko, H. Goronkin, R. Tsui: Solid State Electr. **32**, 1887 (1989)
[10.18] D.J. Day, M. Trudeau, S.P. McAlister, C.M. Hurd: Appl. Phys. Lett. **52**, 2034 (1988)
[10.19] W. Knap, M. Jesewski, J. Lusakovski, W. Kuszko: Solid State Electr. **31**, 813 (1988)
[10.20] K.Karpierz, J. Lusakowski, W. Knap: Acta Physica Polonica **A75**, 207 (1989)
[10.21] C. Backhouse, V.A. Samuilov, L. Young: Appl. Phys. Lett. **60**, 2906 (1992)
[10.22] C. Backhouse, L. Young: Solid State Electr. **35**, 1601 (1992)
[10.23] W. Walukiewicz, J. Lagowski, H.C. Gatos: Appl. Phys. Lett. **43**, 192 (1983)
[10.24] S. Maimon, S.E. Schacham: J. Appl. Phys. **75**, 2035 (1994)
[10.25] D. Richter, A.J. Dianoux, W. Petry, J. Teixeira (eds.): *Dynamics of Disordered Materials* (Springer-Verlag, Berlin 1989)
[10.26] R. Schilling: In *Nonlinear Dynamics in Solids*, ed. by H. Thomas (Springer-Verlag, Berlin 1992)
[10.27] Chao Chen, V.A. Samuilov, T.M. Veselova: In *Proc. 8th Conf. on Semi-Insulating III-V Materials* (Warsaw, Poland, June 6-10, 1994) in print
[10.28] Chao Chen, V.A. Samuilov, T.M. Veselova: In *Proc. of the Third annual Seminar on Nonlinear Phenomena in Complex Systems*, ed. by G. Krylov, V. Kuvshinov (Polatsk, Belarus, February 14-16, 1994) in print
[10.29] V.A. Samuilov, V.A. Dorosinetz, N.A. Poklonski, V.F. Stelmakh: In *Proc. of the First Int'l Conf. on Electronic Materials. New Materials and New Physical Phenomena for Electronics of the 21st Century* (MRS, Pittsburg, Pensilvania 1989) pp. 233-236
[10.30] V.A. Dorosinets, N.A. Poklonskii, V.A. Samuilov, V.F. Stelmakh: Sov. Phys. Semicond. **22**, 476 (1988)
[10.31] V.A. Samuilov, E.A. Bondarionok, V.P. Bondarenko, A.M. Dorofeev, G.M. Troyanova: In *Proc. of the Third annual Seminar on Nonlinear Phenomena in Complex Systems*, ed. by G. Krylov, V. Kuvshinov (Polatsk, Belarus, February 14-16, 1994) in print
[10.32] V.A. Samuilov, T.M. Veselova: In *Proc. of the second annual Seminar on Nonlinear Phenomena in Complex Systems* (Polatsk, Belarus, February 15-17, 1993) in print

[10.33] V.A. Samuilov, T.M. Veselova: In *Proc. 8th Conf. on Semi-Insulating III-V Materials* (Warsaw, Poland, June 6-10, 1994) in print
[10.34] V.A. Samuilov, E.A. Bondarionok, D. Shulman, D.L. Pulfrey: IEEE Electron Device Letters **13**, 396 (1992)
[10.35] V.A. Samuilov, E.A. Bondarionok, N.A. Poklonski: In *Proc. of the second annual Seminar on Nonlinear Phenomena in Complex Systems* (Polatsk, Belarus, February 15-17, 1993) in print
[10.36] D.L. Pulfrey, D. Shulman, V.A. Samuilov, E. Bondarionok, V. Krasnitski, N. Poklonski, V. Stelmakh: *Inverted Heterojunction Bipolar Device Having Undoped Amorphous Silicon Layer* (US Patent No. 5,285,083, issued February 8, 1994)
[10.37] A.A. Reeder, I.P. Thomas, C. Smith, J. Wittgreffe, D.J. Godfrey, J. Hajto, A.E. Owen, A.J. Snell, A.F. Murray, M.J. Rose, I.S. Osborne, P.G. LeComber: In *Proceedings of MRS'92 Spring Meeting* (MRS, San Francisco 1992)
[10.38] G.M. Martin, J.P. Farges, G. Jacob, J.P. Hallais, G. Poiblaud: J. Appl. Phys. **51**, 2840 (1980)
[10.39] G. Muller, D. Hoffmann, P. Kipfer, F. Mosel: In *Proc. 2nd Int'l Conf. on Indium Phosphide and Related Materials* (Denver, Colorado 1990) pp. 21-24
[10.40] Chao Chen, V.A. Bykovski, M.I. Tarasik: Sov. Phys. Semicond. **28**, 35 (1994)
[10.41] L.L. Kazmerski (ed.): *Polycrystalline and Amorphous Thin Films and Devices* (Academic Press, New York 1980)
[10.42] D.J. Bartelink: In *Proc. Materials Research Society Annual Meeting on Grain Boundaries in Semiconductors*, ed. by H.J. Leamy, G.E. Pike, C.H. Seager (North Holland, New York 1982) pp. 249-260
[10.43] M. Peisl, A.W. Wieder: IEEE Trans. ED-**30**, 1792 (1983)
[10.44] J.Y.W. Seto: J. Appl. Phys. **46**, 5247 (1975)
[10.45] N.C.-C. Lu, L. Gerzberg, C.Y. Lu, J.D. Meindl: IEEE Trans. ED-**30**, 137 (1983)
[10.46] L.T. Canham: Appl. Phys. Lett. **57**, 1046 (1990)
[10.47] R.L. Smith, S.D. Collins: J. Appl. Phys **71**, R1 (1992)
[10.48] P.C. Searson, J.M. Macaulay: Nanotechnology **3**, 188 (1992)
[10.49] A.G. Cullis, L.T. Canham: Nature **353**, 335 (1991)
[10.50] J.C. Vial, S. Billat, A. Bsiesy, G. Fishman, F. Gaspard, R. Herino, M. Ligeon, F. Madeore, I. Mihalcescu, F. Muller, R. Romestain: Physica B **185**, 583 (1993)
[10.51] A. Richter, P. Steiner, F. Kozlowski, W. Lang: IEEE Electron Device Letters **12**, 691 (1991)
[10.52] F. Namavar, H.P. Maruska, N.M. Kalkhoran: Appl. Phys. Lett. **60**, 2514 (1992)
[10.53] Z. Chen, G. Bosman, R. Ochoa: Appl. Phys. Lett. **62**, 708 (1993)
[10.54] P. Steiner, F. Koslowski, W. Lang: IEEE Electron Device Letters **14**, 317 (1994)
[10.55] J. Wang, F.-L. Zhang, W.-C. Wang, J.-B. Zheng, X.-Y. Hou, X. Wang: J. Appl. Phys. **75**, 1070 (1994)
[10.56] H.P. Maruska, F. Namavar, N.M. Kalkhoran: Appl. Phys. Lett. **61**, 1338 (1992)
[10.57] H.P. Maruska, F. Namavar, N.M. Kalkhoran: Appl. Phys. Lett. **63**, 45 (1993)
[10.58] C. Cadet, D. Deresmes, D. Vuillaume, D. Stievenard: Appl. Phys. Lett. **64**, 2827 (1994)
[10.59] X. Chen, B. Henderson, K.P. O'Donnel: Appl. Phys. Lett. **60**, 2672 (1992)

[10.60] X. Chen, D. Uttamchandani, D. Sander, K.P. O'Donnel: Physica B **183**, 603 (1993)
[10.61] C.C. Yen, K.Y.J. Hsu, L.K. Samanta, P.C. Chen, H.L. Hwang: Appl. Phys. Lett. **62**, 1617 (1993)
[10.62] B. Willing, J.C. Maan: Phys. Rev. Lett., to be published
[10.63] G.N. Maracas, W. Porod, D.A. Johnson, D.K. Ferry, H. Goronkin: Physica B **134**, 276 (1985)
[10.64] F.-J. Niedernostheide, B.S. Kerner, H.-G. Purwins: Phys. Rev. B **46**, 7559 (1992)
[10.65] U. Rau, J. Peinke, J. Parisi, K. Karpierz, J. Lusakovski, W. Knap: Phys. Lett. A **152**, 356 (1991)
[10.66] W. Pres, B. Flannery, S. Teukovsky, W. Vetterling: *Numerical Recipies in C - The Art of Scientific Computing* (Univ. Press, Cambridge 1988)
[10.67] D.J. Day, M. Trudeau, S.P. McAlister, C.M. Hurd: Appl. Phys. Lett. **52**, 2034 (1988)
[10.68] T. Takanohashi, K. Nakai, K. Nakagima: Japan. J. Appl. Phys. **27**, L113 (1988)
[10.69] D.L. Look: Solid State Commun. **33**, 237 (1980)
[10.70] L.D. Landau: Akad. Nauk Doklady **44**, 339 (1944)
[10.71] D. Ruelle, F. Takens: Commum. Math. Phys. **20**, 167 (1971); **23**, 344 (1971)
[10.72] P. Grassberger, I. Procaccia: Phys. Rev. Lett. **50**, 346 (1983)
[10.73] S. Bumeliene, K. Pyragas, A. Cenys: Sov. Phys. Semicond. **24**, 1509 (1990)
[10.74] A. Cenys, K. Pyragas: Phys. Lett. A **129**, 227 (1988)
[10.75] N.G. Zhdanova, M.S. Kagan: Sov. Phys. Semicond. **15**, 99 (1981)
[10.76] N.G. Zhdanova, M.S. Kagan, S.G. Kalashnikov: Sov. Phys. Semicond. **17**, 1182 (1983)
[10.77] N.C.-C. Lu, L. Gerzberg, C.Y. Lu, J.D. Meindl: IEEE Trans. ED-**28**, 818 (1981)
[10.78] A.J. Madenach, J. Werner, F.J. Stutzler: In *Proc. 18th IEEE Photovoltaic Specialists Conf.* (Las Vegas, Nevada, October 21-25, 1985) pp. 1088-1093
[10.79] P.T. Landsberg, M.S. Abrahams: J. Appl. Phys. **55**, 4284 (1984)
[10.80] E. Schöll: J. Appl. Phys. **60**, 1434 (1986)
[10.81] S. Newhouse, D. Ruelle, F. Takens: Commun. Math. Phys. **64**, 35 (1978)
[10.82] V.L. Bonch-Bruevich: Dokl. AN SSSR **278**, 335 (1984)
[10.83] R.D. Peters: Phys. Lett. A **174**, 216 (1993)
[10.84] H.C. de Graaf, M.T.M. Huybers: J. Appl. Phys. **54**, 2504 (1983)
[10.85] N.S. Minaev, N.A. Poklonski, V.F. Stelmakh, V.D. Tkatchev: Sov. Phys. Semicond. **8**, 1076 (1974)
[10.86] H. Haken: *Synergetics. An Introduction* (Springer, Berlin 1978)
[10.87] A. Cenis, G. Lasiene, K. Pyragas: Solid State Electr. **35**, 975 (1992)

11 Technical Applications of a 2-D Opto-electronic P-N-P-N Winner-Take-All Array

C.V. Radehaus and H. Willebrand

University of Colorado at Boulder, Depart. of Electrical and Computer Engineering, Boulder, Colorado 80309-0425, USA

Abstract. In this contribution we review a 2-D Winner-Take-All (WTA) optoelectronic array-network the basic element of which is a bistable p-n-p-n thyristor structure with a light sensitive nonlinear S-shaped current-voltage characteristic. After giving a review of the operation of the basic element, we describe the functionality of the WTA array. We will show that this array can be used in optical signal and image processing such as minimum detection, correlators and fuzzy logic applications.

11.1 Introduction

In optical signal and image processing, in optical communication systems, and in optical computing, it is very often a necessary task to know the location of the highest or lowest light intensity within a given light distribution. In technical applications the standard way to accomplish this is to use a pixel sensor array (e.g., a CCD-array) with an integrated readout system. The serial data stream containing the information about the light input value of each pixel is digitized, and the calculation of the maximum or minimum pixel value is performed by using a digital computer. This is a serial data processing procedure and may impose a bottleneck in time critical applications. Another more direct way specially well suited for time critical applications is to use optoelectronic Winner-Take-All (WTA) networks. Contrary to the standard approach described above, the information processing is done here by an array of light sensitive elements forming the WTA-array. There are two strategies how to built this type of light sensitive elements: one is to use elaborate electronic circuitry with transistors forming differential amplifiers and separate threshold and memory elements similar to those described in [11.1–3] (but these have no optical input). The other one is to use simpler electronic circuitry with more "intelligent" nonlinear elements such as photothyristors. In the following we review an optoelectronic Winner-Take-All (WTA) network that is able to find the location of the maximum for 1-D [11.4] and 2-D [11.5] light intensity distribution by intrinsic parallel processing. Opposite to WTA-networks using electronic circuitry as described above, this network has an extreme simple design by using only compact p-n-p-n structures forming photothyristors with intrinsic and variable thresholding and memory capabilities. To determine the location of the maximum the array uses the analog information

Springer Proceedings in Physics, Vol. 79 **Nonlinear Dynamics and Pattern Formation in Semiconductors and Devices**
Editor: F.-J. Niedernostheide

of the pixel values and in this way the network acts like an analog parallel computer. Each single element in the array has a light-sensitive, *S*-shaped, bistable *I-V* characteristic which causes multistability of the array. Especially, we will see that the operation of the array is based on this system multistability.

The following section presents some basic optical and electrical features of single p-n-p-n elements as well as for the whole array. Next we will discuss maximum detection experiments done with a 13 $\times$ 13 GaAs-AlGaAs p-n-p-n array, the fabrication of which is described in [11.6]. The third section describes applications of the WTA network. Especially, we will discuss the integration of the WTA-structure into set-ups for extrema detection, optical correlators, fuzzy logic systems and optoelectronic neural nets. Final conclusions are drawn in the last section.

11.2 The Winner-Take-All Network

Before discussing the operation of the p-n-p-n WTA array, lets have a closer look at the function of a single p-n-p-n optical thyristor element. Figure 11.1 shows the p-n-p-n layer sequence.

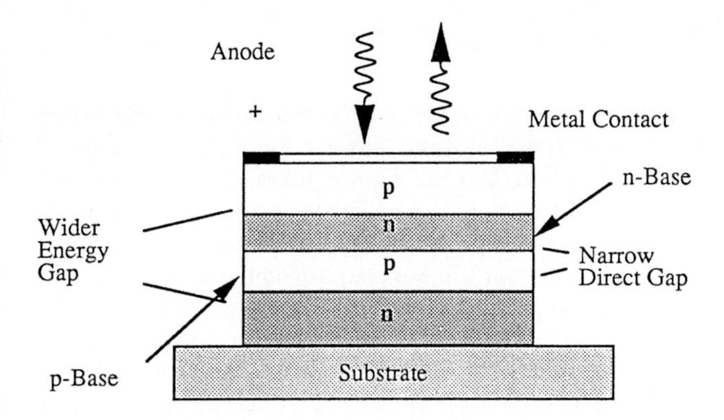

Fig. 11.1. p-n-p-n layer sequence of the optical thryristor

The p-n-p-n structure is grown on a substrate material and has and a metal contact with a light transparent window on the anode side. The structure has all features of an optoelectronic switch, because the middle junction is light sensitive and so we can call this device an optical thyristor. During operation the middle junction is reversed biased and the two outer junctions are forward biased. The device has an on- and an off- state and in the non-conductive off-state it does not emit light, while in the on-state it emits light. The switching between the off- and on-state can be done either by changing the bias voltage or by enlightening the middle junction in its spectral range of sensitivity. In order to illustrate the operation of a single element a schematic band diagram is shown in Fig. 11.2.

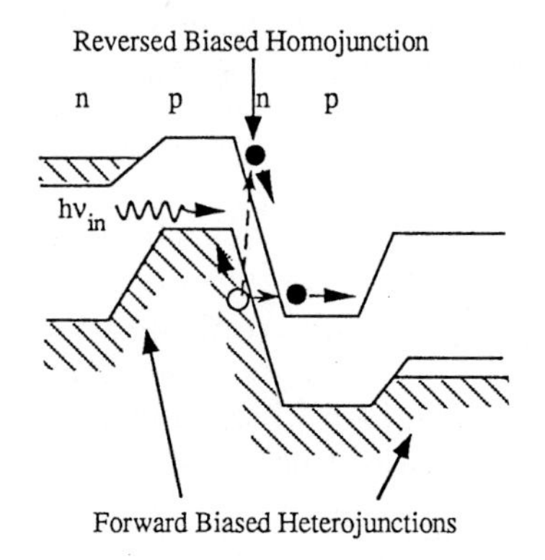

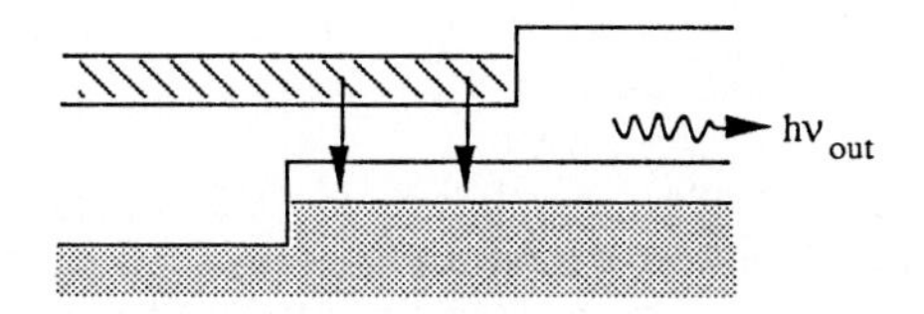

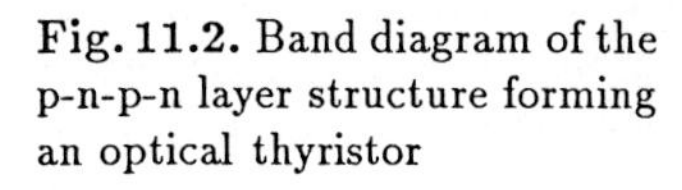
Fig. 11.2. Band diagram of the p-n-p-n layer structure forming an optical thyristor

The switching behaviour of the optical thyristor can be seen in the schematic *I-V* characteristic shown in Fig. 11.3 for the illuminated and the non illuminated case.

In the forward biased mode of operation the inner homojunction of the structure is reversed biased and the outer heterojunctions are slightly forward biased (see Fig. 11.2). Without illumination, the breakdown takes place when a certain critical voltage V_z is reached (see Fig. 11.3). This is due to the Zener effect or to punch-through of the depletion layer to one of the heterojunctions. The breakdown voltage V_z is in the order of a few volts, but depends of course on the special design of the structure. By illuminating the inner junction of the structure, the breakdown voltage is decreased. Assuming the loadline AB in Fig. 11.3, illumination causes the device to switch from the high voltage, low current state (point A) to the low voltage and high current state. Especially this behaviour endows the device with threshold capabilities, e.g., by controlling the intensity of the external background illumination. If the load line is flat (high impedance) the switching of the device can be performed optically either by increasing or decreasing the illumination. During the breakdown process we have double injection and the carriers are flooding the inner two layers. Due to this process the reverse bias is compensated and radiative recombination of the injected carriers can be observed. This process causes the light emission from the structure in its turn-on state. The increase of the optical output power versus the current through an optical thyristor with a mesa of $110\,\mu m \times 55\,\mu m$ and an output window of $100\,\mu m \times 35\,\mu m$ is shown in Fig. 11.4. As a function of current, the integrated light output is quadratic for low current values, while it becomes linear for higher current values.

Once the device is in the on-state, it will remain there as long as a voltage of sufficiently high amplitude is applied. By applying pulses, that will keep the

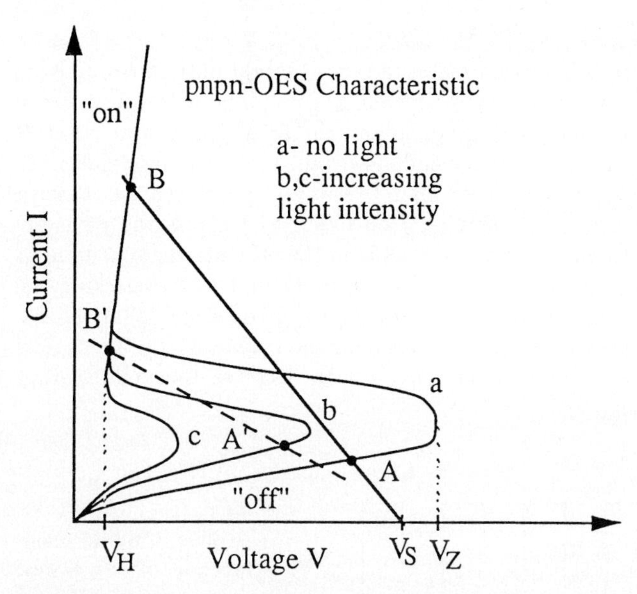

Fig. 11.3. $I(V)$ characteristic of the thyristor switch. Curve (a) shows the device characteristic without external illumination; (b) and (c) illustrate the change of the characteristic with illumination. The load line AB shows an optoelectronic bistability and the load line $A'B'$ an all-optically driven bistability

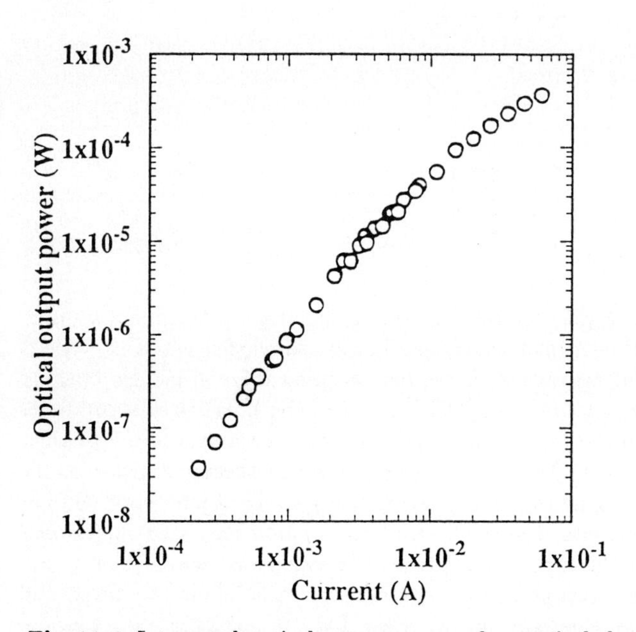

Fig. 11.4. Integrated optical output power of an optical thyristor as a function of the current. The mesa structure was 110 μm × 55 μm and the optical output window was 100 μm × 35 μm (taken from [11.6])

voltage across the thyristor between the holding voltage V_H and the voltage at point B while the device is in the on-state, the optical light output will pulsate between its value at V_H and the value corresponding to point B (see Fig. 11.3). While the light emission at V_H is negligible, and the light emission at point B can be higher by a few orders of magnitude (see Fig. 11.4), the modulation of the optical output signal can be rather high. We would like to remark, that in the on-state the difference in the voltage drop (V_B - V_H) across the device is very small and the light emission pulsates, while in the off-state there is no light emission and the voltage difference V_A and V_H is much higher. By checking the mode of operation it is possible to interrogate the on- or off-state of the device. This characteristic feature also displays the memory capability of the optical thyristor. To turn off the device it is necessary to decrease the voltage drop below the holding voltage V_H.

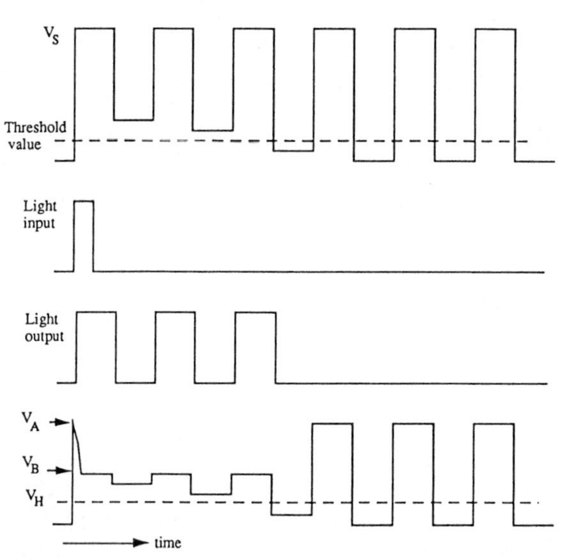

Fig. 11.5. Typical pulse sequence illustrating the memory function of the optical thyristor

Figure 11.5 shows a typical pulse waveshape illustrating the memory and switching function of the optical thyristor. The pulse sequence starts with a situation that an external voltage V_S is applied, keeping the voltage drop across the thyristor at a certain voltage V_A (off-state situation). When a short light pulse hits the photosensitive part of the thyristor structure, the system switches from the off- to the on-state and the voltage across the thyristor drops to its on-state value V_B. As long as the voltage drop stays above the holding voltage V_H, the output light pulsates, indicating that the system has been in the on-state after switching on the input light pulse. The system is switched off when V_S is decreased and the voltage across the thyristor falls below the threshold voltage V_H. Subsequently the light output stops. Even when V_S increases again and passes the threshold voltage, the light output remains zero, indicating that the system is now in the off-state.

Up to now we have discussed the operation of a single thyristor element. In the network array the elements are connected in parallel in order to form a network able to perform the WTA operation. In Fig. 11.6 two variants of a network array can be seen. Both array configurations are connected to a voltage source V_S by a common load resistor. The difference between these two configurations lies in the local coupling of the elements, which becomes obvious in Fig. 11.6b. The resistor network, indicated by R_2 and R_3, causes an electrical coupling of the neighbouring array elements. Besides this electrical coupling there is also an optical coupling due to the penetration of light emission from one element into the neighbouring one. We will now discuss the significance of the coupling on the operation of the network array.

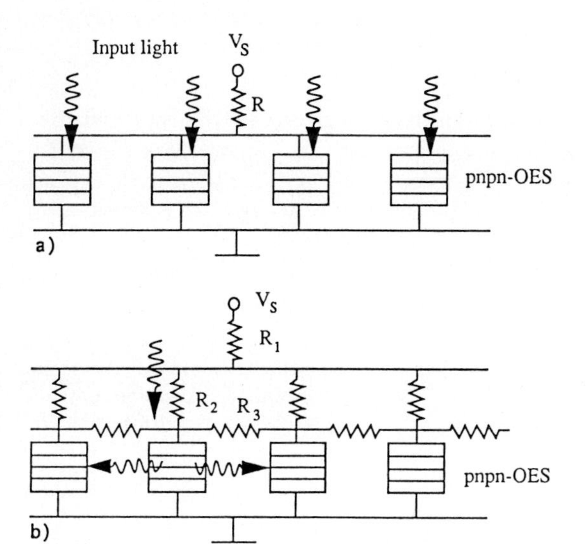

Fig. 11.6. Two possible configurations of the WTA network. In (a) only global coupling is realized by the common load resistor R. In (b) local electrical and optical coupling between adjacent elements can take place

If one of the elements in Fig. 11.6a, all biased through the common resistor R, is illuminated and switches from the off- to the on-state, the voltage across all other elements will drop to the voltage V_B. V_B is below the breakdown voltage of the optical thyristors and due to this fact only one element will stay in the on-state and emit light. In the sequel we will call the action of the load resistor as global inhibition, because it acts on all elements in the same way and limits the total current flow through the network. The operation of the device is that of an "early bird" type winner-take-all network, because only the element that switches first from the off- to the on-state remains the winner. Even when all elements are illuminated at the same time by a light pulse, the competition between the elements will result in the turn on process of only one element. In an array with nearly exactly the same electrical parameters of each element, the device that gets the highest light intensity will win while all other elements will be losers [11.7].

When the same experiments are performed by using the network shown in Fig. 11.6b, the decision process is somewhat different. When one element switches from the off- to the on-state, the voltage drop across the other elements does not drop to the value of the switched element, but is only reduced. This means more light is needed to cause the remaining elements to switch. The electrical interconnection between adjacent elements thus leads to the phenomena of local inhibition. As can be seen in Fig. 11.6b, the elements can also interact by their emitted light fields. In this configuration the light emitted by the switched on element can illuminate the neighbouring element. This light bias causes the reduction of the light intensity necessary to trigger the thyristor. Hence, this way of coupling activates the process of turning on the optical thyristors by an enhancement of their sensitivity. When only adjacent elements are coupled by the radiation field penetrating in free space, this interaction is a local process. By using optical fibers or waveguides, the coupling can become a nonlocal process [11.4].

11.2.1 Maximum Identification

The identification of the maximum intensity of a given light density distribution is an important task in technical applications. The capability of the WTA network to solve this problems can be demonstrated by using the simple experimental set-up shown in Fig. 11.7. The light beam of a He-Ne laser is directed by

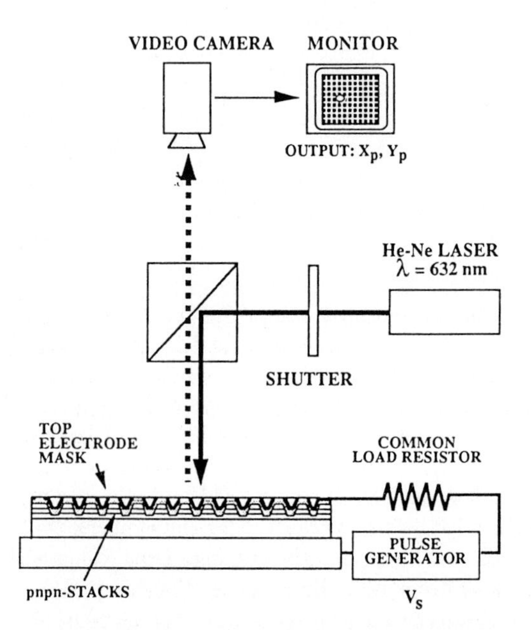

Fig. 11.7. Experimental setup for the demonstration of the maximum identification process (taken from [11.5])

a beam splitter on the surface of the array. By defocusing the laser beam, the diameter and the intensity of the laser spot on the surface of the array can be controlled. When the laser beam profile has a Gaussian intensity distribution, the highest intensity is located just in the center of the laser spot. The shutter is used to turn the input light on and off. The driving voltage is generated by a pulse generator. The video camera, focused onto the array surface, picks up the input light excitation and the optical response of the WTA network. The result of the measurement is displayed on a monitor. The WTA network is an array of the type shown in Fig. 11.6a, which means that there is no local electrical coupling between adjacent detector elements. For the experiments an array of 13×13 p-n-p-n GaAs-AlGaAs double-heterojunction optical thyristors grown by MBE processing have been used. The structure of the thyristors consists of p- and n- layers of various thicknesses and dopant concentrations. For the parallel connection of the thyristor elements in the network array an electrical contact is provided by metallization of the silicon nitride layer. Each array element has a dimension of $20\,\mu\text{m} \times 30\,\mu\text{m}$, and the light input/output window on the top of the structure is $18\,\mu\text{m} \times 20\,\mu\text{m}$ (for details see [11.8]).

To prepare the device to detect an input signal, the detector is biased by a voltage V_S, that keeps the voltage drop across the array just below the breakdown voltage V_B. When the input signal is switched on, the light induced current can be observed. After a dynamic collective competition for the available current, the detector element that gets the highest intensity of the projected light intensity distribution, switches from the off- to the on-state. For this switching process it is of course important that the input intensity exceeds at least in one spot, the threshold light intensity L_T necessary to trigger the switch on process. After the input signal is switched off, only the winning detector element stays in the on state, while all other remain in the off-state. As has been already discussed in Sect. 11.2, the only possibility to reset the WTA network is to switch off the bias voltage V_S. Otherwise the decision process is memorized. As a figure of merit, the decision process itself takes less than 100 ns, while the recovery time of the network after switching off the bias voltage is about 500 ns. The switching performance of the array depends of course on the detector design and the performance of the driving circuit. Recently, it was shown, that an optical switch based on the p-n-p-n structure can operate at 16 MHz with 250 fJ optical input energy [11.8].

The typical result of a maximum identification performed with the WTA network is shown in Fig. 11.8. The two photographs in the upper part of this figure show the location and the shape of the input signal on the surface of the detector array. Due to the Gaussian profile of the input laser beam, the intensity maximum is in the center of the light spots. The two photographs below show the results of the maximum identification process. In both cases only one element is turned on after the bias voltage is switched off. These spots indicate the maximum of the input light intensity distribution. The detection process is widely independent of the local and overall input intensities. For the detection process it is only important that the input intensity exceeds at least at one pixel the required light power for switching the detector element from the off- to the on-state. The value of this threshold light input is in the $10\,\mu\text{W}$ range, and depends on the maximum value of the bias voltage V_S, the common load resistor, the rise time (dV/dt effect) and of course on the duration of the light exposure. To perform the switching process, the exposure time at $10\,\mu\text{W}$ input power, should not be shorter than $3\,\mu\text{s}$. This corresponds to a switching energy of 30 pJ.

Fig. 11.8. Experimental results on the performance of the WTA network. The incident light input signal on the surface of the detector array is shown in the upper part. The pictures below show the result of the maximum identification after the input light is switched off

11.3 Applications of the WTA Network

In the previous section we discussed some general features of the WTA network array. The following section deals with concrete aspects of technical applications of the WTA network. Especially, we will introduce an experimental set-up for a Vander Lugt correlator, for fuzzy logic computation and minimum detection of 2D light fields. Finally, we will discuss the realization of an optical neural network with p-n-p-n elements.

11.3.1 Vander Lugt Correlator with WTA Maximum Identification

The Vander Lugt correlator is used for optical pattern recognition [11.9]. The principle of the pattern recognition process is based on the comparison of the input pattern with a set of stored reference patterns. In combination with a WTA maximum detector an effective device for pattern recognition can be realized. Figure 11.9 shows the schematic setup of a Vander Lugt correlator with an integrated WTA network.

The optical input S is Fourier transformed by using a lens. The output $F(S)$ is projected onto a holographic filter bank, that is used to store the conjugate complex Fourier transform $F(R)^*$ of the reference patterns R and to calculate the convolution function $F(S) \cdot F(R)^*$. Subsequently, the inverse Fourier transform of this signal is calculated and at the output of the second Fourier lens we obtain the correlation function $S \otimes R$. This signal is projected on the WTA network and the network finally does the decision process. We would like to remark, that the

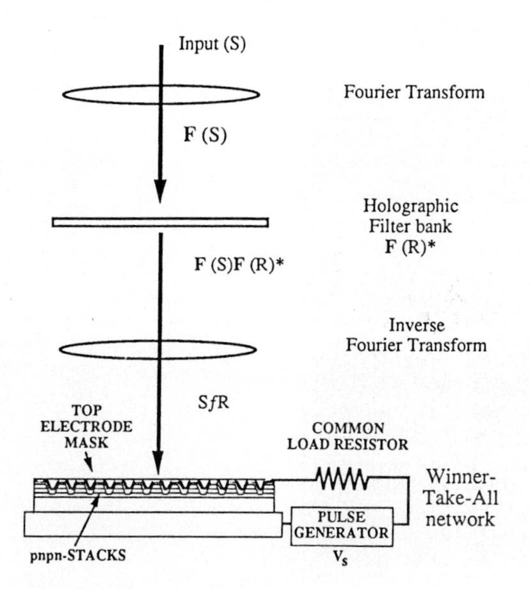

Fig. 11.9. Vander Lugt correlator with WTA network

whole processing is translation invariant and that the phase information will get lost. For the illustration of the pattern recognition procedure, described above in a very abstract way, some simple examples are shown in Fig. 11.10.

The reference patterns, the Fourier transform of which is stored in the holographic filter bank, are shown on the left hand side in Fig. 11.10. The next column shows the input patterns. The correlator output signal is an intensity modulated spot pattern. In case the input pattern exactly corresponds to one of the stored reference patterns, only one bright spot will appear at the spatial position that represents the position of the reference pattern. When the input signal has some properties in common with the reference patterns, the output signal consists of multiple spots with different intensity values. This situation is given in the last two input patterns. The circle, the rectangle and the ellipse are somewhat similar, and their representation in the reference space is not very well separated. Anyway, the spot with the highest intensity indicates the reference pattern which is closest to the input pattern. The task of the winner WTA network is to find the spot with the highest intensity. The network is of the array type discussed in Fig. 11.6a. The position of the maximum is done either optically or electrically by an xy-addressed readout system. Hence, from the spatial position of the maximum on the WTA array the input pattern can be found.

11.3.2 WTA Networks and Fuzzy Logic

During the past years Fuzzy Logic has become very popular to solve problems in control applications. Contrary to binary or Boolean Logic, the sets of elements, e. g., $X = \{x_1, x_2\}$ is not restricted to a binary presentation with elements like (0,1) or (1,0), but also elements with rational numbers like (0.2,0.9) or (0.8, 0.1) are allowed. While the first are Non-Fuzzy subsets, the later are called Fuzzy subsets. Especially when the system behaviour can not be described by

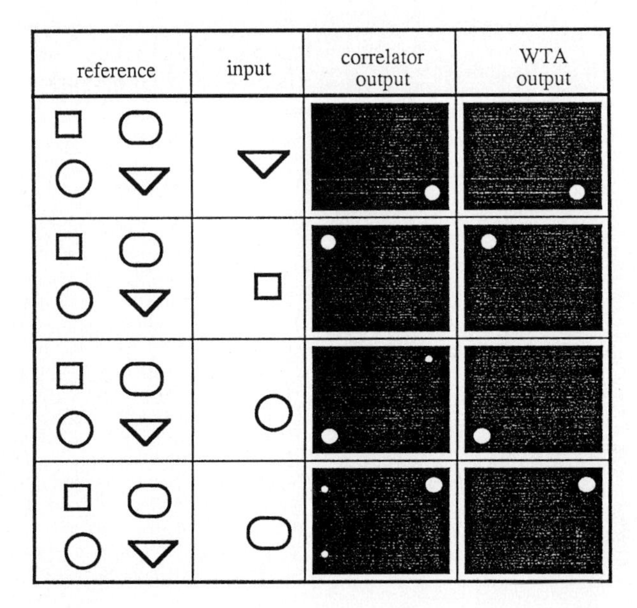

Fig. 11.10. Typical examples of pattern recognition processing with the correlator-WTA network

two logical states, the use of Fuzzy Logic is very useful, because it allows the description of the system by intermediate states. As an example the statement a container is filled with liquid can be answered exactly by boolean statements, only when the container is full or empty. The statement that the container is half filled with liquid, does not exist and makes no sense in Boolean Logic but in Fuzzy Logic.

Like in Boolean Logic logical operations are defined between the elements of Fuzzy subsets. For example:

Boolean logic

- *"or" operation*
  $\{x_1\} \cup \{x_2\} = (1,0) \vee (0,1) = (max\{1,0\}, max\{0,1\}) = (1,1)$
- *"and" operation*
  $\{x_1\} \cap \{x_2\} = (1,0) \wedge (0,1) = (min\{1,0\}, min\{0,1\}) = (0,0)$

Fuzzy logic

- *"or" operation*
  $$\{x_1\} \cup \{x_2\} = (0.2,0.9) \vee (0.8,0.1) = (max\{0.2,0.9\}, max\{0.8,0.1\}) = (0.8,0.9)$$
- *"and" operation*
  $$\{x_1\} \cap \{x_2\} = (0.2,0.9) \wedge (0.8,0.1) = (min\{0.2,0.9\}, min\{0.8,0.1\}) = (0.2,0.1)$$

In the following we will show how the WTA maximum detector can be used in Fuzzy Logic computation. As an example we will discuss the implementation of the Fuzzy "or" operation. The task is to find the maximum of a certain set of function $f_1 \vee ... \vee f_8 = max\{f_1(x), ..., f_8(x)\} = f(x)$.

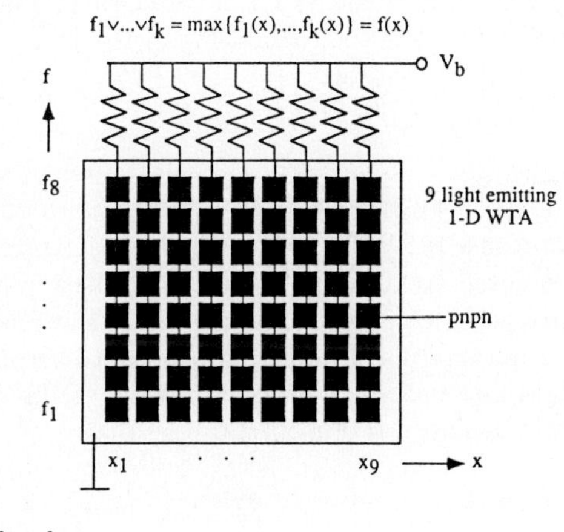

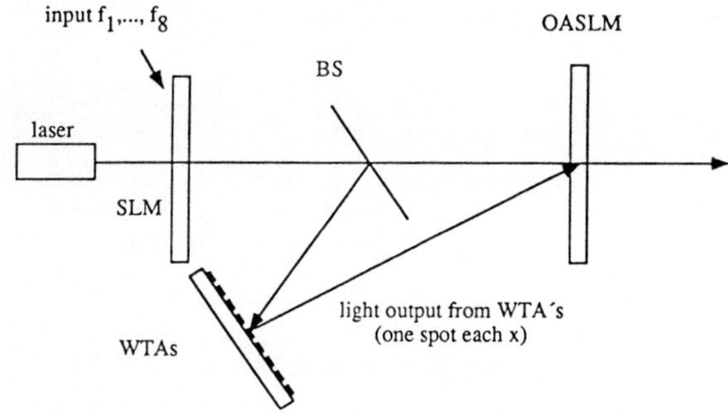

Fig. 11.11. Fuzzy "or" operation with a WTA network

In Fig. 11.11 the experimental set-up for solving this problem is shown. The input functions $f_1, ..., f_8$ are generated by using a laser ($\lambda = 632\,\text{nm}$) and a ferroelectric liquid crystal spatial light modulator (SLM). A beam splitter (BS) incorporated into the set-up deflects a part of the output light on the surface of a WTA array, while another part of the output light is projected on a liquid crystal optical addressed spatial light modulator (OASLM). An OASLM is a device that switches the transmission characteristic when visible light of a certain wavelength is projected on its surface.

The WTA network shown in the upper part of Fig. 11.11 consists of 9×8 elements. Each row represents one of the functions $f_1, ..., f_8$ and each column corresponds to the spatial discretization $x_1, ..., x_9$ of the input functions. The network is constructed in such a way that each column has a common load resistor and therefore the maximum identification is done separately in each column. When the input light pattern is switched on, the WTA network finds the intensity maximums of the complex input light pattern for each position in space. Finally, in each column there will be one detector element in the switch-on state and emit light. Thus the network calculates the Fuzzy "or" operation of the input functions. The light output of the WTA network is projected onto the surface of the OASLM and it switches in the light transmission mode at the illuminated positions. Consequently, the output light pattern represents the function f(x).

In the following section we will see that the WTA network can be used for minimum identification. This will also allow the calculation of the Fuzzy "and" operation.

11.3.3 Minimum Identification

For the minimum identification process the WTA network is used in combination with an OASLM [11.10], that is used to turn the intensity of the input light pattern from maxima into minima and vice versa. After this inversion process the WTA network identifies the maximum of the output light pattern corresponding to the minimum in the input light distribution. The experimental set-up for the minimum identification procedure is shown in Fig. 11.12.

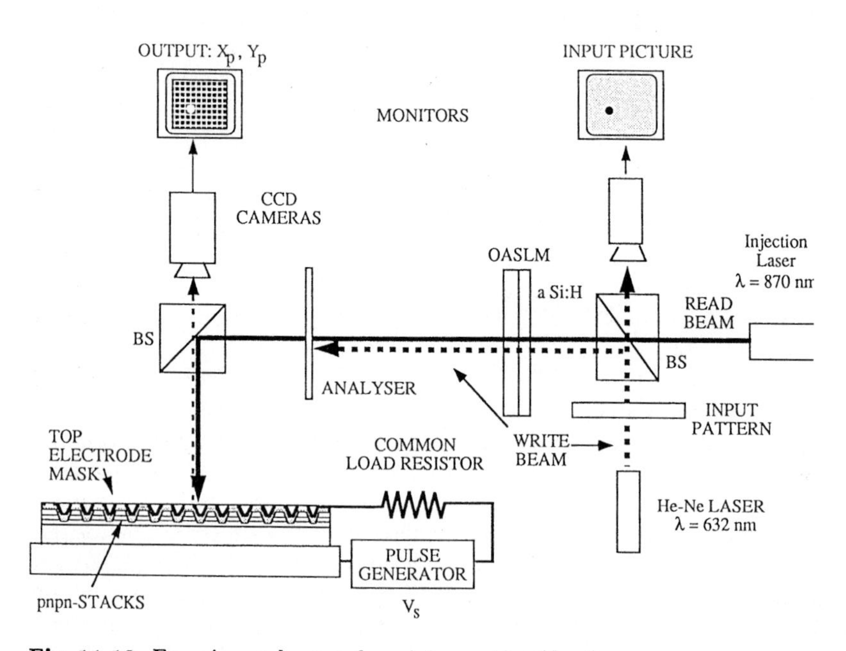

Fig. 11.12. Experimental setup for minimum identification

The OASLM consists of an a-Si:H photo diode layer adjacent to a layer of a ferroelectric liquid crystal (FLC). The liquid crystals are sandwiched between two glass plates coated with transparent Indium Tin Oxides (ITO) electrodes. The thickness of the birefrigend FLC is chosen in such a way, that it acts as half-wave plate. By changing the voltage across the liquid crystal layer, it is possible to adjust the crystal orientation. Initially the liquid crystal is driven into uniform orientation by forward biasing the photo diode. In this case the current flows spatially uniformly through the photo diode and the applied voltage drops entirely and uniformly across the FLC layer. This state is called the off-state of the OASLM. Next the photodiode is reverse-biased by applying a negative voltage. The current flows now only in the illuminated areas of the photodiode and the negative voltage appears only across adjacent regions of the FLC, rotating them in the on-state and leaving the others in the off-state. To ensure a proper function it is necessary to put the OASLM between crossed polarizers.

The adjustment is done in such a way that with incident input light from the write beam, the read beam's polarization is parallel to one of the optical axes of the liquid crystal. Thus it is not rotated and blocked by the analyzer situated just behind the OASLM. In the off-state (no incident light), the crystal orientation is rotated by 45 degree with respect to the on-state. Therefore, the polarization of the read beam incident on region of the FLC in the off-state is rotated by 90 degree. When the readout beam passes the analyzer, only the light which passed through off-state regions of the FLC is transmitted, and in this way the intensity inversion of the input image takes place.

As can be seen in Fig. 11.12, the input light pattern is generated by a He-Ne laser beam ($\lambda = 632\,\mathrm{nm}$). The input light passes through a glass plate with the input pattern. By using a beam splitter a part of the input light is focused on the photodiode area of the OASLM, while another part is picked up by a CCD camera and displayed on a video monitor for visual observation of the input light pattern. An injection laser ($\lambda = 870\,\mathrm{nm}$, $P = 30\,\mathrm{mW}$) is used to generate the read beam. The read beam passes the OASLM and is directed onto the surface of the WTA network via a beam splitter. The WTA array is of that type shown in Fig. 11.6a. After the WTA network has completed the decision process, the result of the minimum identification can be observed by switching off the read beam. The result is displayed on a second video monitor connected to a CCD camera that is focused onto the surface of the WTA array.

Figure 11.13 presents some typical results on minimum identification. In Fig. 11.13a the input patterns, projected onto the photo diode area of the OASLM and recorded with a CCD camera, are shown. For making the recognition of the absolute minimum intensity more easily to the reader, we have simply chosen black spots. The corresponding intensity-inverted images generated by the OASLM are shown in Fig. 11.13b. The responses of the WTA network after the decision process and switching off the read beam, are shown in Fig. 11.13c. As can be seen only one array element is switched on, the spatial position of which corresponds to the absolute minimum light intensity of the input image. The timing of the minimum identification process is the same as for the maximum detection process described in Sect. 11.2.1.

11.3.4 Optical Neural Network

As we have seen in Fig. 11.6b, the resistor network incorporated into the array causes an interaction between single elements of the device. The possibility to

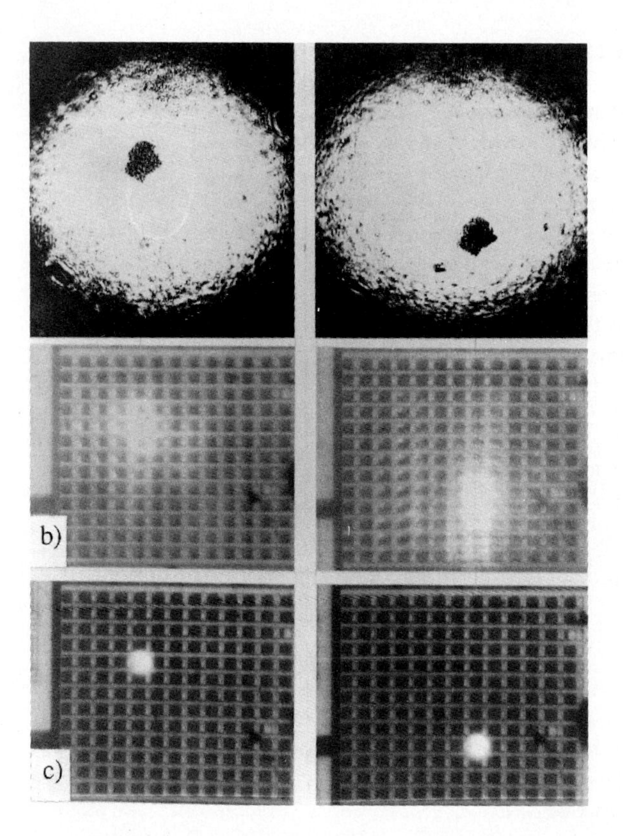

Fig. 11.13. Experimental results on minimum identification with the WTA network. a) shows the inputlight patterns and b) the corresponding intensity-inverted patterns projected onto the surface of the WTA array; c) shows the output light response of the WTA network after the decision process is finished and the read beam is turned off

connect these elements in an appropriate way is a necessary requirement for the realization of an optical neural network. Each element of the device can act as a simple neural switch, which is either in the off- or on-state, depending on the external light stimulation. For making the adaptation and learning possible, it is necessary to adjust the interconnection weights. By replacing the resistor network with photoconductors or phototransistors, one can optically and individually adjust the weights of the interconnections. The interconnection could be done by optical fibers or waveguides, which may be adjustable either electrically or optically to allow a more complex operation of the network. The technical realization of this type of optical neural networks is closely related to the progress in the fabrication of optoelectronic semiconductor chips with integrated waveguide structures.

11.4 Concluding Remarks

We have shown that a 2-D array of optoelectronic p-n-p-n switches can be used to solve the problems of maximum and minimum detection of optical fields. These two kinds of operations are of fundamental interest for optical pattern recognition and optical fuzzy logic computing. Due to the intelligent way of sensing and preprocessing, the system provides a substantial relief to the data conversion and input requirements of standard digital electronics. Furthermore, contrary to the standard approaches to solve these problems, the system uses parallel processing of input data, which results in a higher processing bandwith. However, for building an effective and compact detector device, it will be necessary to use VLSI technology.

Acknowledgments. The authors would like to thank J. I. Pankove, T. A. Rohlev from the University of Colorado at Boulder and G. Borghs, P. Heremans, M. Kuijk, R. A. Vounckx from the Interuniversity Microelectronic Center, Belgium for the kind cooperation. Especially, H. W. likes to thank the DAAD for giving the personal support for his staying at the University of Colorado in Boulder.

References

[11.1] J. Lazzaro, S. Ryckebusch, M. Mahowald, C. Mead: In *Neural Information Processing Systems*, ed. by D. Touretzky, (Morgan Kaufmann, San Mateo, Californien 1989) pp. 703-711

[11.2] K. Wagner, T. Slagle: Optical Computing **6**, of 1991 OSA Technical Digest Series (Optical Society of America, Washington. D.C. 1991) paper WA5

[11.3] M.A. Neifeld, D. Psaltis: Optical Computing **6**, of 1991 OSA Technical Digest Series (Optical Society of America, Washington. D.C. 1991) paper WA4

[11.4] J.I. Pankove, C. Radehaus, K. Wagner: Electron. Lett. **26**, 349 (1990)

[11.5] C.V. Radehaus, J.I. Pankove, M. Kuijk, P. Heremans, G. Borghs: Appl. Opt. **31**, 6303 (1992)

[11.6] P. Heremans, M. Kuijik, R.A. Vounckx, C.V. Radehaus, J.I. Pankove, G. Borghs: IEEE Trans. ED-**39**, 2248 (1992)

[11.7] J.I. Pankove, C. Radehaus: Optoelectronics-Devices and Technology **5**, 311 (1990)

[11.8] P. Heremans, M. Kuijk, R. Vounckx, G. Borghs: Appl. Phys. Lett. **65**, 19 (1994)

[11.9] A. Vander Lugt: *Optical Signal Processing* (Wiley and Sons Inc., New York 1992)

[11.10] B. Landreth, G. Moddel: Appl. Optics **31**, 3937 (1992)

Springer Proceedings in Physics

Managing Editor: H. K. V. Lotsch

1 *Fluctuations and Sensitivity in Nonequilibrium Systems*
Editors: W. Horsthemke and D. K. Kondepudi
2 *EXAFS and Near Edge Structure III*
Editors: K. O. Hodgson, B. Hedman, and J. E. Penner-Hahn
3 *Nonlinear Phenomena in Physics*
Editor: F. Claro
4 *Time-Resolved Vibrational Spectroscopy*
Editors: A. Laubereau and M. Stockburger
5 *Physics of Finely Divided Matter*
Editors: N. Boccara and M. Daoud
6 *Aerogels* Editor: J. Fricke
7 *Nonlinear Optics: Materials and Devices*
Editors: C. Flytzanis and J. L. Oudar
8 *Optical Bistability III*
Editors: H. M. Gibbs, P. Mandel, N. Peyghambarian, and S. D. Smith
9 *Ion Formation from Organic Solids (IFOS III)*
Editor: A. Benninghoven
10 *Atomic Transport and Defects in Metals by Neutron Scattering*
Editors: C. Janot, W. Petry, D. Richter, and T. Springer
11 *Biophysical Effects of Steady Magnetic Fields*
Editors: G. Maret, I. Kiepenheuer, and N. Boccara
12 *Quantum Optics IV*
Editors: J. D. Harvey and D. F. Walls
13 *The Physics and Fabrication of Microstructures and Microdevices*
Editors: M. J. Kelly and C. Weisbuch
14 *Magnetic Properties of Low-Dimensional Systems*
Editors: L. M. Falicov and J. L. Morán-López
15 *Gas Flow and Chemical Lasers*
Editor: S. Rosenwaks
16 *Photons and Continuum States of Atoms and Molecules*
Editors: N. K. Rahman, C. Guidotti, and M. Allegrini
17 *Quantum Aspects of Molecular Motions in Solids*
Editors: A. Heidemann, A. Magerl, M. Prager, D. Richter, and T. Springer
18 *Electro-optic and Photorefractive Materials*
Editor: P. Günter
19 *Lasers and Synergetics*
Editors: R. Graham and A. Wunderlin
20 *Primary Processes in Photobiology*
Editor: T. Kobayashi
21 *Physics of Amphiphilic Layers*
Editors: J. Meunier, D. Langevin, and N. Boccara
22 *Semiconductor Interfaces: Formation and Properties*
Editors: G. Le Lay, J. Derrien, and N. Boccara
23 *Magnetic Excitations and Fluctuations II*
Editors: U. Balucani, S. W. Lovesey, M. G. Rasetti, and V. Tognetti
24 *Recent Topics in Theoretical Physics*
Editor: H. Takayama
25 *Excitons in Confined Systems*
Editors: R. Del Sole, A. D'Andrea, and A. Lapiccirella
26 *The Elementary Structure of Matter*
Editors: J.-M. Richard, E. Aslanides, and N. Boccara
27 *Competing Interactions and Microstructures: Statics and Dynamics*
Editors: R. LeSar, A. Bishop, and R. Heffner
28 *Anderson Localization*
Editors: T. Ando and H. Fukuyama
29 *Polymer Motion in Dense Systems*
Editors: D. Richter and T. Springer
30 *Short-Wavelength Lasers and Their Applications*
Editor: C. Yamanaka
31 *Quantum String Theory*
Editors: N. Kawamoto and T. Kugo
32 *Universalities in Condensed Matter*
Editors: R. Jullien, L. Peliti, R. Rammal, and N. Boccara
33 *Computer Simulation Studies in Condensed-Matter Physics: Recent Developments*
Editors: D. P. Landau, K. K. Mon, and H.-B. Schüttler
34 *Amorphous and Crystalline Silicon Carbide and Related Materials*
Editors: G. L. Harris and C. Y.-W. Yang
35 *Polycrystalline Semiconductors: Grain Boundaries and Interfaces*
Editors: H. J. Möller, H. P. Strunk, and J. H. Werner
36 *Nonlinear Optics of Organics and Semiconductors*
Editor: T. Kobayashi
37 *Dynamics of Disordered Materials*
Editors: D. Richter, A. J. Dianoux, W. Petry, and J. Teixeira
38 *Electroluminescence*
Editors: S. Shionoya and H. Kobayashi
39 *Disorder and Nonlinearity*
Editors: A. R. Bishop, D. K. Campbell, and S. Pnevmatikos
40 *Static and Dynamic Properties of Liquids*
Editors: M. Davidović and A. K. Soper
41 *Quantum Optics V*
Editors: J. D. Harvey and D. F. Walls
42 *Molecular Basis of Polymer Networks*
Editors: A. Baumgärtner and C. E. Picot
43 *Amorphous and Crystalline Silicon Carbide II: Recent Developments*
Editors: M. M. Rahman, C. Y.-W. Yang, and G. L. Harris
44 *Optical Fiber Sensors*
Editors: H. J. Arditty, J. P. Dakin, and R. Th. Kersten
45 *Computer Simulation Studies in Condensed-Matter Physics II: New Directions*
Editors: D. P. Landau, K. K. Mon, and H.-B. Schüttler

Springer-Verlag and the Environment